U0360446

李宁 著

奇妙的 Java

神奇代码漫游之旅

清華大學出版社
北 京

内容简介

本书以 Java 语言为中心，探索其在当代技术领域中的广泛应用。Java 因其跨平台能力、健壮性和安全性，在操作系统、GUI、动画制作、多媒体处理等多个领域中占据了核心地位。本书旨在展现 Java 如何应对各种技术挑战，包括但不限于办公自动化、设备控制、数据加密与解密，以及文件压缩等。

本书通过实战案例详细介绍了 Java 在上述技术领域的应用，内容覆盖全面，从基础的文件系统操作、GUI 设计技巧，到高级的多媒体处理、安全加密技术，以及 ChatGPT 的集成和应用。特别强调了实用插件开发、办公自动化技巧和对设备的控制方法，每个章节都旨在通过具体案例教授技术实现，使读者能够深入理解并实践 Java 编程。

本书面向具备 Java 基础的开发者，包括但不限于软件开发工程师、计算机专业的学生和教师。无论读者是寻求解决特定编程问题的实用方法，还是希望通过学习 Java 在不同领域的应用来拓宽知识面和技能，本书都提供了丰富的资源和灵感。对于那些渴望深入了解 Java 及其在多种技术场景中应用的开发者，本书将是一份宝贵的学习材料。

图书在版编目（CIP）数据

奇妙的 Java : 神奇代码漫游之旅 / 李宁著 . -- 北京 : 清华大学出版社 , 2025. 5.
ISBN 978-7-302-68659-0

Ⅰ. TP312.8

中国国家版本馆 CIP 数据核字第 20250ES360 号

责任编辑： 曾　珊　薛　阳
封面设计： 傅瑞学
责任校对： 申晓焕
责任印制： 刘海龙

出版发行： 清华大学出版社
网　　址： https://www.tup.com.cn，https://www.wqxuetang.com
地　　址： 北京清华大学学研大厦 A 座　　**邮　　编：** 100084
社 总 机： 010-83470000　　**邮　　购：** 010-62786544
投稿与读者服务： 010-62776969，c-service@tup.tsinghua.edu.cn
质量反馈： 010-62772015，zhiliang@tup.tsinghua.edu.cn
课件下载： https://www.tup.com.cn，010-83470236

印 装 者： 三河市铭诚印务有限公司
经　　销： 全国新华书店
开　　本： 186mm × 240mm　　**印　　张：** 25.5　　**字　　数：** 573 千字
版　　次： 2025 年 6 月第 1 版　　**印　　次：** 2025 年 6 月第 1 次印刷
印　　数： 1 ～ 1500
定　　价： 98.00 元

产品编号：107347-01

前言

PREFACE

欢迎来到《奇妙的Java：神奇代码漫游之旅》，本书将带领你踏上一段奇幻的Java之旅，探索代码的神奇力量。Java作为一门健壮而灵活的编程语言，已经成为众多领域的首选。无论你是初学者还是有一定经验的开发者，本书都将为你打开一扇通向Java神奇世界的大门。

在本书中，我们将探索Java在各个领域的应用。从控制操作系统到图形用户界面的构建，再到办公自动化，我们将一起探索Java的无限潜力。无论你是想构建高效的企业级应用，还是通过图像处理和Office应用开发展现创造力，本书都会为你提供全面而实用的指导。

我们将学习如何利用Java在操作系统中执行各种任务，从文件和目录的管理到获取系统信息和显示系统窗口。我们还将探索JavaFX的使用，以及如何创建窗口、设计布局、添加组件和实现交互功能。对人工智能和聊天机器人感兴趣的读者，本书还将向你展示如何解锁ChatGPT的神奇力量，并让它成为你的编程助手。如果希望增强IDE的能力，那么开发Intellij IDEA插件将是你必备的技能，而本书的任务就是让读者拥有这项技能。

除了探索以上领域之外，本书还会教你处理音频、图像和视频，从音乐播放器到视频编辑，从图像处理到动画制作，让你体验到代码创造的魅力。此外，你还将学习如何读写Excel、Word、PowerPoint和PDF文档，以及如何处理文本数据、加密解密信息和进行数学计算。

在本书的每一章，你都将遇到丰富的实例和项目，通过实际的代码演示和练习，提升你的编程技能和解决问题的能力。无论你是希望学习新的技术，还是希望加深对Java的理解，本书都将成为你的指南和伙伴。

无论你是想成为一名职业开发者，还是对编程充满热情的爱好者，我相信《奇妙的Java：神奇代码漫游之旅》将成为你宝贵的学习资料。让我们一起踏上这段奇幻之旅，发现Java世界的无限可能！

编者

2024年11月

目录

CONTENTS

第 1 章　文件系统……1

1.1　打开文件夹……1
1.2　获取文件和目录的属性……3
1.3　使用 JNA 获取 UNIX-like 系统中所有用户的信息……8
1.4　使用 JNA 获取 Windows 系统中所有用户的信息……12
1.5　修改文件的属性……14
1.5.1　修改文件的创建时间、访问时间和修改时间……14
1.5.2　修改文件权限……16
1.5.3　修改文件的所有者和组……18
1.5.4　修改 Windows 的访问控制列表……21
1.6　文件和目录的相关操作……23
1.6.1　创建文件和目录……23
1.6.2　删除文件和目录……26
1.6.3　复制文件和目录……28
1.6.4　重命名文件和目录……32
1.6.5　搜索文件和目录……33
1.6.6　创建快捷方式……35
1.7　回收站……37
1.7.1　将文件和目录放入回收站……37
1.7.2　清空回收站中的文件和目录……41
1.7.3　恢复回收站中的文件和目录……43
1.8　小结……49

第 2 章　驾驭 OS……50

2.1　Windows 注册表……50
2.1.1　读取值的数据……50
2.1.2　读取所有的键……52

2.1.3 读取所有的键和值 ······ 53
2.1.4 添加键和值 ······ 55
2.1.5 重命名键 ······ 57
2.1.6 重命名值 ······ 58
2.2 让程序随 OS 一起启动 ······ 60
2.2.1 将应用程序添加进 macOS 登录项 ······ 60
2.2.2 将应用程序添加进 Windows 启动项 ······ 62
2.2.3 将应用程序添加进 Linux 启动项 ······ 63
2.3 获取系统信息 ······ 64
2.3.1 使用 Java 标准库获取系统信息 ······ 64
2.3.2 使用 OSHI 库获取系统信息 ······ 66
2.4 显示系统窗口 ······ 70
2.4.1 显示 macOS 中的系统窗口 ······ 70
2.4.2 显示 Windows 中的系统窗口 ······ 74
2.4.3 显示 Linux 中的系统窗口 ······ 77
2.5 打开文件夹 ······ 79
2.5.1 打开 macOS 文件夹与回收站 ······ 79
2.5.2 打开 Windows 文件夹与回收站 ······ 80
2.5.3 打开 Linux 文件夹与回收站 ······ 81
2.6 小结 ······ 81

第 3 章 GUI 工具包——JavaFX ······ 82

3.1 Java 支持哪些 GUI 框架 ······ 82
3.2 安装和配置 JavaFX ······ 83
3.3 创建窗口 ······ 86
3.4 布局 ······ 88
3.5 常用组件 ······ 91
3.6 列表组件 ······ 94
3.7 下拉列表组件 ······ 96
3.8 表格组件 ······ 98
3.9 树状组件 ······ 103
3.10 菜单 ······ 106
3.11 对话框 ······ 109
3.12 使用 CSS 设置 JavaFX 组件样式 ······ 114
3.13 项目实战：自由绘画 ······ 116
3.14 项目实战：旋转图像 ······ 119

3.15 小结 ··· 122

第 4 章 有趣的 GUI 技术 ··· 123

4.1 特殊窗口 ··· 123
4.1.1 五角星窗口 ··· 123
4.1.2 异形美女机器人窗口 ··· 127
4.1.3 半透明窗口 ··· 129
4.2 在屏幕上绘制曲线 ··· 131
4.3 控制状态栏 ··· 134
4.3.1 在状态栏上添加图标 ··· 134
4.3.2 添加 Windows 10 风格的 Toast 消息框 ··· 136
4.4 小结 ··· 138

第 5 章 动画 ··· 139

5.1 属性动画 ··· 139
5.2 缓动动画 ··· 142
5.3 正弦波动画 GIF ··· 145
5.4 动画 GIF ··· 150
5.4.1 使用静态图像生成动画 GIF 文件 ··· 150
5.4.2 播放动画 GIF ··· 152
5.5 模拟物理运动 ··· 154
5.6 CSS 动画 ··· 158
5.7 小结 ··· 160

第 6 章 音频 ··· 161

6.1 音乐播放器 ··· 161
6.2 录音机 ··· 165
6.3 音频格式转换 ··· 169
6.4 音频分析和处理 ··· 171
6.4.1 PCM 格式的 WAV 文件 ··· 171
6.4.2 转换为 PCM 格式的 WAV 文件 ··· 172
6.4.3 音高检测 ··· 173
6.4.4 麦克风音频中的音高 ··· 177
6.4.5 添加噪声 ··· 178
6.4.6 降低噪声 ··· 181
6.5 音频编辑 ··· 183

6.5.1 音频裁剪 ······ 183
6.5.2 音频合并 ······ 184
6.5.3 音频混合 ······ 185
6.6 小结 ······ 187

第 7 章 图像与视频 ······ 188

7.1 OpenCV 基础 ······ 188
7.1.1 编译 OpenCV 源代码 ······ 188
7.1.2 在 Java 中使用 OpenCV ······ 192
7.1.3 发布 OpenCV 应用 ······ 192
7.2 获取视频信息 ······ 193
7.3 播放视频 ······ 196
7.4 截屏 ······ 199
7.4.1 截取屏幕 ······ 199
7.4.2 截取 Web 页面 ······ 201
7.5 拍照 ······ 205
7.6 制作录屏视频 ······ 208
7.7 格式转换 ······ 209
7.7.1 图像格式转换 ······ 210
7.7.2 使用 FFmpeg 转换视频格式 ······ 211
7.8 视频编辑 ······ 212
7.8.1 视频裁剪 ······ 212
7.8.2 视频合并 ······ 215
7.8.3 混合音频和视频 ······ 219
7.8.4 提取视频中的音频 ······ 224
7.8.5 制作画中画视频 ······ 226
7.9 小结 ······ 228

第 8 章 图像特效 ······ 229

8.1 图像滤镜 ······ 229
8.2 缩放图像和拉伸图像 ······ 235
8.3 生成圆形头像 ······ 236
8.4 翻转图像 ······ 239
8.5 调整图像的亮度、对比度、饱和度和锐度 ······ 242
8.6 在图像上添加和旋转文字 ······ 246
8.7 混合图像 ······ 248

8.8 油画……251
8.9 波浪扭曲……252
8.10 挤压扭曲……254
8.11 3D 浮雕效果……256
8.12 小结……259

第 9 章 视频特效……260

9.1 旋转视频……260
9.2 镜像视频……262
9.3 变速视频……264
9.4 为视频添加水印……266
9.5 缩放和拉伸视频……268
9.6 视频 3D 透视变换……269
9.7 视频局部高斯模糊……273
9.8 视频转场淡入淡出动画……276
9.9 向视频中添加动态图像……278
9.10 将视频转换为动画 GIF……283
9.11 为视频添加字幕……285
9.12 将彩色视频变为灰度视频……287
9.13 小结……288

第 10 章 IntelliJ IDEA 插件……289

10.1 IntelliJ IDEA 插件简介……289
10.2 IntelliJ IDEA 插件的开发步骤……290
10.3 plugin.xml 文件解读……292
10.4 统计代码行数的插件……293
10.5 圆形头像插件……297
10.6 视频转动画 GIF 插件……301
10.7 绘图插件……305
10.8 小结……310

第 11 章 代码魔法：释放 ChatGPT 的神力……311

11.1 走进 ChatGPT……311
11.1.1 AIGC 概述……311
11.1.2 AIGC 的落地案例……313
11.1.3 ChatGPT 概述……313

11.1.4 ChatGPT vs New Bing ······ 314
11.1.5 ChatGPT Plus，史上最强 AI ······ 315
11.1.6 有了 ChatGPT，程序员真的会失业吗 ······ 316
11.2 注册和登录 ChatGPT ······ 317
11.3 升级为 ChatGPT Plus 账户 ······ 319
11.4 让 ChatGPT 帮你写程序 ······ 319
11.5 将 ChatGPT 嵌入自己的程序：OpenAI API ······ 320
11.6 用 Java 访问 OpenAI API ······ 321
11.7 编程魔匣 ······ 323
11.8 聊天机器人 ······ 325
11.9 小结 ······ 326

第 12 章 读写 Office 文档 ······ 327

12.1 Apache POI 简介 ······ 327
12.2 下载和引用 Apache POI 库 ······ 327
12.3 读写 Excel 文档 ······ 328
12.3.1 生成 Excel 表格 ······ 328
12.3.2 读取 Excel 表格 ······ 331
12.3.3 为 Excel Cell 指定公式 ······ 333
12.4 读写 Word 文档 ······ 335
12.4.1 创建 Word 文档 ······ 335
12.4.2 插入图像 ······ 337
12.4.3 插入页眉、页脚和页码 ······ 338
12.4.4 统计 Word 文档生成云图 ······ 340
12.5 PowerPoint 文档 ······ 343
12.5.1 创建 PowerPoint 文档 ······ 343
12.5.2 将图片批量插入 PPT 中 ······ 346
12.6 小结 ······ 348

第 13 章 读写 PDF 文档 ······ 349

13.1 Apache PDFBox 简介 ······ 349
13.2 生成简单的 PDF 文档 ······ 350
13.3 在 PDF 文档中插入图像和表格 ······ 353
13.4 加密 PDF 文档 ······ 356
13.5 在 PDF 文档上绘制图表 ······ 358
13.6 小结 ······ 361

第 14 章 控制软件和设备 362

14.1 浏览器 362
14.2 鼠标和键盘 364
14.2.1 录制键盘和鼠标的动作 364
14.2.2 回放键盘和鼠标的动作 368
14.3 剪贴板 370
14.4 小结 370

第 15 章 加密与解密 371

15.1 MD5 摘要算法 371
15.2 SHA 摘要算法 372
15.3 Base64 编码和解码 374
15.4 DES 加密和解密 375
15.5 AES 加密和解密 377
15.6 RSA 加密和解密 379
15.7 小结 381

第 16 章 文件压缩与解压 382

16.1 zip 格式 382
16.1.1 压缩成 zip 文件 382
16.1.2 解压 zip 文件 385
16.2 7z 格式 387
16.2.1 压缩成 7z 格式 387
16.2.2 解压 7z 文件 390
16.2.3 设置 7z 文件的密码 392
16.3 小结 393

第1章 文件系统

文件系统是现代操作系统最重要的组成部分之一。然而，操作文件系统通常使用操作系统自带的管理工具，例如，Windows 中使用资源管理器操作文件和目录，macOS 中对应的工具是 Finder（中文名叫“访达”）。但在很多场景中，需要使用程序来完成同样的工作，所以本章主要介绍如何通过 Java 来管理文件和目录。

1.1 打开文件夹

使用 Java 打开文件夹需要使用 java.awt.Desktop 类，该类是 Java AWT（Abstract Window Toolkit）的一部分，提供了一个跨平台的方式来与主机桌面交互。通过使用 Desktop 类，可以执行一些与主机桌面环境交互的操作，如打开文件、浏览 URL 或启动邮箱客户端等。以下是 java.awt.Desktop 类的一些主要特点和方法。

- 检查支持性

public static boolean isDesktopSupported()：检查当前平台是否支持 Desktop 类。

- 获取 Desktop 实例

public static Desktop getDesktop()：获取当前 Desktop 实例。

- 打开、编辑和打印文件

（1）public void open(File file) throws IOException：打开一个文件，该文件用默认的关联应用程序打开。

（2）public void edit(File file) throws IOException：用默认的关联应用程序编辑文件。

（3）public void print(File file) throws IOException：用系统默认的打印机打印文件。

- 浏览 URI

public void browse(URI uri) throws IOException：使用系统默认的浏览器打开指定的 URI。

- 邮件支持

（1）public void mail() throws IOException：启动用户默认的邮件客户端。

（2）public void mail(URI mailtoURI) throws IOException：启动用户默认的邮件客户端，

以处理给定的 mailtoURI。

❑ 检查动作的支持性

public boolean isSupported(Desktop.Action action)：检查指定的动作是否受当前平台和 Desktop 实例的支持。

这个类是在 Java 6（即 JDK 1.6）中引入的，旨在提供一种简单、跨平台的方式来与用户的桌面环境交互。通过 Desktop 类，开发人员可以在不依赖特定平台或外部库的情况下实现与桌面的基本交互。例如，可以使用 Desktop 类来打开用户的系统中的文件和目录，或者使用系统默认的浏览器浏览网页。

下面的例子使用 open 方法打开当前文件夹，该方法会用默认的关联程序打开当前文件夹，例如，在 macOS 中会使用 Finder 打开文件夹，在 Windows 中会使用资源管理器打开文件夹。

代码位置：src/main/java/file/system/OpenFolder.java

```
package file.system;
import java.awt.Desktop;
import java.io.File;
import java.io.IOException;
public class OpenFolder {
    public static void main(String[] args) {
        openFolder(".");                            // 将路径更改为想要打开的文件夹的路径
    }
    public static void openFolder(String folderPath) {
        // 确保 Desktop API 支持
        if (!Desktop.isDesktopSupported()) {
            System.err.println("Desktop API 不受支持 ");
            return;
        }
        Desktop desktop = Desktop.getDesktop();
        // 确保可以打开文件夹
        if (!desktop.isSupported(Desktop.Action.OPEN)) {
            System.err.println(" 打开文件夹操作不受支持 ");
            return;
        }
        File folder = new File(folderPath);
        // 确保文件夹存在并且确实是一个文件夹
        if (!folder.exists() || !folder.isDirectory()) {
            System.err.println(" 文件夹不存在或不是一个有效的文件夹 ");
            return;
        }
        try {
            desktop.open(folder);
        } catch (IOException e) {
            e.printStackTrace();
        }
    }
}
```

这段代码是跨平台的，可以在 macOS、Windows 和 Linux 中成功打开当前文件夹，例如，在 macOS 下打开当前文件夹的效果如图 1-1 所示。

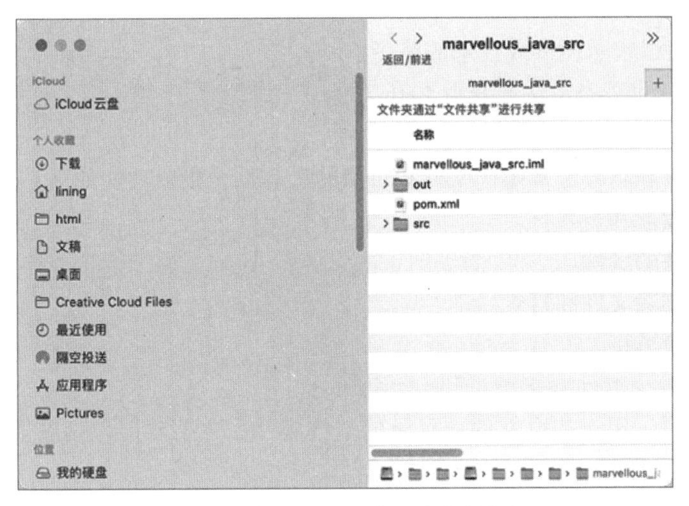

图 1-1　在 macOS 下打开当前文件夹

1.2　获取文件和目录的属性

文件和目录的属性有很多，常用的属性包括文件的尺寸、最新访问时间、最新编辑时间、文件模式等。这些属性需要使用不同的 API 获取，例如文件尺寸需要使用 BasicFileAttributes 类、最新访问时间需要使用 BasicFileAttributes.lastAccessTime() 方法，下面将详细介绍获取文件和目录属性的相关 API。

❑ java.io.File

这是 Java 中最基本的文件操作类。它表示文件和目录路径的抽象表示形式，主要功能如下：

（1）创建新文件或目录。

（2）删除文件或目录。

（3）获取和设置文件或目录的属性（例如是否可读、可写等）。

（4）列出目录的内容。

❑ java.io.FileWriter

这是一个用于写入字符流到文件的类，主要功能如下：

（1）创建一个新的文件并写入数据。

（2）向现有文件追加数据。

❑ java.io.IOException

这是输入输出操作期间可能抛出的异常类。

❑ java.nio.file.Path 和 java.nio.file.Paths

在新的 NIO（New I/O）框架中，Path 是文件和目录路径的抽象表示，而 Paths 类提供了方法来操作路径。

（1）Path：可以表示文件或目录的路径。

（2）Paths：提供静态方法来获取 Path 对象。

❑ java.nio.file.Files

这是一个提供静态方法来操作文件和目录的工具类，主要功能如下：

（1）文件的读取和写入。

（2）获取文件的属性。

（3）创建和删除文件和目录。

（4）创建符号链接。

❑ java.nio.file.attribute.BasicFileAttributes

这是一个接口，用于获取文件或目录的基本属性，主要功能包括获取文件的大小、创建时间、最后修改时间等。

❑ PosixFileAttributeView、PosixFileAttributes 和 PosixFilePermission

这 3 个类和接口都属于 java.nio.file.attribute 包，这些类和接口用于在支持 POSIX[①]文件属性的文件系统（如 Linux、macOS 等）中操作文件属性，主要功能如下：

（1）PosixFileAttributeView：允许读取和更新 POSIX 定义的文件属性。

（2）PosixFileAttributes：提供对 POSIX 文件属性的访问。

（3）PosixFilePermission：枚举定义了 POSIX 权限模型中的各种权限。

在前面的内容中提到了 Java NIO，这是从 Java 1.4 版本开始引入的一个 I/O API，用于替代标准的 Java I/O API。NIO 提供了与传统 I/O 不同的 I/O 工作方式，并增加了一些新的功能。以下是 NIO 的主要特点和功能。

❑ 缓冲区（Buffer）

（1）在 NIO 中，所有数据都是通过缓冲区处理的。缓冲区实质上是一个容器，它允许数据在读入和写出时进行缓存。

（2）NIO 包含多种类型的缓冲区，如 ByteBuffer、CharBuffer、IntBuffer 等，分别用于处理不同类型的数据。

❑ 通道（Channel）

通道是 NIO 中的一个核心概念，可以视为流的替代品。与流不同的是，通道可以同时进行读写操作。

（1）通道通常与缓冲区一起使用，数据从通道读入缓冲区或从缓冲区写入通道。

---

① POSIX（Portable Operating System Interface of UNIX）是一个由 IEEE 定义的为各种 UNIX-like 操作系统提供的接口标准。其中，对文件系统的定义是该标准的一个重要部分。当我们谈论 POSIX 文件系统，实际上是指遵循 POSIX 标准的文件系统的特定行为和特性。

（2）主要的通道实现有 FileChannel、SocketChannel、ServerSocketChannel 等。

- 非阻塞 I/O

在 NIO 中，可以使通道进入非阻塞模式。这意味着线程可以在没有数据可以读取或写入的情况下从该通道返回，从而线程可以执行其他任务。

- 选择器（Selector）

选择器是 NIO 中的一个独特组件，允许单个线程监控多个通道的事件（如连接、数据到达等）。这意味着使用单个线程可以监听多个客户端通道，这种方式通常称为“事件驱动的 I/O”或“异步 I/O”。

- 内存映射文件

NIO 提供了内存映射文件的功能，允许将文件或文件的部分映射到内存中，这样可以直接在内存中读写文件，而不需要复制数据到中间缓冲区。

- 文件锁定

NIO 支持文件级的锁定，允许程序锁定文件的部分或全部，以确保其他程序不能同时修改它。

- 字符集 Coder

NIO 提供了字符集编解码器，用于字符和字节之间的转换。

Java NIO 主要用于构建高性能的 I/O 密集型应用，如文件系统、网络通信等。它特别适用于需要处理成千上万并发连接的场景，例如高并发的服务器。与传统的 Java I/O API 相比，NIO 提供了更高级、更灵活的处理 I/O 的方法。

Java 的标准库中并没有直接提供获取文件模式的方法，但我们可以自己编写一个 permissionsToMode 方法来实现这个功能。该方法将 POSIX 文件权限集转换为其等效的整数模式表示。在 POSIX 系统（如 Linux、UNIX、macOS 等）中，文件权限通常表示为 3 组三个数字（如 755、644 等），其中每个数字代表一个八进制值，该值表示一组特定的读、写和执行权限。

以下介绍权限到模式转换的基本原理。

- 权限位

（1）r（读）：4。

（2）w（写）：2。

（3）x（执行）：1。

- 权限组

用户 (u): 文件的所有者的权限。

组 (g): 文件的组的权限。

其他 (o): 所有其他用户的权限。

每个组的权限是通过将相应的读、写和执行权限值相加来计算的。例如，如果一个文件的所有者有读、写和执行权限，那么该组的值为 $r + w + x = 4 + 2 + 1 = 7$。

整个文件的权限模式由 3 组权限组成，例如：

（1）用户：rwx (7)。

（2）组：r-x (5)。

（3）其他：r-- (4)。

最终文件的模式是 754。

现在，要实现 permissionsToMode 方法，可以按照以下步骤操作。

（1）为每个权限组（用户、组、其他）定义一个初始值为 0 的变量。

（2）对于每个权限组，检查其是否包含读、写和执行权限，并相应地增加该组的值。

（3）最后，将这三个值组合成一个三位数，得到完整的模式。

使用 java.nio.file.attribute.PosixFilePermission 枚举，可以轻松检查每个权限和组的存在，并按照上述逻辑计算模式。

需要注意的是，为了在 Java 中实现这一功能，可能需要使用 java.nio.file.attribute.PosixFileAttributeView 和相关的 API 来获取文件的 POSIX 权限。关于 permissionsToMode 方法的具体实现，见本节后面的案例。

下面的例子获取了文件的多个属性，包括文件尺寸、文件最新访问时间、文件最新编辑时间、是否为可执行文件、是否为只读文件、文件模式等。

代码位置：src/main/java/file/attributes/FileAttributesDemo.java

```
package file.attributes;

import java.io.File;
import java.io.FileWriter;
import java.io.IOException;
import java.nio.file.Files;
import java.nio.file.Path;
import java.nio.file.Paths;
import java.nio.file.attribute.*;
import java.util.Set;

public class FileAttributesDemo {
    public static void main(String[] args) throws IOException {
        // 创建一个临时目录
        String tempDirName = "temp";
        File tempDir = new File(tempDirName);
        tempDir.mkdir();

        // 在临时目录中创建一个文件
        File tempFile = new File(tempDir, "test.txt");
        try (FileWriter writer = new FileWriter(tempFile)) {
            writer.write("test");
        }
        // 获取临时目录中的文件属性信息
        Path tempFilePath = Paths.get(tempDirName, "test.txt");
        BasicFileAttributes tempFileAttributes = Files.readAttributes
                            (tempFilePath, BasicFileAttributes.class);
```

```
        System.out.println("Temp File size: " + tempFileAttributes.size());

        // 输出其他文件属性信息
        FileTime lastAccessTime = tempFileAttributes.lastAccessTime();
        FileTime lastModifiedTime = tempFileAttributes.lastModifiedTime();
        System.out.println("Last access time: " + lastAccessTime.toMillis() / 1000.0);
        System.out.println("Last modification time: " +
lastModifiedTime.toMillis() / 1000.0);
        System.out.println("Absolute path: " + tempFilePath.toAbsolutePath());

// 由于 Java 不直接提供文件模式的信息，所以可以使用 File 类来获取文件是否可执行、可读、可写的信息
        System.out.println("Is executable: " + Files.isExecutable
(tempFilePath));
        System.out.println("Is readable: " + Files.isReadable(tempFilePath));
        System.out.println("Is writable: " + Files.isWritable(tempFilePath));


        // 获取文件模式 (file mode)
        PosixFileAttributeView view = Files.getFileAttributeView
(tempFilePath, PosixFileAttributeView.class);
        PosixFileAttributes attrs = view.readAttributes();
        Set<PosixFilePermission> permissions = attrs.permissions();
        // 将权限转换为数字表示的文件模式
        System.out.println("Temp File mode: " + permissionsToMode(permissions));
        // 创建一个符号链接
        Path link = Paths.get("test_link.txt");
        Files.createSymbolicLink(link, tempFilePath);

        // 获取符号链接指向的文件的属性信息
        BasicFileAttributes linkedFileAttributes = Files.readAttributes
(link, BasicFileAttributes.class);
        System.out.println("Link File size: " + linkedFileAttributes.size());

        // 获取符号链接本身的属性信息
        BasicFileAttributes linkAttributes = Files.readAttributes
(link, BasicFileAttributes.class, java.nio.file.LinkOption.NOFOLLOW_LINKS);
        System.out.println("Link size: " + linkAttributes.size());

        // 删除符号链接
        Files.delete(link);

        // 删除临时文件和目录
        tempFile.delete();
        tempDir.delete();
    }
    // 将 POSIX 文件权限集转换为其等效的整数模式表示
    private static String permissionsToMode(Set<PosixFilePermission>
permissions) {
        int mode = 0;
        if (permissions.contains(PosixFilePermission.OWNER_READ)) mode |= 0400;
        if (permissions.contains(PosixFilePermission.OWNER_WRITE)) mode |= 0200;
```

```
        if (permissions.contains(PosixFilePermission.OWNER_EXECUTE)) mode |= 0100;
        if (permissions.contains(PosixFilePermission.GROUP_READ)) mode |= 0040;
        if (permissions.contains(PosixFilePermission.GROUP_WRITE)) mode |= 0020;
        if (permissions.contains(PosixFilePermission.GROUP_EXECUTE)) mode |= 0010;
        if (permissions.contains(PosixFilePermission.OTHERS_READ)) mode |= 0004;
        if (permissions.contains(PosixFilePermission.OTHERS_WRITE)) mode |= 0002;
        if (permissions.contains(PosixFilePermission.OTHERS_EXECUTE)) mode |= 0001;
        return Integer.toOctalString(mode);
    }
}
```

运行程序，会在终端输出如下内容：

```
Temp File size: 4
Last access time: 1.698461199513E9
Last modification time: 1.698461199513E9
Absolute path: /marvellous_java_src/temp/test.txt
Is executable: false
Is readable: true
Is writable: true
Temp File mode: 644
Link File size: 4
Link size: 13
```

由于 permissionsToMode 方法是基于 POSIX 文件权限模型的，所以该程序只能在 Linux、UNIX、macOS 等 POSIX 兼容的操作系统中运行。这些系统使用特定的权限位来表示文件或目录的读、写和执行权限。

而在 Windows 操作系统下，文件权限模型与 POSIX 不同。Windows 使用访问控制列表（ACL）来表示文件和目录的权限，它提供了更详细和细粒度的权限控制。因此，尝试在 Windows 上使用基于 POSIX 的 permissionsToMode 方法或其他 POSIX 文件权限相关功能可能不会正常工作或根本不会有任何效果。

如果是在 Java 中处理 Windows 文件权限，可能需要使用其他的 Java NIO API，如 AclFileAttributeView，以适应 Windows 的 ACL 模型。

1.3 使用 JNA 获取 UNIX-like 系统中所有用户的信息

要在 Java 中获取系统用户信息，需要依赖于一些第三方库，如 JNA（Java Native Access）。但是，这在跨平台环境中可能并不总是适用，因为它依赖于平台特定的 API。

JNA（Java Native Access）是一个 Java 库，它为 Java 程序提供了简化的方式来调用本地共享库的函数（例如，在 Windows 上的“.dll”文件，在 Linux/UNIX 上的“.so”文件）而不需要写 JNI（Java Native Interface）代码。JNA 的主要优势在于其简洁性和易用性。使用 JNA，可以直接从 Java 代码中访问 C 库，而不必为每个函数编写 JNI 包装器。

JNA 通过使用 Java 的反射 API 在运行时动态生成本地方法调用，从而消除了 JNI 的复杂性。JNA 使用 Java 的接口定义来描述 C 库中的函数和结构。这些接口提供了一种声明本地库的方法和结构的方式，这与 C 语言的头文件有些相似。

例如，下面的代码加载了一个 C 库并将其绑定到一个 Java 接口上：

```
CLibrary INSTANCE = (CLibrary) Native.load("c", CLibrary.class);
```

对这行代码的解释如下。

（1）CLibrary：这是自己定义的 Java 接口，用于描述 C 库中可供 Java 调用的函数。这个接口可能包含一些方法声明，它们对应于 C 库中的函数。

（2）Native.load("c", CLibrary.class)：Native.load 是 JNA 库中的方法，用于加载并连接本地库。"c" 是要加载的库的名称。在这里，它指的是标准 C 库，这在大多数 UNIX-like 系统上是 libc.so，在 macOS 上是 libSystem.B.dylib。如果要加载其他的库，只需替换 "c" 为相应的库名称即可。CLibrary.class 是希望将库绑定到的 Java 接口的类对象。

（3）(CLibrary)：这是一个类型转换，它告诉 Java，希望得到的是 CLibrary 接口的一个实例，而不是一个通用的 Object。

最终，INSTANCE 变量将持有 C 库的一个实例，可以通过该实例调用 C 库中的函数。

总的来说，使用 JNA 可以使 Java 程序员轻松地调用本地 C/C++ 代码，而无须深究 JNI 的复杂性。

由于 JNA 是第三方库，所以在使用之前需要安装 JNA 库，读者可以在 JNA github 的主页下载相关的 jar 文件，JNA github 主页的地址为：https://github.com/java-native-access/jna。

进入主页，下载 jna-5.13.0.jar 文件和 jna-jpms-5.13.0.jar 文件即可。如果读者直接在命令行编译并运行 Java 程序，需要将这两个 jar 文件的绝对路径添加到 CLASSPATH 环境变量中。如果读者在 IDE 中编译并运行 Java 程序，需要在 IDE 中引用这两个 jar 文件。这里以 IntelliJ IDEA[①]为例（本书其他的案例也在 IntelliJ IDEA 中编写和测试）。在工程根目录下创建一个 libs 子目录，然后将这两个 jar 文件复制到 libs 目录中，如图 1-2 所示。

然后在工程右键菜单中选择 Open Module Settings 菜单项，打开 Project Structure 对话框（这个对话框可以用多种方式打开，这个方式是其中之一），然后单击左侧的 Libraries，接下来单击中间上方的“+”按钮，在弹出的对话框中选择相应的 jar 文件即可，选择相应的 jar 文件后，会在列表中添加相应的项，如图 1-3 所示。最后单击 OK 按钮完成对 jar 文件的引用。

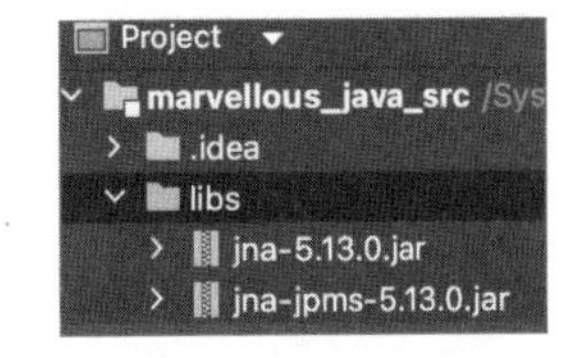

图 1-2　将 JNA 库文件复制到 IntelliJ IDEA 中

① IntelliJ IDEA 是由 JetBrains 公司开发的一款集成开发环境（IDE），特别针对 Java 开发。不过，通过插件，它也支持许多其他编程语言和框架。IntelliJ IDEA 有商业版和社区版，其中社区版是免费的。本书的案例可以完全使用社区版编译和运行。

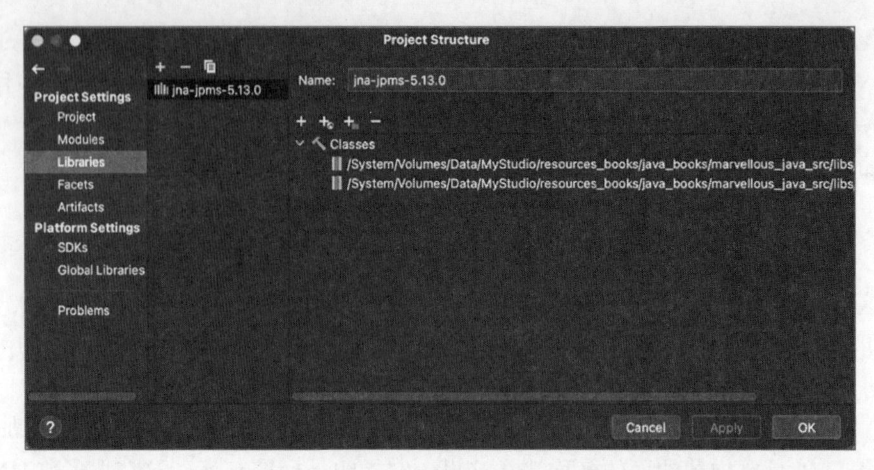

图 1-3　引用 jar 文件

下面的例子使用 JNA 获取 UNIX-like 系统中所有的用户名和相关信息。

代码位置：src/main/java/users/GetUsers.java

```java
package users;
import com.sun.jna.Library;
import com.sun.jna.Native;
import com.sun.jna.Structure;
import java.util.Arrays;
import java.util.List;

public class GetUsers {
    public interface CLibrary extends Library {
        CLibrary INSTANCE = (CLibrary) Native.load("c", CLibrary.class);

        // 定义 passwd 结构体，用于获取用户信息
        class Passwd extends Structure {
            public String pw_name;                  // 用户名
            public String pw_passwd;                // 用户密码
            public int pw_uid;                      // 用户 ID
            public int pw_gid;                      // 用户组 ID
            public String pw_gecos;
            public String pw_dir;                   // 用户家目录
            public String pw_shell;                 // 用户的默认 shell

            @Override
            protected List<String> getFieldOrder() {
                return Arrays.asList("pw_name", "pw_passwd", "pw_uid", "pw_gid",
"pw_gecos", "pw_dir", "pw_shell");
            }
        }
        void setpwent();
        Passwd getpwent();
        void endpwent();
```

```
    }
    public static void main(String[] args) {
        // 初始化获取用户信息
        CLibrary.INSTANCE.setpwent();
        CLibrary.Passwd user;
        // 遍历所有用户信息
        while ((user = CLibrary.INSTANCE.getpwent()) != null) {
            // 打印用户名和用户 ID
            System.out.println(user.pw_name + ": " + user.pw_uid);
        }
        // 结束获取用户信息
        CLibrary.INSTANCE.endpwent();
    }
}
```

运行程序，会在终端输出如下信息（部分内容）：

```
_appstore: 33
_mcxalr: 54
_appleevents: 55
_geod: 56
_devdocs: 59
_sandbox: 60
```

在这段代码中，定义了一个 Passwd 接口，这个接口是与 UNIX-like 系统中的 Passwd 结构体对应的。这个结构体通常在 pwd.h 头文件中定义，在不同的 UNIX-like 系统中，pwd.h 文件的位置不同，例如，如果在 macOS 系统中安装了 Command Line Tools 或 Xcode，那么 pwd.h 文件的路径如下：

```
/Library/Developer/CommandLineTools/SDKs/MacOSX.sdk/usr/include/pwd.h
```

打开 pwd.h 文件会看到如下的代码片段，其中有一个 passwd 结构体。Passwd 接口就是 passwd 结构体在 JNA 中的映射。

```
struct passwd {
  char    *pw_name;                  /* user name */
  char    *pw_passwd;                /* encrypted password */
  uid_t   pw_uid;                    /* user uid */
  gid_t   pw_gid;                    /* user gid */
  __darwin_time_t pw_change;         /* password change time */
  char    *pw_class;                 /* user access class */
  char    *pw_gecos;                 /* Honeywell login info */
  char    *pw_dir;                   /* home directory */
  char    *pw_shell;                 /* default shell */
  __darwin_time_t pw_expire;         /* account expiration */
};
#include <sys/cdefs.h>
```

```
__BEGIN_DECLS
... ...
struct passwd     *getpwent(void);
void               setpwent(void);
void               endpwent(void);
__END_DECLS
```

定义 Passwd 接口，需要了解如下几点。

（1）Passwd 接口的名称可以随便起，如称 UserInformation、UserDetails 都没有问题。

（2）如果 Passwd 中的字段和方法与 passwd 结构体中的成员与函数对应，那么名称必须相同，而且参数个数和数据类型也必须对应。

（3）Passwd 接口并不需要定义 passwd 结构体中的全部成员和相关函数，用哪一个定义哪一个。

（4）由于 C 结构体的字段是在内存中连续分配的，所以字段的顺序非常关键。如果 Java 类的字段顺序与原生 C 结构体的字段顺序不一致，那么当 JNA 尝试读取或写入这个结构体时，数据可能会被放到错误的位置。所以需要使用 getFieldOrder 方法让 Java 中的字段顺序与 C 结构体中的成员顺序保持一致。

1.4 使用 JNA 获取 Windows 系统中所有用户的信息

在 1.3 节给出的例子只能获取 UNIX-like 系统中的用户信息，并不能获取 Windows 系统中的用户信息。要想获取 Windows 系统中的用户信息，需要引用 JNA 中的另外两个 jar 文件：jna-platform-5.13.0.jar 和 jna-platform-jpms-5.13.0.jar。这两个 jar 文件包含了一些可跨平台的特性。读者可以从 1.3 节给出的 JNA Github 主页中下载这两个 jar 文件。

与获取 Windows 系统中用户信息相关的 API 的解释如下。

（1）Advapi32Util 类提供了许多有用的方法，允许 Java 程序与 Windows 系统中的 Advapi32.dll 库进行交互。这个库包含了与账户、用户组和安全相关的 API。

（2）Advapi32Util.getCurrentUserGroups() 方法获取的是当前 Windows 系统中用户所属的所有用户组。这些用户组信息存储在 Windows 系统的安全设置中，并决定了用户的许多权限。

（3）每个 Advapi32Util.Account 对象都代表一个用户组。它包含了用户组的名称（name）和安全标识符（sidString）。安全标识符是 Windows 系统用来识别用户、用户组和计算机账户的唯一值。

（4）Shell32Util.getFolderPath() 方法是与 Windows 系统的 Shell32.dll 库交互的。它允许程序获取特定的系统文件夹路径。这里我们用它来获取当前用户的主目录，获得的路径通常是 C:\Users\<用户名>。

下面的例子使用 JNA API 输出的 Windows 系统中所有用户的信息，以及当前用户的 Home 目录。

代码位置：src/main/java/users/GetWindowsUsers.java

```
// 导入必要的 JNA 库和 Windows API 相关的类
import com.sun.jna.platform.win32.Advapi32Util;
import com.sun.jna.platform.win32.WinNT;
import com.sun.jna.platform.win32.Shell32Util;
import com.sun.jna.platform.win32.ShlObj;

public class GetWindowsUsers {

    public static void main(String[] args) {
        // WinNT.HANDLEByReference 是一个指针或句柄的引用，通常用于 Windows API 函数中，
        // 当函数需要返回或修改一个句柄时使用
        // 这里我们创建一个句柄的引用，以备后用，但在此示例中我们实际上没有使用它
        WinNT.HANDLEByReference hToken = new WinNT.HANDLEByReference();

        // 使用 Advapi32Util.getCurrentUserGroups() 方法获取当前 Windows 系统中用户所属
        // 的所有用户组
        // 返回的是一个 Account 数组，每个 Account 对象都代表一个用户组
        Advapi32Util.Account[] accounts = Advapi32Util.getCurrentUserGroups();

        // 遍历 Account 数组，打印每个用户组的名称和 SID（安全标识符）
        for (Advapi32Util.Account account : accounts) {
            System.out.println("Name: " + account.name);       // 用户组的名称
            System.out.println("SID: " + account.sidString); // 用户组的安全标识符
        }

        // 使用 Shell32Util.getFolderPath() 方法获取当前 Windows 系统中用户的主目录
        // ShlObj.CSIDL_PROFILE 是一个常量，代表用户的主目录
        String homeDir = Shell32Util.getFolderPath(ShlObj.CSIDL_PROFILE);

        // 打印当前用户的主目录
        System.out.println("Home Directory: " + homeDir);
    }
}
```

运行程序，会输出如图 1-4 所示的内容（在读者的机器上输出的结果可能会有所差异）。

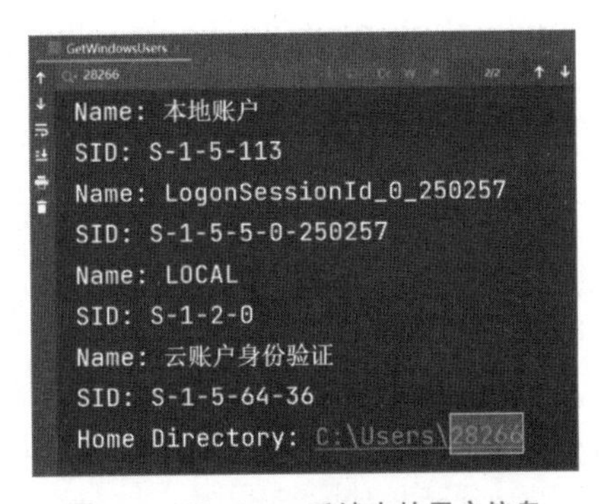

图 1-4　Windows 系统中的用户信息

1.5 修改文件的属性

本节会介绍修改文件的一些常用属性，这些属性包括创建时间、访问时间、修改时间、文件权限、文件所有者和组以及 Windows 的访问控制列表（Access Control List，ACL）。

1.5.1 修改文件的创建时间、访问时间和修改时间

使用 java.nio.file.attribute.BasicFileAttributeView.setTimes 方法可以修改文件或目录的时间戳。具体地说，该方法允许用户单独修改文件的创建时间、访问时间和编辑时间。下面是 setTimes 方法的原型：

```
void setTimes(FileTime lastModifiedTime,
              FileTime lastAccessTime,
              FileTime createTime)
       throws IOException
```

参数含义如下。

（1）lastModifiedTime：文件内容上次修改的时间。如果传入 null，则表示不更改这个时间。

（2）lastAccessTime：文件上次被访问的时间。注意，这并不意味着文件内容已被修改，只是表示文件被访问或读取了。同样，如果传入 null，则表示不更改这个时间。

（3）createTime：这是文件的创建时间。但需要注意的是，不是所有的文件系统都支持创建时间的概念。例如，POSIX 系统通常不支持真正的创建时间。此参数在大多数情况下应该传入 null，除非确实知道正在使用的文件系统支持并需要更改创建时间。

如果这些时间戳超出了文件系统所支持的范围或精度，那么会向上或向下四舍五入为最接近的值。

该方法可能抛出 IOException，这表示在尝试设置时间戳时出现了问题。可能的原因有许多，例如文件可能不存在、没有适当的权限或是其他 I/O 相关的问题。

下面的例子使用 setTimes 方法修改了文件的创建时间、访问时间和修改时间。

代码位置：src/main/java/file/attributes/FileTimeModifier.java

```
package file.attributes;

import java.io.IOException;
import java.nio.file.*;
import java.nio.file.attribute.BasicFileAttributeView;
import java.nio.file.attribute.FileTime;
import java.time.Instant;

public class FileTimeModifier {
    public static void main(String[] args) {
```

```
        Path filePath = Paths.get("test.txt");                        // 替换为文件路径

        Instant createTime = Instant.parse("2022-10-30T10:15:30.00Z");
        // 设置创建时间
        Instant accessTime = Instant.parse("2022-11-10T10:20:30.00Z");
        // 设置访问时间
        Instant modifiedTime = Instant.parse("2022-12-25T10:25:30.00Z");
        // 设置修改时间

        try {
            BasicFileAttributeView attributes = Files.getFileAttributeView(filePath,
                          BasicFileAttributeView.class);
            attributes.setTimes(FileTime.from(modifiedTime), FileTime.from
(accessTime),
                          FileTime.from(createTime));
            System.out.println("时间已成功修改！");
        } catch (IOException e) {
            System.out.println("修改时间出现问题：");
            e.printStackTrace();
        }
    }
}
```

运行程序之前，要保证当前目录存在 test.txt 文件。如果输出“时间已成功修改！”，就表明成功修改了 test.txt 文件的 3 个时间。

在前面的程序中，时间的格式使用了 2022-10-30T10:15:30.00Z，这是一个遵循 ISO 8601 日期和时间的表示法。其中 T 用于分隔日期和时间部分；Z 表示 UTC（协调世界时），当日期和时间后面跟着 Z 时，表示这个日期和时间是 UTC 时区的。也就是说，它没有时区偏移。所以，2022-10-30T10:15:30.00Z 可以解释为：2022 年 10 月 30 日，UTC 时区的上午 10 点 15 分 30 秒。

下面是在 Windows、macOS 和 Linux 系统下查询文件的创建时间、访问时间和修改时间的命令。

- Windows：

查询 test.txt 的创建时间：

```
dir /T:C test.txt
```

查询 test.txt 的访问时间：

```
dir /T:A test.txt
```

查询 test.txt 的修改时间：

```
dir /T:W test.txt
```

- macOS：

查询 test.txt 的创建时间：

```
date -r $(stat -f "%B" test.txt) "+%Y-%m-%d %H:%M:%S"
```

查询 test.txt 的访问时间：

```
date -r $(stat -f "%a" test.txt) "+%Y-%m-%d %H:%M:%S"
```

查询 test.txt 的修改时间：

```
date -r $(stat -f "%m" test.txt) "+%Y-%m-%d %H:%M:%S"
```

- Linux：

查询 test.txt 的创建时间（请注意，不是所有的文件系统都支持记录创建时间，例如 ext4 支持但是 ext3 不支持）：

```
stat --format="Birth: %w" test.txt
```

查询 test.txt 的访问时间：

```
stat --format="Access: %x" test.txt
```

查询 test.txt 的修改时间：

```
stat --format="Modify: %y" test.txt
```

1.5.2 修改文件权限

setPosixFilePermissions 是 java.nio.file.Files 类中的一个静态方法，用于更新文件的 POSIX 文件权限。

该方法的原型如下：

```
public static Path setPosixFilePermissions(Path path, Set<PosixFilePermission>
perms) throws IOException
```

参数含义如下。

（1）path：要修改其权限的文件或目录的路径。它是 java.nio.file.Path 类型的对象。

（2）perms：要设置的权限集。它是 java.nio.file.attribute.PosixFilePermission 枚举值的集合。这些枚举值描述了 POSIX 样式的文件权限，例如读取、写入和执行权限。

setPosixFilePermissions 方法返回更新后权限的文件的路径。通常，返回的路径与传递的路径相同。

setPosixFilePermissions 方法可能会抛出以下异常。

（1）IOException：如果发生 I/O 错误。

（2）UnsupportedOperationException：如果关联的文件系统不支持 POSIX 文件权限。

（3）SecurityException：如果安全管理器存在并且它拒绝写入文件权限。

PosixFilePermission 枚举中的可能值及其含义如下。

（1）OWNER_READ：拥有者有读取权限。

（2）OWNER_WRITE：拥有者有写入权限。

（3）OWNER_EXECUTE：拥有者有执行权限。

（4）GROUP_READ：所属组有读取权限。

（5）GROUP_WRITE：所属组有写入权限。

（6）GROUP_EXECUTE：所属组有执行权限。

（7）OTHERS_READ：其他用户有读取权限。

（8）OTHERS_WRITE：其他用户有写入权限。

（9）OTHERS_EXECUTE：其他用户有执行权限。

使用这些枚举值，可以创建一个 Set<PosixFilePermission>，然后传递给 setPosixFilePermissions 方法来更新文件或目录的权限。

下面的例子演示了如何修改文件的权限。

代码位置：src/main/java/file/attributes/ModifyFilePermissions.java

```
package file.attributes;

import java.nio.file.Files;
import java.nio.file.Path;
import java.nio.file.Paths;
import java.nio.file.attribute.PosixFilePermission;
import java.util.Set;
import java.util.HashSet;

public class ModifyFilePermissions {
    public static void main(String[] args) {
        try {
            // 定义文件路径
            Path filePath = Paths.get("example.txt");

            // 更改文件权限，使所有用户都可以读取和写入该文件
            Set<PosixFilePermission> permissions1 = new HashSet<>();
            permissions1.add(PosixFilePermission.OWNER_READ);
            permissions1.add(PosixFilePermission.OWNER_WRITE);
            permissions1.add(PosixFilePermission.GROUP_READ);
            permissions1.add(PosixFilePermission.GROUP_WRITE);
            permissions1.add(PosixFilePermission.OTHERS_READ);
            permissions1.add(PosixFilePermission.OTHERS_WRITE);
```

```
            Files.setPosixFilePermissions(filePath, permissions1);

            // 更改文件权限，使文件所有者可以读取、写入和执行该文件，其他用户只能读取该文件
            Set<PosixFilePermission> permissions2 = new HashSet<>();
            permissions2.add(PosixFilePermission.OWNER_READ);
            permissions2.add(PosixFilePermission.OWNER_WRITE);
            permissions2.add(PosixFilePermission.OWNER_EXECUTE);
            permissions2.add(PosixFilePermission.GROUP_READ);
            permissions2.add(PosixFilePermission.OTHERS_READ);
            Files.setPosixFilePermissions(filePath, permissions2);

        // 更改文件权限，使文件所有者和文件所属组可以读取和写入该文件，其他用户不能访问该文件
            Set<PosixFilePermission> permissions3 = new HashSet<>();
            permissions3.add(PosixFilePermission.OWNER_READ);
            permissions3.add(PosixFilePermission.OWNER_WRITE);
            permissions3.add(PosixFilePermission.GROUP_READ);
            permissions3.add(PosixFilePermission.GROUP_WRITE);
            Files.setPosixFilePermissions(filePath, permissions3);

        } catch (Exception e) {
            e.printStackTrace();
        }
    }
}
```

注意，上面的程序只能在 UNIX-like 系统（macOS、Linux 等）上运行。

1.5.3 修改文件的所有者和组

修改文件的所有者和组，可以直接执行 chown 命令，完整的命令行如下：

```
chown 272:4 example.txt
```

其中 272 是用户的 ID，4 是用户组的 ID。在 Java 中，可以使用 Runtime.getRuntime().exec(command) 方法执行外部命令。

下面的例子通过 chown 命令修改了 example.txt 文件的用户和用户组。

代码位置：src/main/java/file/attributes/ChangeFileOwnerAndGroup.java

```
package file.attributes;
public class ChangeFileOwnerAndGroup {
    public static void main(String[] args) {
        try {
            // 定义文件路径
            String filePath = "example.txt";
            // 手工设置用户 ID 和组 ID
            String userId = "272";                          // 请设置为正确的用户 ID
            String groupId = "4";                           // 请设置为正确的组 ID
```

```
            // 使用 chown 命令来更改文件的所有者和组
            String command = "chown " + userId + ":" + groupId + " " + filePath;
            Process process = Runtime.getRuntime().exec(command);
            // 等待命令执行完成
            process.waitFor();
            System.out.println(" 文件所有者和组已更改 ");
        } catch (Exception e) {
            e.printStackTrace();
        }
    }
}
```

如果执行上面的代码，抛出了 Operation not permitted 异常，说明执行的用户不是 root，需要使用 sudo 命令执行 ChangeFileOwnerAndGroup 类，现在进入 src/main/java 目录，然后执行下面的命令：

```
sudo java file/attributes/ChangeFileOwnerAndGroup
```

注意：如果直接通过 Java 命令运行 ChangeFileOwnerAndGroup，当前目录不再是原来的目录了，所以要确保新的当前目录存在 example.txt 文件，或者使用 example.txt 文件的绝对路径。

使用 Java 设置文件的所有者和用户组，还有另外一种方式，就是通过 Files.setOwner 方法设置文件的所有者，通过 PosixFileAttributeView.setGroup 方法设置的实现代码如下。

代码位置：src/main/java/file/attributes/ChangeFileOwnerAndGroup1.java

```
package file.attributes;

import java.nio.file.Files;
import java.nio.file.Path;
import java.nio.file.Paths;
import java.nio.file.attribute.GroupPrincipal;
import java.nio.file.attribute.PosixFileAttributeView;
import java.nio.file.attribute.UserPrincipal;
import java.nio.file.attribute.UserPrincipalLookupService;

public class ChangeFileOwnerAndGroup1 {

    public static void main(String[] args) {
        try {
            // 定义文件路径
            Path path = Paths.get("example.txt");
            // 获取 UserPrincipalLookupService 来查找用户和组
            UserPrincipalLookupService lookupService =
                    path.getFileSystem().getUserPrincipalLookupService();
            // 设置新的所有者
```

```
                UserPrincipal owner = lookupService.lookupPrincipalByName("_appinstalld");
                // 替换指定的用户名
                Files.setOwner(path, owner);
                // 设置新的组
                PosixFileAttributeView fileAttributeView = Files.getFileAttributeView
                                    (path, PosixFileAttributeView.class);
                GroupPrincipal group = lookupService.lookupPrincipalByGroupName
("staff");
                // 替换为指定的组名
                fileAttributeView.setGroup(group);
            } catch (Exception e) {
                e.printStackTrace();
            }
        }
    }
```

执行上面的代码同样需要 root 权限，需要使用下面的命令行才可以设置文件的所有者和用户组。

```
sudo java file/attributes/ChangeFileOwnerAndGroup1
```

注意：本节提供的两段代码只能在 UNIX-like 系统中运行。而且在运行代码之前，需要先修改代码中的所有者 ID、用户组 ID、所有者名称和用户组名称，要确保这些数据在自己的系统中存在。

如果读者不知道系统中有哪些用户，可以使用下面的命令获取用户和用户组的相关信息。

- macOS

获取所有用户的名称和 ID：

```
dscl . -list /Users UniqueID
```

获取所有组的名称和 ID：

```
dscl . -list /Groups PrimaryGroupID
```

- Linux

获取所有用户的名称和 ID：

```
awk -F':' '{ print $1, $3 }' /etc/passwd
```

获取所有组的名称和 ID：

```
awk -F':' '{ print $1, $3 }' /etc/group
```

1.5.4 修改 Windows 的访问控制列表

访问控制列表是一个由多个访问控制项组成的列表，用于定义资源的访问权限。有两种类型的 ACL：DACL 和 SACL。DACL 定义了哪些用户或组可以或不可以做什么操作；SACL 定义了哪些操作应当被审计。大多数情况下，除非特殊说明，否则 ACL 通常是指 DACL。

在 Windows 中，可以使用命令行工具 icacls 修改 ACL。icacls 用于显示和修改文件和目录的访问控制列表。它是从 Windows Vista 开始提供的，作为 cacls 命令的替代品。

- icacls 的主要用途

（1）显示文件或目录的 ACL。

（2）替换文件或目录的 ACL。

（3）修改文件或目录的 ACL，例如添加或删除访问规则。

（4）保存文件或目录的 ACL 到一个文本文件。

（5）从一个之前保存的文本文件恢复文件或目录的 ACL。

- 常用参数

（1）/T：递归地处理指定的目录和其子目录。

（2）/C：在处理多个文件时，即使发生错误也继续。

（3）/L：在符号链接本身上操作，而不是在其目标上。

（4）/Q：安静模式，不显示被修改的文件名。

（5）/P user:perm：为指定的用户添加 / 替换权限。

（6）/R user：撤销指定用户的权限。

（7）/GRANT user:perm：明确为用户授予权限。

（8）/DENY user:perm：明确拒绝用户的权限。

（9）/REMOVE user：删除所有匹配用户的访问控制项（ACE）。

（10）/REVOKE user：撤销指定用户的明确的允许或拒绝的权限。

- 权限说明

（1）R：只读。

（2）W：写入。

（3）C：更改（写入）。

（4）F：完全控制。

- 示例

为指定用户添加完全控制权限：

```
icacls "C:\path\to\file_or_folder" /grant User:F
```

递归地为指定用户添加只读权限到一个文件夹和其所有子项：

```
icacls "C:\path\to\folder" /T /grant User:R
```

删除指定用户的所有权限：

```
icacls "C:\path\to\file_or_folder" /remove User
```

保存文件的 ACL 到一个文本文件：

```
icacls "C:\path\to\file_or_folder" /save ACLFile.txt
```

从文本文件恢复文件的 ACL：

```
icacls "C:\path\to\file_or_folder" /restore ACLFile.txt
```

下面的例子使用 icacls 工具为 demo.png 添加 Everyone 用户并设置读写权限。

代码位置：src/main/java/file/attributes/WindowsFilePermission.java

```
package file.attributes;
import java.io.IOException;
public class WindowsFilePermission {
    public static void main(String[] args) {
        String filePath = "d:\\data\\demo.png";
        addEveryoneReadWritePermission(filePath);
    }
    /**
     * 使用 icacls 命令为文件添加 Everyone 用户并设置读写权限
     * @param filePath 文件的完整路径
     */
    public static void addEveryoneReadWritePermission(String filePath) {
        try {
            // 调用 icacls 命令为文件设置权限
            Process process = new ProcessBuilder("cmd.exe", "/c", "icacls",
                    filePath, "/grant", "Everyone:(R,W)").start();
            process.waitFor();
            int exitValue = process.exitValue();
            if (exitValue == 0) {
                System.out.println("成功为文件" + filePath + "添加 Everyone 用户并设
                                置读写权限。");
            } else {
                System.err.println("设置文件权限失败，退出代码：" + exitValue);
            }
        } catch (IOException | InterruptedException e) {
            e.printStackTrace();
        }
    }
}
```

这段代码必须在 Windows 系统下运行，在运行之前，确保 d:\data 目录存在 demo.png 文件。运行代码之前，demo.png 文件的“安全”页面信息如图 1-5 所示。

设置后的“安全”页面如图 1-6 所示。

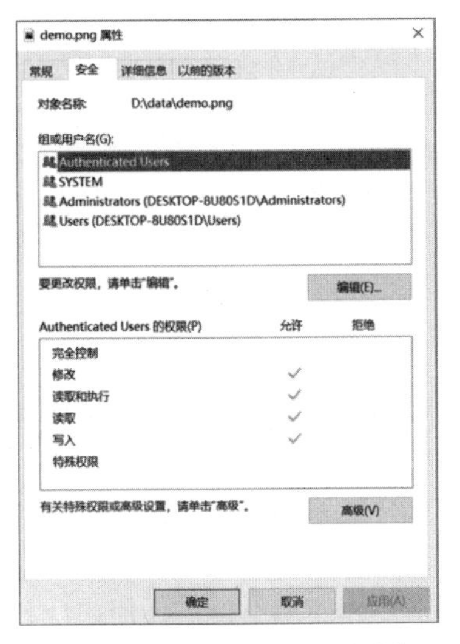

图 1-5　demo.png 设置前的“安全”页面

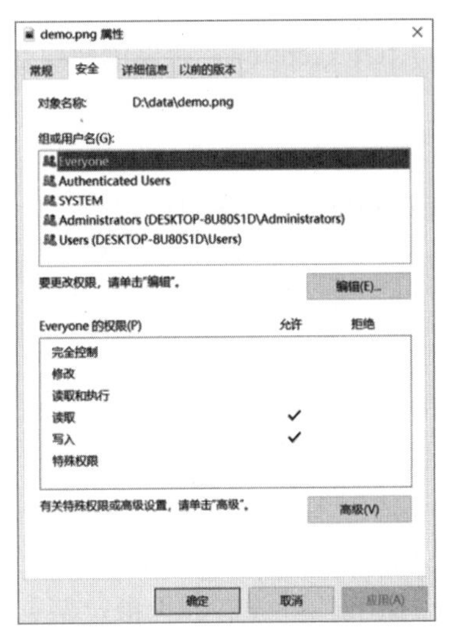

图 1-6　demo.png 设置后的“安全”页面

很明显，设置后的“安全”页面多了 Everyone 用户，而且是读写权限。

1.6　文件和目录的相关操作

本节会介绍一些常用的文件和目录操作，如创建文件和目录、删除文件和目录、复制文件和目录、重命名文件和目录、搜索文件和目录、创建快捷方式。

1.6.1　创建文件和目录

使用 java.nio.file.Files 类中的 createDirectories 方法和 createFile 方法可以创建目录和文件，下面是对这两个方法的详细介绍。

createDirectories 方法的原型如下。

```
public static Path createDirectories(Path dir, FileAttribute<?>... attrs)
throws IOException
```

功能：此方法用于创建目录，包括任何必要但不存在的父目录。如果目录因为已经存在或其他原因而无法创建，则会抛出异常。

参数的详细含义如下。

（1）dir：要创建的目录的路径。

（2）attrs：可选参数，指定新创建的目录的文件属性。如果不提供任何属性，将使用文件系统的默认属性。

（3）返回值：返回一个 Path 对象，表示新创建的目录（或已存在的目录）的路径。

可能抛出的异常如下。

（1）IOException：如果创建目录的过程中出现 I/O 错误。

（2）UnsupportedOperationException：如果指定的属性不受支持。

（3）FileAlreadyExistsException：如果目录无法创建，因为已经存在一个非目录文件。

createFile 方法的原型如下。

```
public static Path createFile(Path path, FileAttribute<?>... attrs) throws
IOException
```

功能：此方法用于创建一个新的空文件，如果文件因为已经存在或因其他原因而无法创建，则会抛出异常。

参数的详细含义如下。

（1）path：要创建的文件的路径。

（2）attrs：可选参数，指定新创建的文件的文件属性。如果不提供任何属性，将使用文件系统的默认属性。

（3）返回值：返回一个 Path 对象，表示新创建的文件的路径。

可能抛出的异常如下。

（1）IOException：如果创建文件的过程中出现 I/O 错误。

（2）UnsupportedOperationException：如果指定的属性不受支持。

（3）FileAlreadyExistsException：如果文件无法创建，因为已经存在一个文件。

这两个方法都可以接收一个或多个 FileAttribute 对象作为参数，以指定新创建的目录或文件的属性。这允许在创建时设置例如文件权限或所有者等属性。如果不指定这些属性，将使用文件系统的默认属性。在大多数情况下，可能不需要提供任何文件属性，因此这些参数是可选的。

下面的例子完整地演示了如何创建目录以及多层目录，并且创建新文件，同时写入一个字符串。

代码位置：src/main/java/file/DirectoryFileCreator.java

```
package file;

import java.io.IOException;
import java.nio.file.FileAlreadyExistsException;
import java.nio.file.Files;
import java.nio.file.Path;
import java.nio.file.Paths;
```

```
public class DirectoryFileCreator {

    public static void main(String[] args) {
        try {
            // 创建目录
            createDirectory("my_directory");

            // 创建文件并写入内容
            createFile("my_directory/my_file.txt", "Hello World!");

            // 创建多层目录
            createDirectory("my_directory/subdirectory/subsubdirectory");

            // 创建文件并写入内容
            createFile("my_directory/subdirectory/subsubdirectory/my_file.txt",
"世界你好！");
        } catch (IOException e) {
            e.printStackTrace();
        }
    }

    /**
     * 创建目录的方法
     * @param dirPath 目录的路径
     * @throws IOException 如果创建目录过程中出现错误
     */
    private static void createDirectory(String dirPath) throws IOException {
        Path path = Paths.get(dirPath);
        if (Files.notExists(path)) {
            Files.createDirectories(path);
        }
    }

    /**
     * 创建文件并写入内容的方法
     * @param filePath 文件的路径
     * @param content 写入文件的内容
     * @throws IOException 如果创建文件或写入内容过程中出现错误
     */
    private static void createFile(String filePath, String content) throws
IOException {
        Path path = Paths.get(filePath);
        try {
            Files.createFile(path);
        } catch (FileAlreadyExistsException e) {
            // 文件已存在，无须创建
        }
        Files.write(path, content.getBytes());
    }
}
```

运行程序，就会在当前目录创建子目录和文件，如图 1-7 所示。

图 1-7 创建的子目录和文件

1.6.2 删除文件和目录

可以使用 java.nio.file.Files 类的 delete 方法删除目录和文件，不过 delete 方法只能删除空目录，如果删除非空目录，会抛出 DirectoryNotEmptyException 异常，所以如果要删除非空目录，就需要使用遍历非空目录中的所有子目录，然后删除非空目录中所有子目录和文件后，让非空目录变成空目录后，才能使用 delete 方法删除这个目录。

为了方便，可以编写一个 deleteNonEmptyDirectory 方法，用于删除非空目录。该方法通过 java.nio.file.Files 类的 walkFileTree 方法递归遍历所有的子目录和文件，这个方法的原型如下。

```
public static Path walkFileTree(Path start, FileVisitor<? super Path> visitor)
throws IOException
```

参数含义如下。

（1）start：要遍历的起始路径，也就是要删除的非空目录的路径。

（2）visitor：一个 FileVisitor 实例，它指定了如何访问每个文件和目录。

FileVisitor 是一个接口，它定义了在遍历文件树时如何处理文件和目录。我们通过创建 SimpleFileVisitor 的匿名子类来提供 FileVisitor 实例，并覆盖了如下两个方法。

（1）visitFile(Path file, BasicFileAttributes attrs)：此方法在访问单个文件时被调用。这里调用 Files.delete 方法来删除每个文件。

（2）postVisitDirectory(Path dir, IOException exc)：此方法在访问完一个目录的所有文件和子目录后被调用。这里调用 Files.delete 方法来删除目录。

deleteNonEmptyDirectory 方法的工作方式是从指定的非空目录开始，首先删除所有文件，然后递归地删除所有子目录。每个子目录都是在删除其内容后被删除的，从而确保整个目录树被完全删除。

下面的例子删除了文件、空目录和非空目录。

代码位置：src/main/java/file/DirectoryFileDeleter.java

```
package file;

import java.io.IOException;
```

```
import java.nio.file.*;
import java.nio.file.attribute.BasicFileAttributes;
public class DirectoryFileDeleter {
    public static void main(String[] args) {
        try {
            // 永久删除文件
            deleteFile("file.txt");
            // 永久删除空目录
            deleteEmptyDirectory("empty_directory");
            // 永久删除非空目录
            deleteNonEmptyDirectory("non_empty_directory");
        } catch (IOException e) {
            e.printStackTrace();
        }
    }
    /**
     * 删除文件的方法
     * @param filePath 文件的路径
     * @throws IOException 如果删除文件过程中出现错误
     */
    private static void deleteFile(String filePath) throws IOException {
        Path path = Paths.get(filePath);
        if (Files.exists(path)) {
            Files.delete(path);
        }
    }
    /**
     * 删除空目录的方法
     * @param dirPath 空目录的路径
     * @throws IOException 如果删除空目录过程中出现错误
     */
    private static void deleteEmptyDirectory(String dirPath) throws IOException {
        Path path = Paths.get(dirPath);
        if (Files.exists(path)) {
            Files.delete(path);
        }
    }
    /**
     * 删除非空目录的方法
     * @param dirPath 非空目录的路径
     * @throws IOException 如果删除非空目录过程中出现错误
     */
    private static void deleteNonEmptyDirectory(String dirPath) throws
IOException {
        Path path = Paths.get(dirPath);
        if (Files.exists(path)) {
            Files.walkFileTree(path, new SimpleFileVisitor<Path>() {
                @Override
                public FileVisitResult visitFile(Path file, BasicFileAttributes
attrs) throws IOException
                {
```

```
                    Files.delete(file);
                    return FileVisitResult.CONTINUE;
                }
                @Override
                public FileVisitResult postVisitDirectory(Path dir,
IOException exc) throws
                IOException {
                    Files.delete(dir);
                    return FileVisitResult.CONTINUE;
                }
            });
        }
    }
}
```

运行程序之前，在当前目录创建一个 file.txt 文件、一个空目录（empty_directory）和一个非空目录（non_empty_directory），然后运行程序，会看到这文件和目录都被删除了。

1.6.3 复制文件和目录

使用 java.nio.file.Files 类的 copy 方法可以直接复制文件，但 Files 类并没有提供直接复制目录的方法，所以需要自己使用 Files.walkFileTree 方法实现，该方法用于遍历指定目录中所有的文件和子目录（包括根目录自身）。copy 方法和 walkFileTree 方法的详细解释如下。

❑ copy 方法

原型如下。

```
public static Path copy(Path source, Path target, CopyOption... options) throws
IOException
```

参数含义如下。

（1）source：源文件的路径。

（2）target：目标文件的路径。

（3）options：复制选项，可以设置的值包括 StandardCopyOption.REPLACE_EXISTING（如果目标文件已经存在，就替换它）、StandardCopyOption.COPY_ATTRIBUTES（复制文件的属性，如修改时间、访问时间、权限等）和 StandardCopyOption. ATOMIC_MOVE（将文件的移动操作作为一个原子文件系统操作去执行）。

❑ walkFileTree 方法

原型如下。

```
public static Path walkFileTree(Path start, FileVisitor<? super Path> visitor)
throws IOException
```

参数含义如下。

（1）start：开始遍历的目录的路径。

（2）visitor：一个实现了 FileVisitor 接口的对象，用于访问遍历过程中遇到的每个文件和目录。

下面的例子在当前目录创建了若干文件和目录，然后将这些文件和目录复制到另一个位置。

代码位置：src/main/java/file/FileAndDirectoryOperations.java

```
package file;
import java.io.IOException;
import java.nio.charset.Charset;
import java.nio.file.*;
import java.nio.file.attribute.BasicFileAttributes;

public class FileAndDirectoryOperations {
    public static void main(String[] args) {
        // 创建目录
        createDirectories(Paths.get("my_directory"));
        // 创建文件并写入内容
        createFile(Paths.get("my_file.txt"), "Hello World!");
        // 创建多层目录
        createDirectories(Paths.get("my_directory", "subdirectory",
"subsubdirectory"));
        // 在多层目录中创建文件并写入内容
        createFile(Paths.get("my_directory", "subdirectory", "subsubdirectory",
"my_file.txt"), "世界你好！");
        // 将 'my_file.txt' 文件复制到 'destination.txt'
        copyFile(Paths.get("my_file.txt"), Paths.get("destination.txt"));
        // 将 'my_directory' 目录复制到 'destination_directory'
        copyDirectory(Paths.get("my_directory"), Paths.get
("destination_directory"));
    }
    public static void createDirectories(Path path) {
        try {
            Files.createDirectories(path);
            System.out.println("目录 " + path + " 已创建");
        } catch (IOException e) {
            System.err.println("创建目录时出错：" + e.getMessage());
        }
    }
    public static void createFile(Path path, String content) {
        try {
            Files.write(path, content.getBytes(Charset.forName("UTF-8")),
          StandardOpenOption.CREATE, StandardOpenOption.TRUNCATE_EXISTING);
            System.out.println("文件 " + path + " 已创建并写入内容");
        } catch (IOException e) {
            System.err.println("创建文件时出错：" + e.getMessage());
        }
```

```
    }

    public static void copyFile(Path source, Path destination) {
        try {
            Files.copy(source, destination, StandardCopyOption.
REPLACE_EXISTING);
            System.out.println("文件 " + source + " 已复制到 " + destination);
        } catch (IOException e) {
            System.err.println("复制文件时出错: " + e.getMessage());
        }
    }

    public static void copyDirectory(Path source, Path destination) {
        try {
            Files.walkFileTree(source, new CopyDirVisitor(source,
destination, StandardCopyOption.REPLACE_EXISTING));
            System.out.println("目录 " + source + " 已复制到 " + destination);
        } catch (IOException e) {
            System.err.println("复制目录时出错: " + e.getMessage());
        }
    }
    static class CopyDirVisitor extends SimpleFileVisitor<Path> {
        private final Path fromPath;
        private final Path toPath;
        private final CopyOption copyOption;

        CopyDirVisitor(Path fromPath, Path toPath, CopyOption copyOption) {
            this.fromPath = fromPath;
            this.toPath = toPath;
            this.copyOption = copyOption;
        }

        @Override
        public FileVisitResult preVisitDirectory(Path dir, BasicFileAttributes
attrs) throws IOException {
            Path targetPath = toPath.resolve(fromPath.relativize(dir));
            if(!Files.exists(targetPath)) {
                Files.createDirectory(targetPath);
            }
            return FileVisitResult.CONTINUE;
        }

        @Override
        public FileVisitResult visitFile(Path file, BasicFileAttributes attrs)
throws IOException {
            Files.copy(file, toPath.resolve(fromPath.relativize(file)), copyOption);
            return FileVisitResult.CONTINUE;
        }
    }
}
```

运行程序，会在终端输出如下内容：

```
目录 my_directory 已创建
文件 my_file.txt 已创建并写入内容
目录 my_directory/subdirectory/subsubdirectory 已创建
文件 my_directory/subdirectory/subsubdirectory/my_file.txt 已创建并写入内容
文件 my_file.txt 已复制到 destination.txt
目录 my_directory 已复制到 destination_directory
```

在上面的代码中，CopyDirVisitor 实现了 FileVisitor 接口，用于遍历文件系统的目录结构，并对遍历到的每个文件和目录执行特定的操作。在给出的 Java 代码中，CopyDirVisitor 类被设计用于复制目录及其子目录和文件。

下面详细介绍 CopyDirVisitor 类及其方法的实现原理。

CopyDirVisitor 类的构造方法如下。

```
CopyDirVisitor(Path fromPath, Path toPath, CopyOption copyOption)
```

构造方法接收 3 个参数：fromPath 表示源目录路径，toPath 表示目标目录路径，copyOption 表示复制操作的选项(例如是否替换现有文件)。这些参数在类的实例中被保存，以便在后续的方法调用中使用。

- preVisitDirectory 方法

```
public FileVisitResult preVisitDirectory(Path dir, BasicFileAttributes attrs)
throws IOException
```

这个方法在遍历到每个目录之前被调用。它使用 Files.createDirectory 方法创建目标目录（如果目标目录不存在）。方法体中，toPath.resolve(fromPath.relativize(dir)) 计算了目标目录的完整路径，然后检查这个目录是否已经存在，如果不存在，则创建它。

- visitFile 方法

```
public FileVisitResult visitFile(Path file, BasicFileAttributes attrs) throws
IOException
```

这个方法在遍历到每个文件时被调用。它使用 Files.copy 方法复制文件。方法体中，toPath.resolve(fromPath.relativize(file)) 计算了目标文件的完整路径，然后调用 Files.copy 方法将源文件复制到目标路径。

CopyDirVisitor 类通过实现 FileVisitor 接口，为 Files.walkFileTree 方法提供了操作文件和目录的具体实现。这使得 CopyDirVisitor 类能够在遍历文件系统目录结构时，按照指定的方式复制文件和目录，实现了目录复制的功能。

1.6.4 重命名文件和目录

使用 java.nio.file.Files 类中的 move 方法可以移动或重命名文件和目录[①]。copy 方法的原型如下。

```
public static Path move(Path source, Path target, CopyOption... options) throws
IOException
```

参数含义如下。

（1）source：源文件或目录的路径。这是要被移动或重命名的文件或目录。

（2）target：目标文件或目录的路径。这是源文件或目录移动或重命名后的新路径。

（3）options：一个可变参数，它指定了移动操作的选项。CopyOption 是一个接口，它有两个实现类：StandardCopyOption 和 LinkOption。这些选项可以影响 move 方法的行为。

下面的例子使用 move 方法将 my_file.txt 文件重命名为 new_file.txt，将非空目录 my_directory 重命名为 new_directory。

代码位置：src/main/java/file/RenameOperations.java

```
package file;
import java.io.IOException;
import java.nio.file.*;
public class RenameOperations {
    public static void main(String[] args) {
        // 将 my_file.txt 文件重命名为 my_file.txt
        rename(Paths.get("my_file.txt"), Paths.get("my_file.txt"));
        // 将 my_directory 目录重命名为 new_directory
        rename(Paths.get("my_directory"), Paths.get("new_directory"));
    }
    /**
     * 重命名文件或目录
     * @param source 源文件或目录的路径
     * @param target 新的文件或目录的路径
     */
    public static void rename(Path source, Path target) {
        try {
            // 使用 Files.move 方法实现重命名
            Files.move(source, target, StandardCopyOption.REPLACE_EXISTING);
            System.out.println(source + " 已重命名为 " + target);
        } catch (IOException e) {
            System.err.println("重命名时出错：" + e.getMessage());
```

① move 方法在移动非空目录时，根据不同操作系统平台以及 JRE 的版本不同，可能表现会有所不同。但通常在同一个逻辑磁盘内移动非空目录没有问题。例如，在 Windows 中的 D 盘内移动非空目录。但如果跨盘符移动非空目录，可能会抛出异常，如在 Windows 中从 D 盘将非空目录移动到 E 盘。

```
            }
        }
    }
```

在运行程序之前，应该确保当前目录存在 my_file.txt 文件和 my_directory 目录。现在运行程序，会在终端输出如下内容：

```
my_file.txt 已重命名为 my_file.txt
my_directory 已重命名为 new_directory
```

1.6.5　搜索文件和目录

在前面曾多次使用 Files.walkFileTree 方法，该方法的主要功能是递归遍历指定目录中的所有子目录和文件。所以可以用该方法获取想要的文件和目录名。

通过 FileVisitor.visitFile 方法可以获取当前扫描到的文件名，通过 FileVisitor.postVisitDirectory 方法可以获取当前扫描到的目录名。

下面的例子使用 Files.walkFileTree 方法递归扫描当前目录中所有的 .class 文件，并将扫描到的 .class 文件名输出到终端。

代码位置：src/main/java/file/FileSearcher.java

```
package file;
import java.io.File;
import java.io.IOException;
import java.nio.file.FileVisitOption;
import java.nio.file.FileVisitResult;
import java.nio.file.FileVisitor;
import java.nio.file.Files;
import java.nio.file.Path;
import java.nio.file.Paths;
import java.nio.file.attribute.BasicFileAttributes;
import java.util.ArrayList;
import java.util.EnumSet;
import java.util.List;
import java.util.regex.Pattern;

public class FileSearcher {
    // 搜索与pattern匹配的文件和目录
    public static List<String> find(String pattern, String path) {
        Pattern compiledPattern = Pattern.compile(
                pattern.replace(".", "\\.")
                        .replace("?", ".")
                        .replace("*", ".*"));
        List<String> result = new ArrayList<>();
        try {
```

```
            Files.walkFileTree(Paths.get(path), EnumSet.noneOf(FileVisitOption.
class),
                                        Integer.MAX_VALUE,
                    new FileVisitor<Path>() {
                        @Override
                        public FileVisitResult preVisitDirectory(Path dir,
BasicFileAttributes attrs) {
                            return FileVisitResult.CONTINUE;
                        }
                        // 扫描文件
                        @Override
                        public FileVisitResult visitFile(Path file,
BasicFileAttributes attrs) {
                            if (compiledPattern.matcher(file.getFileName().
toString()).matches()) {
                                result.add(file.toString());
                            }
                            return FileVisitResult.CONTINUE;
                        }

                        @Override
                        public FileVisitResult visitFileFailed(Path file,
IOException exc) {
                            return FileVisitResult.CONTINUE;
                        }
                        // 扫描目录
                        @Override
                        public FileVisitResult postVisitDirectory(Path dir,
IOException exc) {
                            if (compiledPattern.matcher(dir.getFileName().
toString()).matches()) {
                                result.add(dir.toString());
                            }
                            return FileVisitResult.CONTINUE;
                        }
                    });
        } catch (IOException e) {
            e.printStackTrace();
        }
        return result;
    }
    public static void main(String[] args) {
        // 搜索指定路径中所有的文本文件
        List<String> result = find("*.class", ".");
        for (String file : result) {
            System.out.println(file);
        }
    }
}
```

运行程序，会在终端输出如下内容（这里只展示部分内容）：

```
./out/production/marvellous_java_src/file/DirectoryFileCreator.class
./out/production/marvellous_java_src/file/DirectoryFileDeleter.class
./out/production/marvellous_java_src/file/DirectoryFileDeleter$1.class
./out/production/marvellous_java_src/file/FileSearcher$1.class
./out/production/marvellous_java_src/file/FileAndDirectoryOperations.class
./out/production/marvellous_java_src/users/GetUsers$CLibrary.class
```

1.6.6　创建快捷方式

在 Windows、macOS 和 Linux 中，要使用不同的方式创建快捷方式。

- 创建 Windows 快捷方式

在 Windows 系统中，可以使用 powershell 命令创建快捷方式。

```
powershell "$s= (New-Object -COM
WScript.Shell).CreateShortcut('D:\\data\\demo.lnk');$s.TargetPath=
'D:\\data\\demo.png';$s.Save()"
```

上面的命令为 d:\data\data.png 文件创建了 d:\data\demo.lnk 快捷方式文件。Windows 中的快捷方式文件的扩展名是 lnk。双击 demo.lnk 文件，就可以查看 demo.png 图像文件的内容。

注意，在执行 CreateShortcut 方法和 Save 方法时，左侧圆括号与方法名之间不能有空格，否则会抛出异常。

- 创建 macOS 快捷方式

在 macOS 系统中，需要创建一个简单的 bash 脚本来实现快捷方式功能。可以使用 Java 的标准 API 来创建一个新文件，并向其中写入内容。可以直接使用 sh 命令执行 bash 脚本文件。

- 创建 Linux 快捷方式

在 Linux 系统中，需要创建一个桌面入口文件（.desktop 文件）来实现快捷方式功能。下面是一个标准的 Linux 快捷方式文件：

```
[Desktop Entry]
Type=Link
URL=/root/MyStudio/Java/java_knowledge/photo.jpg
Name=photo.jpg
```

下面的例子定义了一个 createShortcut 方法，该方法接收两个参数：target 和 shortcutLocation。target 参数表示快捷方式指向的目标文件路径，而 shortcutLocation 参数表示快捷方式文件的保存路径。在方法内部，分别调用了 createWindowsShortcut 方法、createMacShortcut 方法和 createLinuxShortcut 方法创建了 Windows 快捷方式、macOS 快捷方式和 Linux 快捷方式。

代码位置：src/main/java/file/ShortcutCreator.java

```
package file;
import java.io.*;
import java.nio.file.*;
import java.nio.file.attribute.PosixFilePermissions;
public class ShortcutCreator {
    public static void createShortcut(String target, String shortcutLocation)
throws IOException {
        String os = System.getProperty("os.name").toLowerCase();
        if (os.contains("win")) {
            // 在 Windows 下创建快捷方式
            createWindowsShortcut(target, shortcutLocation);
        } else if (os.contains("mac")) {
            // 在 macOS 下创建快捷方式
            createMacShortcut(target, shortcutLocation);
        } else {
            // 在 Linux 下创建快捷方式
            createLinuxShortcut(target, shortcutLocation);
        }
    }
    private static void createWindowsShortcut(String target, String
shortcutLocation) throws IOException {
        // 构建 PowerShell 命令
        String command = String.format(
                "powershell \"$s=(New-Object -COM WScript.Shell).
CreateShortcut('%s');$s.TargetPath='%s';$s.Save()\"",
                shortcutLocation.replace("\\", "\\\\"),        // 转义反斜杠
                target.replace("\\", "\\\\")                   // 转义反斜杠
        );
        System.out.println(command);
        // 使用 ProcessBuilder 执行命令
        ProcessBuilder processBuilder = new ProcessBuilder("cmd.exe", "/c",
command);
        processBuilder.redirectErrorStream(true);       // 合并错误流和输出流
        Process process = processBuilder.start();       // 启动进程

        // 等待进程完成
        try {
            int exitCode = process.waitFor();
            if (exitCode != 0) {
                throw new IOException("Failed to create shortcut, exit code: "
+ exitCode);
            }
        } catch (InterruptedException e) {
            throw new IOException("Interrupted while waiting for process to
finish", e);
        }
    }
```

```
    private static void createMacShortcut(String target, String
shortcutLocation) throws IOException {
        // 获取目标文件的绝对路径
        Path targetPath = Paths.get(target).toAbsolutePath();
        // 创建快捷方式文件
        try (BufferedWriter writer = Files.newBufferedWriter(Paths.get
(shortcutLocation))) {
            writer.write("#!/bin/sh\nopen \"" + targetPath + "\"\n");
        }
        // 设置快捷方式文件的权限
        Files.setPosixFilePermissions(Paths.get(shortcutLocation),
PosixFilePermissions.fromString("rwxr-xr-x"));
    }
    private static void createLinuxShortcut(String target, String
shortcutLocation) throws IOException {
        // 获取目标文件的绝对路径
        Path targetPath = Paths.get(target).toAbsolutePath();
        // 创建快捷方式文件
        try (BufferedWriter writer = Files.newBufferedWriter(Paths.get
(shortcutLocation))) {
            writer.write("[Desktop Entry]\nType=Link\nURL=" + targetPath +
"\nName=" + targetPath.getFileName() + "\n");
        }
    }
    public static void main(String[] args) throws IOException {
        // 创建 macOS 快捷方式
        createShortcut("/Volumes/backup/data.xlsx", "/Volumes/backup/data.sh");
    }
}
```

1.7　回收站

本节会详细介绍如何操作 Windows、macOS 和 Linux 平台的回收站[①]，包括将文件和目录放入回收站、清除回收站中的内容以及恢复回收站中的文件和目录。

1.7.1　将文件和目录放入回收站

使用 java.awt.Desktop.moveToTrash 方法可以将文件放入回收站（对目录也同样有效），该方法需要传入一个 File 类的实例，用来描述要放入回收站的文件的原路径。具体的实现代码如下：

① 回收站在不同平台的叫法不同，如在 macOS 中称废纸篓，其实都是一回事。为了方便，本书统一称为回收站。

代码位置：src/main/java/trash/MoveToTrash.java

```
package trash;
import java.awt.*;
import java.io.File;
public class MoveToTrash {
    public static void main(String[] args) {
        File file = new File("D:\\data\\file.txt");      // 替换为想要删除的文件的路径
        moveToTrash(file);
    }

    public static void moveToTrash(File file) {
        if (file.exists()) {
            try {
                Desktop desktop = Desktop.getDesktop();
                boolean result = desktop.moveToTrash(file);
                if (result) {
                    System.out.println("File moved to trash successfully!");
                } else {
                    System.err.println("Failed to move file to trash");
                }
            } catch (UnsupportedOperationException e) {
                System.err.println("Move to trash operation is not supported on
the current platform");
            }
        } else {
            System.err.println("File does not exist");
        }
    }
}
```

尽管上面的代码是跨平台的，但在某些平台，如 Linux 中，可能会不好使，例如，在某些 Linux 版本中可能会显示如下的提示：

```
Move to trash operation is not supported on the current platform
```

为了让程序尽可能支持不同的操作系统平台，可以直接调用平台本身的命令行程序，或者直接调用当前平台的 API，将文件放入回收站。下面是不同平台的处理方式。

- Windows

使用 JNA 库来调用 Windows 的 Shell API 函数 SHFileOperation，这个函数提供了执行各种文件操作的能力，包括复制、移动、删除文件等。在这种情况下，我们配置 SHFileOperation 函数来将文件移动到回收站，而不是永久删除它。

- macOS

macOS 的回收站目录是 ~/.Trash，所以可以直接使用 mv 命令将文件移动到回收站目录。

❑ Linux

Linux 的回收站目录是 ~/.local/share/Trash/files，所以与 macOS 类似，可以使用 mv 命令将文件移动到回收站目录。

下面的例子使用上述方式将文件放入了回收站目录。

代码位置：src/main/java/trash/MoveToTrash1.java

```
package trash;
import java.awt.*;
import java.io.File;
import java.io.IOException;
import com.sun.jna.*;
import com.sun.jna.platform.win32.*;
import com.sun.jna.platform.win32.ShellAPI.SHFILEOPSTRUCT;
import com.sun.jna.ptr.IntByReference;
public class MoveToTrash1 {
    public static void main(String[] args) {
        File file = new File("D:\\data\\test.txt");      // 替换为想要删除的文件的路径
        moveToTrashWindows(file);
    }
    public static void moveToTrashWindows(File file) {
        if (file.exists()) {
            WString from = new WString(file.getAbsolutePath() + '\0');
            SHFILEOPSTRUCT fileop = new SHFILEOPSTRUCT();
            fileop.wFunc = ShellAPI.FO_DELETE;
            fileop.pFrom = from.toString();
            fileop.fFlags = ShellAPI.FOF_ALLOWUNDO | ShellAPI.FOF_NOCONFIRMATION;

            int result = Shell32.INSTANCE.SHFileOperation(fileop);
            if (result == 0) {
                System.out.println("File moved to trash successfully!");
            } else {
                System.err.println("Failed to move file to trash, error code: "
 + result);
            }
        } else {
            System.err.println("File does not exist");
        }
    }
    public static void moveToTrashMac(File file) {
        try {
            String cmd = "mv " + file.getAbsolutePath() + " ~/.Trash";
            System.out.println(cmd);
            Runtime.getRuntime().exec(new String[]{"/bin/sh", "-c", cmd});
            System.out.println("File moved to trash successfully!");
        } catch (IOException e) {
            e.printStackTrace();
            System.err.println("Failed to move file to trash");
        }
    }
```

```
    public static void moveToTrashLinux(File file) {
        try {
            String cmd = "mv " + file.getAbsolutePath() + " ~/.local/share/Trash
/files";
            System.out.println(cmd);
            Runtime.getRuntime().exec(new String[]{"/bin/sh", "-c", cmd});
            System.out.println("File moved to trash successfully!");
        } catch (IOException e) {
            e.printStackTrace();
            System.err.println("Failed to move file to trash");
        }
    }
}
```

上面代码中比较复杂的是 moveToTrashWindows 方法，下面是对该方法实现原理的详细解释。

1. 检查文件是否存在

```
if (file.exists()) {
    // ...
} else {
    System.err.println("File does not exist");
}
```

在尝试将文件移动到回收站之前，首先检查文件是否存在。如果文件不存在，我们打印一个错误消息并退出方法。

2. 准备 SHFileOperation 函数的参数

```
WString from = new WString(file.getAbsolutePath() + '\0');
SHFILEOPSTRUCT fileop = new SHFILEOPSTRUCT();
fileop.wFunc = ShellAPI.FO_DELETE;
fileop.pFrom = from;
fileop.fFlags = ShellAPI.FOF_ALLOWUNDO | ShellAPI.FOF_NOCONFIRMATION;
```

相关部分解释如下。

（1）WString from：创建一个 WString 对象来存储要删除的文件的路径。WString 是 JNA 用来处理 Windows Unicode 字符串的类。文件路径后面添加了一个 null 字符 (\0)，因为 SHFileOperation 函数要求这样做。

（2）SHFILEOPSTRUCT fileop：创建一个 SHFILEOPSTRUCT 结构体对象来存储 SHFileOperation 函数的参数。SHFILEOPSTRUCT 是 Shell API 中定义的一个结构体，它包含了执行文件操作所需的各种信息。

（3）fileop.wFunc = ShellAPI.FO_DELETE：我们设置 wFunc 字段为 FO_DELETE，指示想要删除文件。

（4）fileop.pFrom = from：设置 pFrom 字段为要删除的文件的路径。

（5）fileop.fFlags：设置 fFlags 字段为 FOF_ALLOWUNDO | FOF_NOCONFIRMATION，表示想要将文件移动到回收站 (FOF_ALLOWUNDO)，并且不想显示删除确认对话框 (FOF_NOCONFIRMATION)。

3. 调用 SHFileOperation 函数

```
int result = Shell32.INSTANCE.SHFileOperation(fileop);
if (result == 0) {
    System.out.println("File moved to trash successfully!");
} else {
    System.err.println("Failed to move file to trash, error code: " + result);
}
```

我们调用 SHFileOperation 函数并传递刚刚准备好的 SHFILEOPSTRUCT 结构体。SHFileOperation 函数执行文件操作，并返回一个错误代码。如果错误代码为 0，则表示操作成功；否则，表示操作失败。

通过这种方式，我们能够利用 Windows 的 Shell API 和 JNA 库在 Java 中将文件移动到回收站。

1.7.2　清空回收站中的文件和目录

下面详细解释 Windows、macOS 和 Linux 平台清除回收站中文件和目录的方式。

- Windows

在 Windows 中可以直接通过命令行工具 powershell.exe 清除回收站中的文件和目录：

```
powershell Clear-RecycleBin -Force
```

- macOS

macOS 回收站的目录是 ~/ .Trash，所以只需要扫描该目录中的所有子目录和文件，然后删除我们感兴趣的文件和目录，或者全部清除即可。

- Linux

Linux 回收站的目录是 ~/ .local/share/Trash/files，所以只需要扫描该目录中的所有子目录和文件，然后删除我们感兴趣的文件和目录，或者全部清除即可。

下面的例子清除了回收站中的所有文件和目录。

代码位置：src/main/java/trash/RecycleBinCleaner.java

```
package trash;
import java.io.IOException;
import java.nio.file.*;
import java.nio.file.attribute.BasicFileAttributes;
import java.util.stream.Stream;
```

```
public class RecycleBinCleaner {
    public static void main(String[] args) {
        emptyRecycleBin();
    }
    public static void emptyRecycleBin() {
        String osName = System.getProperty("os.name").toLowerCase();
        try {
            if (osName.contains("win")) {
                emptyRecycleBinWindows();
            } else if (osName.contains("mac")) {
                emptyRecycleBinMac();
            } else if (osName.contains("nix") || osName.contains("nux") ||
osName.contains("aix")) {
                emptyRecycleBinLinux();
            }
        } catch (IOException e) {
            e.printStackTrace();
        }
    }
    // 清除 Windows 回收站
    public static void emptyRecycleBinWindows() {
        try {
            String command = "powershell Clear-RecycleBin -Force";
            Process process = Runtime.getRuntime().exec(command);
            process.waitFor();
            System.out.println("Recycle bin emptied successfully on Windows!");
        } catch (IOException | InterruptedException e) {
            e.printStackTrace();
            System.err.println("Failed to empty recycle bin on Windows");
        }
    }
    // 清除 macOS 回收站
    public static void emptyRecycleBinMac() throws IOException {
        Path trashPath = Paths.get(System.getProperty("user.home"), ".Trash");
        try (Stream<Path> stream = Files.list(trashPath)) {
            stream.forEach(path -> {
                try {
                    if (Files.isDirectory(path)) {
                        deleteDirectory(path);
                    } else {
                        Files.delete(path);
                    }
                } catch (IOException e) {
                    e.printStackTrace();
                }
            });
        }
    }
    // 清除 Linux 回收站
    public static void emptyRecycleBinLinux() throws IOException {
```

```
        Path trashPath = Paths.get(System.getProperty("user.home"),
".local/share/Trash/files");
        try (Stream<Path> stream = Files.list(trashPath)) {
            stream.forEach(path -> {
                try {
                    if (Files.isDirectory(path)) {
                        deleteDirectory(path);
                    } else {
                        Files.delete(path);
                    }
                } catch (IOException e) {
                    e.printStackTrace();
                }
            });
        }
    }
    // 递归删除 path 指定目录中的所有文件和子目录
    public static void deleteDirectory(Path path) throws IOException {
        Files.walkFileTree(path, new SimpleFileVisitor<Path>() {
            @Override
            public FileVisitResult visitFile(Path file, BasicFileAttributes
attrs) throws IOException {
                Files.delete(file);
                return FileVisitResult.CONTINUE;
            }
            @Override
            public FileVisitResult postVisitDirectory(Path dir,
IOException exc) throws IOException {
                Files.delete(dir);
                return FileVisitResult.CONTINUE;
            }
        });
    }
}
```

1.7.3　恢复回收站中的文件和目录

恢复回收站中的文件的方式可分为如下 3 步。

（1）获取回收站中文件的原始路径。

（2）将回收站中的文件复制到原始路径。

（3）删除回收站中的文件。

其实这个过程与剪切文件的方式类似，只是源目录是回收站目录。由于恢复目录与恢复文件的方式类似，所以本节就只提及恢复文件。

macOS、Linux 和 Windows 在恢复回收站文件的方式上略有不同，但大同小异，不管是使用第三方 API，还是直接使用内建 API，都是遵循前面提到的 3 步。下面分别讲解如何在这 3 个平台恢复回收站中的文件。

❑ 恢复 macOS 回收站中的文件

macOS 回收站的绝对路径是“**/Users/ 用户名 /.Trash**”，其中“用户名”是当前登录的用户名，加上用户名 great，macOS 回收站的绝对路径是“**/Users/great/.Trash**”。在路径下有一个 **.DS_Store** 文件，该文件存储了当前目录的元数据，对于回收站来说，就存储了回收站中所有文件和目录的相关信息，如原始路径，被删除时间等，但由于 .DS_Store 文件的格式苹果公司并未公开，也没有提供任何可以读取 .DS_Store 文件的 API，而且 .DS_Store 文件用的是二进制格式存储。所以通过正常的手段是无法读取 .DS_Store 文件内容的，自然也就无法获取回收站中文件的原始目录了。因此，在 macOS 下恢复回收站中的文件，只能通过 **osascript** 命令了。osascript 是 macOS 上执行 **AppleScript** 的命令行工具。AppleScript 是一种脚本语言，用于自动化 macOS 应用程序的操作。使用 osascript 命令可以在终端中运行 AppleScript 脚本，也可以在脚本中使用 AppleScript 来发送系统通知。以下是一个发送系统通知的例子：

```
osascript -e 'display notification "Hello World!" with title "Greetings"'
```

在终端执行这行命令，将在屏幕右上角显示一个如图 1-8 所示的通知。

图 1-8 使用 osascript 命令弹出的通知

AppleScript 几乎能操作 macOS 中的一切，控制回收站更不在话下。AppleScript 会用接近自然语言（英语）的方式描述如何操作回收站（trash）。本例通过 AppleScript 打开回收站，并模拟键盘按下 Command + Delete 键来恢复回收站中被选中的文件或目录，当然，在做这个操作之前，先要通过 AppleScript 获取回收站顶层的所有文件和目录。下面是完整的 AppleScript 代码。

```
-- 打开 Finder 应用程序
tell application "Finder"
   -- 激活 Finder 窗口
   activate
   -- 获取垃圾桶中已删除文件的数量
   set file_count to count of (trash's items)
   -- 重复以下步骤，直到所有文件都被恢复
   repeat file_count times
        -- 调用 recoverMyFile() 函数来恢复文件
        recoverMyFile() of me
   end repeat
end tell

-- 定义 recoverMyFile() 函数来恢复单个文件
on recoverMyFile()
   -- 打开 System Events 应用程序
   tell application "System Events"
        -- 将 Finder 窗口置于最前面
        set frontmost of process "Finder" to true
        -- 打开垃圾桶窗口并选择第一个文件
```

```
        tell application "Finder"
            open trash
            select the first item of front window
        end tell
        -- 使用键盘快捷键 Command + Delete 来恢复文件
        tell process "Finder"
            key code 51 using command down
            delay 2 -- 延迟 2s
        end tell
    end tell
end recoverMyFile
```

将这段代码保存在 **apple.script** 文件中，然后执行 **osascript apple.script** 即可将回收站中的所有文件和目录放回原处。

在执行 apple.script 文件时，有可能出现下面的错误：

```
execution error: "System Events" 遇到一个错误："osascript" 不允许发送按键。 (1002)
```

这个错误通常出现在使用 macOS 自带的 Script Editor（脚本编辑器）应用程序时，它试图向某些应用程序发送按键信号但被系统阻止。

请使用下面的步骤解决这个问题。

（1）在 System Preferences 中找到“安全性与隐私”，然后切换到“隐私”选项卡。

（2）在左侧菜单中选择“辅助功能”，然后单击右侧的锁形图标以进行更改。

（3）输入管理员密码以解锁更改，并将 Script Editor 从列表中添加到允许应用程序列表中，如图 1-9 所示。

（4）如果问题仍然存在，请尝试退出并重新启动 Script Editor 应用程序。

图 1-9　设置脚本编辑器权限

如果想要用 Java 完成这一切，只需要用 Java 自动生成 apple.script 文件，然后执行该文件即可，或者干脆直接在 Java 中执行这段代码。不过由于这段代码较长，需要分段执行，所以推荐生成 apple.script 文件，然后再用 Java 执行的方式。

❑ 恢复 Linux 回收站中的文件

Linux 回收站的路径是"**~/.local/share/Trash**"，而回收站中每一个文件和目录都在"**~/.local/share/Trash/info**"目录中有一个元数据文件，文件名是 **filename.trashinfo**，其中 filename 表示回收站中的文件或目录名。例如，如果回收站中有一个 abc.txt 文件，那么对应的元数据文件是 abc.txt.trashinfo。

元数据文件是纯文本格式，里面保存了回收站文件中的原始路径，以及被移入回收站的时间，下面就是标准元数据文件的内容：

```
[Trash Info]
Path=/root/software/nginx.zip
DeletionDate=2023-03-30T21:49:37
```

根据元数据文件的内容，可以很容易获取回收站中文件和目录的原始路径，然后可以用相应的 API 将这些回收站中的文件和目录复制回原始目录，然后再删除回收站中对应的文件和目录。

❑ 恢复 Windows 回收站中的文件

Java 中并没有 API 直接恢复 Windows 回收站中的文件，也没有什么第三方库直接完成这个功能。如果非要用 Java 恢复 Windows 回收站中的文件，可以使用其他编程语言做一个脚本（如 Python）或可执行程序，然后使用 Java 来调用。

当然，使用 Java 调用 COM 组件也是一种实现方式，不过过程非常复杂，因此建议读者采用多种语言结合的方式实现这个功能。如果读者想用 Python 完成这个功能，可以参阅《奇妙的 Python：神奇代码漫游之旅》中 1.10.3 节的内容。

下面的例子恢复了 macOS 回收站和 Linux 回收站中的所有文件和目录。

代码位置：src/main/java/trash/TrashRestorer.java

```
package trash;
import java.io.*;
import java.nio.charset.StandardCharsets;
import java.nio.file.*;
import java.util.List;
import java.util.stream.*;

public class TrashRestorer {
    public static void main(String[] args) {
        putBackTrash();
    }
    public static void putBackTrash() {
        String osName = System.getProperty("os.name").toLowerCase();
```

```
        if (osName.contains("mac")) {
            recoverMacOSAllFiles();
        } else if (osName.contains("win")) {
            recoverWindowsAllFiles();
        } else if (osName.contains("nix") || osName.contains("nux") ||
osName.contains("aix")) {
            recoverLinuxAllFiles();
        } else {
            System.err.println("Unsupported operating system.");
        }
    }
    public static void recoverMacOSAllFiles() {
        String appleScript = "tell application \"Finder\"\n" +
                "    activate\n" +
                "    set file_count to count of (trash's items)\n" +
                "    repeat file_count times\n" +
                "        recoverMyFile() of me\n" +
                "    end repeat\n" +
                "end tell\n" +
                "\n" +
                "on recoverMyFile()\n" +
                "    tell application \"System Events\"\n" +
                "        set frontmost of process \"Finder\" to true\n" +
                "        tell application \"Finder\"\n" +
                "            open trash\n" +
                "            select the first item of front window\n" +
                "        end tell\n" +
                "        tell process \"Finder\"\n" +
                "            key code 51 using command down\n" +
                "            delay 2 -- Yes, it's stupid, but necessary :(\n" +
                "        end tell\n" +
                "    end tell\n" +
                "end recoverMyFile";
        Path scriptPath = Paths.get("apple.script");
        try {
            Files.write(scriptPath, appleScript.getBytes());
            Process process = Runtime.getRuntime().exec("osascript "
+ scriptPath.toString());
            process.waitFor();
            System.out.println("Successfully restored all files on macOS!");
        } catch (IOException | InterruptedException e) {
            e.printStackTrace();
            System.err.println("Failed to restore files on macOS");
        } finally {
            try {
                Files.deleteIfExists(scriptPath);
            } catch (IOException e) {
                e.printStackTrace();
            }
        }
    }
```

```
    public static void recoverWindowsAllFiles() {
        System.out.println("不支持还原 Windows 回收站中的文件和目录.");
    }
    public static void recoverLinuxAllFiles() {
        Path recycleBinPath = Paths.get(System.getProperty("user.home"),
".local/share/Trash/files");
        try (Stream<Path> stream = Files.list(recycleBinPath)) {
            stream.forEach(path -> {
                try {
                    String filename = path.getFileName().toString();
                    recoverLinuxFile(filename);
                } catch (IOException e) {
                    e.printStackTrace();
                }
            });
        } catch (IOException e) {
            e.printStackTrace();
        }
    }
    public static void recoverLinuxFile(String filename) throws IOException {
        Path trashInfoPath = Paths.get(System.getProperty("user.home"),
".local/share/Trash/info", filename + ".trashinfo");
        List<String> lines = Files.readAllLines(trashInfoPath);
        for (String line : lines) {
            if (line.startsWith("Path=")) {
                String originalPath = java.net.URLDecoder.decode
(line.substring(5).trim(), StandardCharsets.UTF_8);
                Path trashFilePath = Paths.get(System.getProperty("user.home"),
".local/share/Trash/files", filename);
                if (!Files.exists(Paths.get(originalPath))) {
                    Files.move(trashFilePath, Paths.get(originalPath));
                    System.out.println("Successfully restored " + filename +
" to " + originalPath);
                } else {
                    System.err.println("Error: File " + originalPath +
"already exists");
                }
                break;
            }
        }
    }
}
```

如果在 macOS 版的 IntelliJ IDEA 中第一次运行这个程序，会弹出如图 1-10 所示的对话框。

单击“好”按钮，会弹出如图 1-11 所示的对话框，选中 IntelliJ IDEA 或 IntelliJ IDEA CE 选项即可。

图 1-10　授权对话框

图 1-11　设置 IntelliJ IDEA 的权限

1.8　小结

使用 Java 管理文件和目录有着不小的挑战，因为并不是对文件和目录的每一种操作，Java 都提供了标准的 API，有一些功能需要借助第三方库以及其他编程语言的脚本、命令行工具等才能实现。而且不同操作系统有着不一样的操作方式，甚至部分功能（如恢复回收站中的文件）在某些操作系统（如 Linux）上甚至没有第三方模块可用，这就要求我们自己分析背后的原理，自己从底层实现所有的功能。尽管挑战一直存在，但同时也充满乐趣。

第 2 章 驾 驭 OS

使用 Java 可以借助很多第三方库和系统命令控制操作系统（Operating System，OS），例如，向 Windows 注册表中写入信息，将应用程序添加进启动项，获取系统硬件信息、显示设置窗口、打开文件夹等。本章将介绍一些控制系统的第三方库和系统命令。

2.1 Windows 注册表

Windows 注册表是 Windows 操作系统中的一个重要组成部分，它是一个分层数据库，包含了对 Windows 操作至关重要的数据以及运行在 Windows 上的应用程序和服务。注册表记录了用户安装在计算机上的软件和每个程序的相互关联信息，它包括了计算机的硬件配置，包括自动配置的即插即用的设备和已有的各种设备说明、状态属性以及各种状态信息和数据。

我们可以使用注册表编辑器（**regedit.exe** 或 **regedt32.exe**）、注册表（.reg）文件或运行 Visual Basic 脚本文件等来修改注册表。不过本节要使用 Java 来读写注册表中的数据。

在 Java 中，可以使用 JNA（Java Native Access）直接调用 Windows API 读写 Windows 注册表。

2.1.1 读取值的数据

使用 JNA 读写 Windows 注册表需要导入下面两个类：

```
import com.sun.jna.platform.win32.Advapi32Util;
import com.sun.jna.platform.win32.WinReg;
```

这两个类用于访问 Windows 的特定功能。其中 Advapi32Util 类提供了访问 Windows 高级 API（如注册表操作）的实用方法。WinReg 类则提供了表示 Windows 注册表中的各种根键（如 HKEY_LOCAL_MACHINE）的常量。

使用 Advapi32Util 类的 registryGetStringValue 方法来读取注册表项的值。该方法的原型如下。

```
public static String registryGetStringValue(WinReg.HKEY hKey, String keyPath,
String valueName);
```

参数含义如下。

（1）hKey：枚举类型的参数，表示 Windows 注册表的顶级键（如 HKEY_LOCAL_MACHINE 或 HKEY_CURRENT_USER）。WinReg.HKEY 是 JNA 库中定义的一个特殊的枚举，它包含了所有顶级注册表键的引用。

（2）keyPath：指明了从指定的根键（hKey）开始的路径，用来定位注册表中的项。这个路径是注册表项的完整路径，但不包括根键本身。

（3）valueName：指定在上面给出路径的注册表项中要检索的值的名称。如果指定的值名称存在，此方法将返回它的字符串数据。

registryGetStringValue 方法返回一个 String 类型的值，这个值就是从注册表中检索到的字符串数据。如果指定的键、路径或值名称不存在，JNA 通常会抛出一个异常。

注意： registryGetStringValue 方法只适用于读取字符串类型的数据。如果需要读取整数或二进制数据，应该使用相应的方法，比如 registryGetIntValue 或 registryGetBinaryValue。而且，为了使这些方法工作，需要有适当的权限来访问 Windows 注册表，并且在使用它们时要小心，因为错误地修改注册表可能会导致系统不稳定。

下面的代码读取了 HKEY_LOCAL_MACHINE\SOFTWARE\Microsoft\Windows NT\CurrentVersion 键中 ProductName 的值。

代码位置： src/main/java/winreg/ReadWinregKey.java

```
package winreg;
import com.sun.jna.platform.win32.Advapi32Util;
import com.sun.jna.platform.win32.WinReg;
public class ReadWinregKey {
    public static void main(String[] args) {
        // 确保有读取注册表的权限
        String value = Advapi32Util.registryGetStringValue(
                WinReg.HKEY_LOCAL_MACHINE,                       // 注册表主键
                "SOFTWARE\\Microsoft\\Windows NT\\CurrentVersion",    // 路径
                "ProductName"                                    // 要获取的值的名称
        );
        System.out.println("ProductName value is: " + value);
    }
}
```

运行程序，会在终端输出如下内容：

```
ProductName value is: Windows 10 Pro
```

2.1.2 读取所有的键

这里要讲解一下注册表中的键和值。注册表由键和值组成，其中键是一个包含值的容器。键可以包含子键，而子键可以包含更多的子键和值。值是与键关联的数据。每个值包含名称、类型和数据。图 2-1 是 Themes 键、Themes 键的子键以及 Themes 键中包含的值。在 2.1.1 节中获得的 ProductName 就是一个值，该值属于 CurrentVersion 键。而本节将获取 Themes 键的所有子键，并输出这些子键。

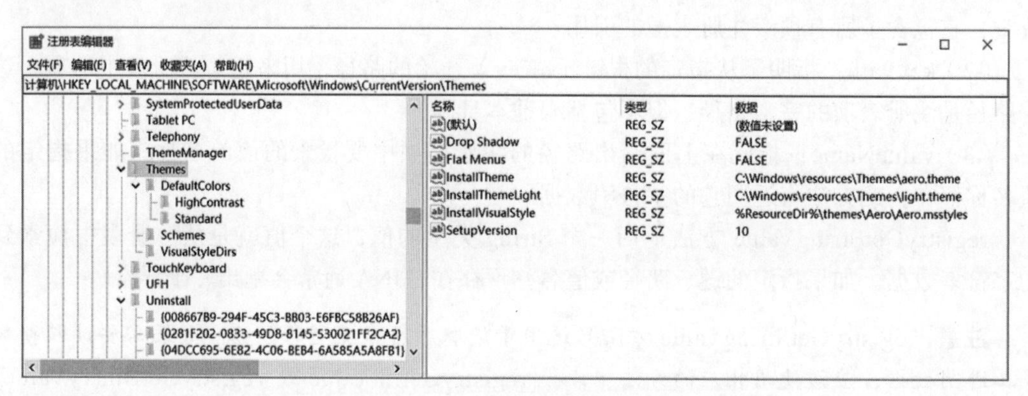

图 2-1 Windows 注册表

registryGetKeys 方法是 Advapi32Util 类的一个静态方法，它是通过 JNA（Java Native Access）库提供的，用于获取 Windows 注册表中一个给定键路径下的所有子键名称。这个方法的原型如下。

```
public static String[] registryGetKeys(WinReg.HKEY root, String keyPath);
```

参数含义如下。

（1）root：表示注册表的一个顶级键。WinReg.HKEY 是一个枚举类型，包含了所有可能的顶级键，例如 HKEY_LOCAL_MACHINE, HKEY_CURRENT_USER 等。这个参数告诉 registryGetKeys 方法从哪个顶级键开始查找。

（2）keyPath：指定了注册表中的一个完整键路径，registryGetKeys 方法将在这个路径下查找所有的直接子键。路径应该是相对于传入的 root 顶级键来说的。

registryGetKeys 方法返回一个 String[] 数组，包含了在指定路径下找到的所有直接子键的名称。如果指定路径下没有子键，或者出现错误，它可能返回 null 或抛出异常。

registryGetKeys 方法通常用于读取注册表信息，用于系统配置或软件配置的读取操作。例如，可以通过这个方法来获取系统中安装的软件列表或其他系统配置信息。

下面的例子枚举了 SOFTWARE\Microsoft\Windows\CurrentVersion\Themes 键以及所有子键，并输出这些键的名称。

代码位置：src/main/java/winreg/ReadWinregAllKeys.java

```
package winreg;
import com.sun.jna.platform.win32.Advapi32Util;
import com.sun.jna.platform.win32.WinReg;
public class ReadWinregAllKeys {
    public static void main(String[] args) {
        // 指定要枚举的顶级键和键路径
        WinReg.HKEY hKey = WinReg.HKEY_LOCAL_MACHINE;
        String keyPath = "SOFTWARE\\Microsoft\\Windows\\CurrentVersion\\
Themes";
        readKey(hKey, keyPath);
    }
    public static void readKey(WinReg.HKEY root, String path) {
        // 使用Advapi32Util提供的方法来访问注册表
        String[] subKeys = Advapi32Util.registryGetKeys(root, path);
        if (subKeys == null) {
            return;
        }
        for (String key : subKeys) {
            // 构建子键的完整路径
            String subKeyPath = path + "\\" + key;
            System.out.println(subKeyPath);
            // 递归枚举当前子键下的所有子键
            readKey(root, subKeyPath);
        }
    }
}
```

执行这段代码，会输出如下内容：

```
SOFTWARE\Microsoft\Windows\CurrentVersion\Themes\DefaultColors
SOFTWARE\Microsoft\Windows\CurrentVersion\Themes\DefaultColors\HighContrast
SOFTWARE\Microsoft\Windows\CurrentVersion\Themes\DefaultColors\Standard
SOFTWARE\Microsoft\Windows\CurrentVersion\Themes\Schemes
SOFTWARE\Microsoft\Windows\CurrentVersion\Themes\VisualStyleDirs
```

对比图2-1可以看出，Themes有5个子键，已经将这5个子键的完整路径全部输出。

2.1.3 读取所有的键和值

读取键可以使用2.1.2节介绍的registryGetKeys方法，通过递归的方式就可以得到指定键下的所有子键。

registryGetValues方法是Advapi32Util类中提供的一个静态方法，它用于获取Windows注册表项下所有值的名称及其对应的数据。

下面是registryGetValues方法的原型。

```
public static Map<String, Object> registryGetValues(WinReg.HKEY hKey, String keyPath);
```

参数含义如下。

（1）hKey：这是一个枚举值，表示 Windows 注册表中的一个顶级键（如 HKEY_LOCAL_MACHINE 或 HKEY_CURRENT_USER）。此参数指定了方法开始查询值的注册表分支的起点。

（2）keyPath：指定了注册表项的路径，相对于前面指定的顶级键。此参数告诉方法在哪个具体的注册表路径下查询值。

registryGetValues 方法返回一个 Map<String, Object> 类型的值，Key 是一个字符串，代表注册表值的名称；Value 是一个对象，它可以是多种类型，如字符串（String）、整数（Integer）、字节数组（byte[]）等，具体类型取决于注册表值的数据类型。

使用 registryGetValues 方法可以方便地获取注册表项中所有值的详细信息，而无须逐一查询每个值。例如想要获取某个软件在注册表中的设置，可以指定该软件设置的路径，然后获取该路径下所有的设置项及其值。

下面的例子用递归的方式读取了 SOFTWARE\Microsoft\Windows\CurrentVersion\Themes 键及所有子键和对应的所有值。

代码位置：src/main/java/winreg/ReadWinregAllKeysValues.java

```
package winreg;
import com.sun.jna.platform.win32.Advapi32Util;
import com.sun.jna.platform.win32.WinReg;
import java.util.Map;
public class ReadWinregAllKeysValues {
    public static void main(String[] args) {
        // 打开注册表的路径
        WinReg.HKEY hKey = WinReg.HKEY_LOCAL_MACHINE;
        String path = "SOFTWARE\\Microsoft\\Windows\\CurrentVersion\\Themes";
        printAllSubkeys(hKey, path);                                // 打印所有子键和值
    }
    public static void printAllSubkeys(WinReg.HKEY root, String path) {
        // 打印当前路径
        System.out.println(path);
        // 获取并打印当前路径下的所有值
        printAllValues(root, path);
        // 获取当前路径下的所有子键
        String[] subKeys = Advapi32Util.registryGetKeys(root, path);
        if (subKeys != null) {
            for (String subKey : subKeys) {
                // 递归调用以打印子键及其值
                printAllSubkeys(root, path + "\\" + subKey);
            }
        }
    }
    public static void printAllValues(WinReg.HKEY root, String path) {
        // 获取当前路径下的所有值及其数据
        Map<String, Object> values = Advapi32Util.registryGetValues(root, path);
        for (Map.Entry<String, Object> entry : values.entrySet()) {
```

```
            // 打印值的名称和数据
            System.out.println(entry.getKey() + " = " + entry.getValue());
        }
    }
}
```

执行这段代码，会输出如下内容（不同的环境，可能输出的内容略有差异）：

```
DefaultColors
HighContrast
ActiveTitle = 7209015
ButtonFace = 0
ButtonText = 16777215
GrayText = 4190783
Hilight = 16771866
... ...
```

2.1.4　添加键和值

在 JNA 库中，Advapi32Util 类提供了一系列用于修改 Windows 注册表的方法。其中 registryCreateKey 方法用于创建或打开键，registrySetStringValue 方法用于设置指定键的字符串值。

registryCreateKey 方法的原型如下。

```
public static void registryCreateKey(WinReg.HKEY hKey, String keyPath);
```

参数含义如下。

（1）hKey：这是一个指向注册表顶级键的枚举值，如 HKEY_LOCAL_MACHINE 或 HKEY_CURRENT_USER。

（2）keyPath：这是一个字符串，指定了从顶级键下要创建或打开的注册表项的路径。

该方法没有返回值。它会在指定路径创建一个新的注册表项，如果注册表项已经存在，则打开它。如果操作成功，注册表项将被创建或打开；如果操作失败，将抛出异常。

registrySetStringValue 方法的原型如下。

```
public static void registrySetStringValue(WinReg.HKEY hKey, String keyPath,
String valueName, String valueData);
```

参数含义如下。

（1）hKey：同上，表示注册表中的一个顶级键。

（2）keyPath：同上，表示注册表项的路径。

（3）valueName：这是一个字符串，表示要设置的值的名称。如果要设置默认值，则此参数应为 " "（空字符串）。

（4）valueData：这是一个字符串，表示要设置的值的数据。

该方法没有返回值。它会在指定的注册表项路径下，为给定的值名称设置字符串数据。如果操作成功，值将被设置；如果操作失败，将抛出异常。

下面的例子为 Software 键添加了一个 MyApp 子键，并设置了该子键的默认值，并为该子键添加了一个名为 myValueData 的值。

代码位置：src/main/java/winreg/AddKeyValue.java

```
package winreg;
import com.sun.jna.platform.win32.Advapi32Util;
import com.sun.jna.platform.win32.WinReg;
public class AddKeyValue {
    public static void main(String[] args) {
        // 定义要添加的键的路径
        String keyPath = "Software\\MyApp";
        // 定义要设置的值的名称和数据
        String valueName = "myValueName";
        String valueData = "myValueData";
        String defaultValue = "myDefaultValue";
        // 打开或创建键
        Advapi32Util.registryCreateKey(WinReg.HKEY_CURRENT_USER, keyPath);
        // 为键设置一个值
        Advapi32Util.registrySetStringValue(WinReg.HKEY_CURRENT_USER, keyPath,
valueName, valueData);
        // 设置键的默认值
        Advapi32Util.registrySetStringValue(WinReg.HKEY_CURRENT_USER, keyPath,
"", defaultValue);
    }
}
```

执行这段代码，会看到 Software 键多了一个 MyApp 子键，以及一个 myValueData 值，如图 2-2 所示。

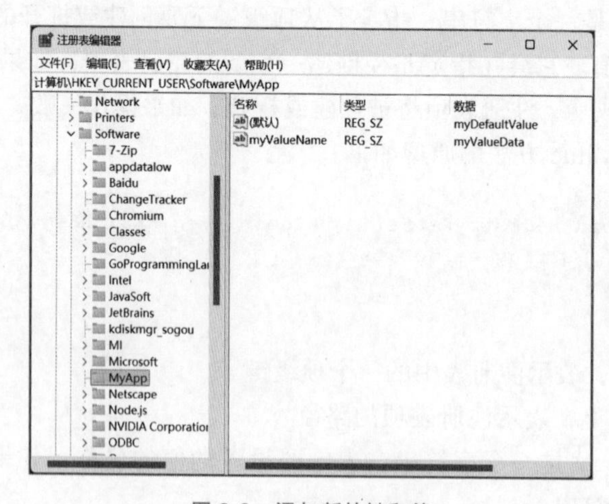

图 2-2　添加新的键和值

2.1.5　重命名键

JNA 并没有提供重命名键的方法，所以要想重命名键，需要按下面的步骤操作。

（1）使用新的键名创建新的键。

（2）获取旧键中的所有数据（包括键默认值、键中所有的值、键中的子键等）。

（3）将旧键中的所有数据复制到新键中。

（4）删除旧键。

完成这些操作的大部分方法在前面已经介绍过，只有一个 registryDeleteKey 方法还未涉及，该方法用于删除键，原型如下。

```
public static void registryDeleteKey(WinReg.HKEY hKey, String keyName);
```

参数含义如下。

（1）hKey：这是一个指向注册表顶级键的枚举值，如 HKEY_LOCAL_MACHINE 或 HKEY_CURRENT_USER。它指定了要删除的键的根部分。

（2）keyName：这是一个字符串，指定了从顶级键下要删除的注册表项的完整路径。

registryDeleteKey 方法没有返回值。当调用该方法时，如果指定的键存在，它将被删除。如果键不存在或方法无法删除键（例如，由于权限问题或键下还有子键存在），则会抛出异常。需要注意的是，删除注册表键是一个不可逆操作，一旦删除，信息将无法恢复。因此，在删除之前，请确保已经正确备份了所有重要数据，并且确认不再需要该键。

下面的例子完整地演示了如何重命名键。

代码位置：src/main/java/winreg/RenameKey.java

```
package winreg;
import com.sun.jna.platform.win32.Advapi32Util;
import com.sun.jna.platform.win32.WinReg;
import java.util.Map;
public class RenameKey {
    public static void renameKey(String oldKeyPath, String newKeyName) {
        String newKeyPath = oldKeyPath.substring(0,
oldKeyPath.lastIndexOf('\\')) + "\\" + newKeyName;
        // 读取旧键的所有值
        Map<String, Object> values =
       Advapi32Util.registryGetValues(WinReg.HKEY_CURRENT_USER, oldKeyPath);
        // 创建新键
        Advapi32Util.registryCreateKey(WinReg.HKEY_CURRENT_USER, newKeyPath);

        // 复制旧键的所有值到新键
        for (Map.Entry<String, Object> valueEntry : values.entrySet()) {
            Advapi32Util.registrySetStringValue(
                    WinReg.HKEY_CURRENT_USER, newKeyPath, valueEntry.getKey(),
(String) valueEntry.getValue());
        }
```

```
        // 获取旧键的所有子键
        String[] subKeys = Advapi32Util.registryGetKeys(WinReg.HKEY_CURRENT_
USER, oldKeyPath);
        if (subKeys != null) {
            for (String subKey : subKeys) {
                // 递归复制子键
                renameKey(oldKeyPath + "\\" + subKey, subKey);
            }
        }
        // 删除旧键
        Advapi32Util.registryDeleteKey(WinReg.HKEY_CURRENT_USER, oldKeyPath);
    }
    public static void main(String[] args) {
        String testKeyPath = "Software\\test_key";
        String newTestKeyName = "new_test_key";

        // 创建测试键
        Advapi32Util.registryCreateKey(WinReg.HKEY_CURRENT_USER, testKeyPath);
        Advapi32Util.registrySetStringValue(WinReg.HKEY_CURRENT_USER,
testKeyPath, "test_value", "test_data");
        // 重命名测试键
        renameKey(testKeyPath, newTestKeyName);
        // 验证新键是否存在
        try {
            String newTestValue = Advapi32Util.registryGetStringValue(
                    WinReg.HKEY_CURRENT_USER, "Software\\" + newTestKeyName,
"test_value");
            assert "test_data".equals(newTestValue);
            System.out.println("测试通过");
        } catch (Exception e) {
            System.out.println("测试失败：" + e.getMessage());
        }

        // 清理测试键
        Advapi32Util.registryDeleteKey(WinReg.HKEY_CURRENT_USER, "Software\\"
+ newTestKeyName);
    }
}
```

执行代码，如果输出如下内容，说明已成功重命名了键。

```
测试通过
```

2.1.6 重命名值

JNA 并没有提供重命名值的方法，所以要想重命名值，需要按下面的步骤操作。

（1）使用新的值名创建新的值。

（2）获取旧值中的所有数据。

（3）将旧值中的所有数据复制到新值中。

（4）删除旧值。

完成这 4 步的大多数方法在前面都已经讲解过，只有 registryDeleteValue 方法还未涉及，该方法的原型如下。

```
public static void registryDeleteValue(WinReg.HKEY hKey, String keyPath,
String valueName);
```

参数含义如下。

（1）hKey：这是一个指向注册表顶级键的枚举值，如 HKEY_LOCAL_MACHINE 或 HKEY_CURRENT_USER。它指定了要操作的键的根部分。

（2）keyPath：这是一个字符串，表示顶级键下的注册表项的完整路径。

（3）valueName：这是一个字符串，指定了要删除的值的名称。

registryDeleteValue 方法没有返回值。当调用该方法时，它会尝试删除在指定路径下的键的指定值。如果指定的值不存在，或者方法无法删除值（例如，由于权限问题），则会抛出异常。

下面的例子完整地演示了如何重命名键。

代码位置：src/main/java/winreg/RenameValue.java

```
package winreg;
import com.sun.jna.platform.win32.Advapi32Util;
import com.sun.jna.platform.win32.WinReg;
public class RenameValue {
    public static void renameValue(String keyPath, String oldName,
String newName) {
        // 获取旧值的数据
        Object valueData = Advapi32Util.registryGetValue(WinReg.HKEY_CURRENT_
USER, keyPath, oldName);

        // 判断值数据的类型，并据此设置新值
        if (valueData instanceof String) {
            // 值是字符串类型
            Advapi32Util.registrySetStringValue(WinReg.HKEY_CURRENT_USER,
keyPath, newName, (String) valueData);
        } else if (valueData instanceof Integer) {
            // 值是整数类型
            Advapi32Util.registrySetIntValue(WinReg.HKEY_CURRENT_USER, keyPath,
newName, (Integer) valueData);
        } else if (valueData instanceof byte[]) {
            // 值是二进制类型
            Advapi32Util.registrySetBinaryValue(WinReg.HKEY_CURRENT_USER,
keyPath, newName, (byte[]) valueData);
        } // ... 处理其他已知的类型
        else {
            throw new IllegalArgumentException("Unknown data type for registry
```

```
value: " + oldName);
        }
        // 删除旧值
        Advapi32Util.registryDeleteValue(WinReg.HKEY_CURRENT_USER, keyPath,
oldName);
    }
    public static void main(String[] args) {
        String keyPath = "Software\\MyTest";
        String oldValueName = "OldValue";
        String newValueName = "NewValue";
        String data = "Hello World";
        // 创建测试键并设置原始值
        Advapi32Util.registryCreateKey(WinReg.HKEY_CURRENT_USER, keyPath);
        Advapi32Util.registrySetStringValue(WinReg.HKEY_CURRENT_USER, keyPath,
oldValueName, data);
        // 重命名值
        renameValue(keyPath, oldValueName, newValueName);
        // 验证新值是否设置成功
        String newData = Advapi32Util.registryGetStringValue(WinReg.HKEY_
CURRENT_USER, keyPath, newValueName);
        if ("Hello World".equals(newData)) {
            System.out.println(" 测试通过 ");
        } else {
            System.out.println(" 测试失败 ");
        }
        // 清理测试数据
        Advapi32Util.registryDeleteValue(WinReg.HKEY_CURRENT_USER, keyPath,
newValueName);
        Advapi32Util.registryDeleteKey(WinReg.HKEY_CURRENT_USER, keyPath);
    }
}
```

执行代码，如果输出如下内容，说明已成功重命名了值。

```
测试通过
```

2.2 让程序随 OS 一起启动

本节主要介绍如何分别在 Windows、macOS 和 Linux 下将应用程序添加进各自的启动项。启动项在不同的操作系统中的叫法有所不同，但作用是一样的，就是在操作系统登录时自动运行启动项中的应用程序。

2.2.1 将应用程序添加进 macOS 登录项

所有被添加进 macOS 登录项的程序，会在 macOS 启动的过程中依次运行。将应用程序添加进登录项有多种方式，其中常规的操作步骤如下。

（1）显示“系统偏好设置”。
（2）单击“用户与群组”。
（3）单击当前用户的名称。
（4）单击左下角的“小锁”，并输入“用户名”和“密码”。
（5）单击“登录项”选项卡。
（6）单击左下角的“+”按钮。
（7）在弹出的窗口中，选择要添加的应用程序。
（8）单击“添加”按钮将选中的应用程序添加进登录项。
按照这 8 步做，就会将 macOS 应用程序添加进登录项，效果如图 2-3 所示。

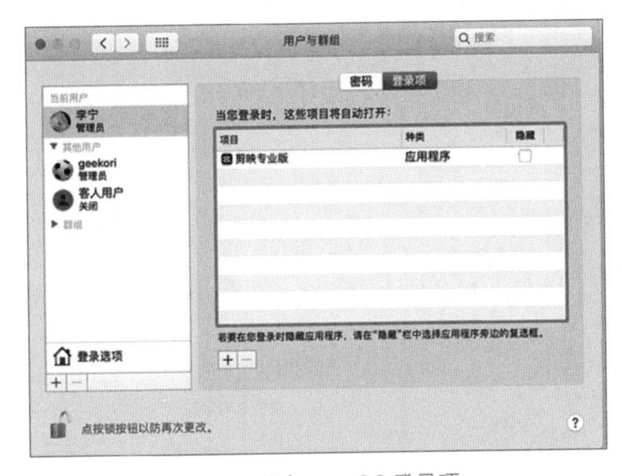

图 2-3　添加 macOS 登录项

用 AppleScript 也可以实现同样的功能。首先创建一个 starter.script 文件，并输入如下内容：

```
tell application "System Events"
    make login item at end with properties {path:"/Applications/WeChat.app",
hidden:false}
end tell
```

这段代码的含义是将 WeChat.app（微信）添加到 macOS 登录项列表的最后一个位置。其中 hidden 属性为 true，在登录时启动该应用程序，将无法看到应用程序的窗口。该输出就是设置图 2-3 所示窗口的“隐藏”列的值。

如果想在 Java 中将应用程序添加进 macOS 登录项，可以使用 Runtime.getRuntime().exec 方法执行前面的 AppleScript，代码如下：

代码位置：src/main/java/startup/MacosStartup.java

```
package harness.os;
import java.io.IOException;
```

```
public class MacosStartup {
    public static void main(String[] args) {
        // 尝试将微信添加到登录项
        addAppToLoginItems("/Applications/WeChat.app");
    }
    public static void addAppToLoginItems(String appPath) {
        // 构建 AppleScript 命令，该命令会向系统事件发送指令，制作一个登录项
        String script = String.format("tell application \"System Events\" to
make login item at end with properties {path:\"%s\", hidden:false}", appPath);
        try {
            // 使用 Runtime 类来执行 AppleScript 命令
            Process process = Runtime.getRuntime().exec(new String[]
{"osascript", "-e", script});
            // 等待命令执行完毕
            process.waitFor();
        } catch (IOException | InterruptedException e) {
            // 打印异常信息
            e.printStackTrace();
        }
    }
}
```

执行代码，会看到“微信”被添加到了 macOS 登录项列表最后的位置，如图 2-4 所示。

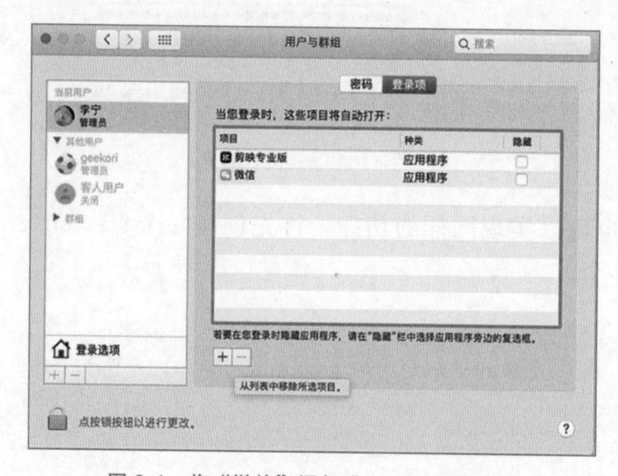

图 2-4　将“微信”添加进 macOS 登录项

2.2.2　将应用程序添加进 Windows 启动项

操作 Windows 启动项需要使用 JNA 向注册表中如下的键中添加新的值。

```
HKEY_CURRENT_USER\Software\Microsoft\Windows\CurrentVersion\Run
```

下面的例子向该键添加了 MyApp 值，数据是 d:\\software\\myapp.exe，如果添加成功，那么在 Windows 登录时，将会自动运行 myapp.exe。

代码位置： src/main/java/startup/WindowsStartup.java

```
package harness.os;
import com.sun.jna.platform.win32.Advapi32Util;
import com.sun.jna.platform.win32.WinReg;
public class WindowsStartup {
    public static void main(String[] args) {
        // 将应用程序添加到启动项
        addToStartup("d:\\software\\myapp.exe");
    }
    public static void addToStartup(String filePath) {
        // 指定注册表的路径
        String keyPath = "Software\\Microsoft\\Windows\\CurrentVersion\\Run";
        // 使用 JNA 的 Advapi32Util 类来访问注册表
        // HKEY_CURRENT_USER 对应 Java 中的 WinReg.HKEY_CURRENT_USER
        Advapi32Util.registrySetStringValue(WinReg.HKEY_CURRENT_USER, keyPath,
"MyApp", filePath);

        // 输出一条消息确认操作已完成
        System.out.println("应用程序已添加到启动项。");
    }
}
```

执行程序，会在Run键中添加如图2-5所示的MyApp值。

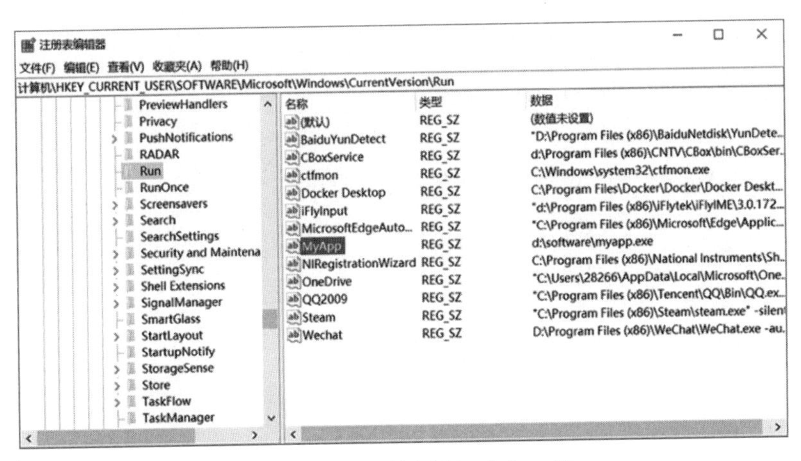

图2-5 向Run键中添加MyApp值

注意： 读者在运行本节程序时，需要将d:\\software\\myapp.exe改成自己机器上存在的可执行文件。

2.2.3 将应用程序添加进Linux启动项

添加Linux启动项的方式非常多，其中比较常用的就是在~/.bashrc文件最后添加一行命令，Linux 在登录时就会按顺序执行 ~/.bashrc 文件中的每一条命令。

下面的例子将 java ~/myapp.py 命令追加到 ~/.bashrc 文件的结尾，当 Linux 登录时，就是执行 myapp.py 脚本文件。

代码位置：src/main/java/startup/LinuxStartup.java

```
package harness.os;
import java.io.BufferedWriter;
import java.io.FileWriter;
import java.io.IOException;
import java.nio.file.Path;
import java.nio.file.Paths;
public class LinuxStartup {
    public static void main(String[] args) {
        // 传入想要添加到启动项的 Java 脚本的路径
        addToStartup("~/myapp.py");
    }
    public static void addToStartup(String filePath) {
        try {
            // 获取用户的 home 目录
            String homeDir = System.getProperty("user.home");
            // 构建 .bashrc 文件的路径
            Path bashrcPath = Paths.get(homeDir, ".bashrc");
            // 使用 BufferedWriter 来向 .bashrc 文件追加内容
            try (BufferedWriter writer = new BufferedWriter(new
FileWriter(bashrcPath.toString(), true)))
            {
                // 写入执行 Java 脚本的命令
                writer.write("java " + filePath + "\n");
            }
            System.out.println("脚本已添加到启动项。");
        } catch (IOException e) {
            // 打印异常信息
            e.printStackTrace();
        }
    }
}
```

2.3 获取系统信息

本节主要介绍如何使用 Java 的标准库以及 oshi 库获取大量的系统信息。

2.3.1 使用 Java 标准库获取系统信息

通过 Java 的标准库可以获取一些系统信息，例如操作系统的名称和版本信息、系统内存信息等。下面是相关 Java 库的详细描述。

❑ 获取操作系统信息

使用 System.getProperty("os.name") 和 System.getProperty("os.version") 可以获取操作系统的名称和版本信息。

❑ 获取内存信息

（1）使用 ManagementFactory.getPlatformMXBean(OperatingSystemMXBean.class) 可获取操作系统的 MXBean，这个 Bean 提供了系统内存的相关信息。

（2）使用 getTotalPhysicalMemorySize() 可获取系统的总物理内存大小。

（3）使用 getFreePhysicalMemorySize() 可获取系统的空闲物理内存大小。

❑ 获取磁盘信息

（1）File.listRoots() 方法返回一个 File 数组，包含了可用的文件系统根。这些根可以被视为磁盘分区。

（2）对于每一个文件系统根，可以获取磁盘名称（绝对路径）、磁盘的总容量。

尽管上述涉及的基本都是 Java 标准库，但 MXBean 对象要使用 com.sun.management.OperatingSystemMXBean 类。这个类不是 Java 标准库的一部分。它属于 com.sun.* 包，这些包通常是 Sun Microsystems 提供的特定于实现的 API，可能在不同版本的 Java 或不同的 JVM 实现中不可用或有所不同。因此，使用这些类可能会导致代码的可移植性问题。

下面的例子会使用 Java 标准 API 获取并输出系统信息。

代码位置：src/main/java/system/info/SystemInfo.java

```
package system.info;
import java.lang.management.ManagementFactory;
import com.sun.management.OperatingSystemMXBean;
import java.io.File;
public class SystemInfo {
    public static void main(String[] args) {
        // 获取操作系统的版本信息
        String os = System.getProperty("os.name") + " " +
System.getProperty("os.version");
        System.out.println("操作系统版本信息: " + os);
        // 获取系统的总内存、已经使用的内存、剩余内存
        OperatingSystemMXBean osBean = ManagementFactory.getPlatformMXBean
(OperatingSystemMXBean.class);
        // 总内存
        long totalPhysicalMemorySize = osBean.getTotalPhysicalMemorySize();
        double totalMem = totalPhysicalMemorySize / (1024.0 * 1024 * 1024);
        System.out.println("系统总内存: " + String.format("%.2f", totalMem) + "GB");

        // 已经使用的内存
        long usedPhysicalMemorySize = totalPhysicalMemorySize -
osBean.getFreePhysicalMemorySize();
        double usedMem = usedPhysicalMemorySize / (1024.0 * 1024 * 1024);
        System.out.println("已经使用的内存: " + String.format("%.2f", usedMem) + "GB");
        // 剩余内存
```

```
        System.out.println("剩余内存: " + String.format("%.2f", totalMem - usedMem) + "GB");
        // 获取磁盘信息
        /* 在 Java 中没有直接的方法来获取磁盘的分区信息
           这里我们获取可用的文件系统根，但并不完全等同于 Python 中的磁盘分区 */
        File[] roots = File.listRoots();
        for (File root : roots) {
            System.out.println("硬盘名称: " + root.getAbsolutePath());
            System.out.println("硬盘容量: " + String.format("%.2f", (double) root.getTotalSpace() / (1024 * 1024 * 1024)) + "GB");
        }
    }
}
```

运行程序，会在终端输出如下内容：

```
操作系统版本信息: macOS X 10.16
系统总内存: 32.00GB
已经使用的内存: 27.28GB
剩余内存: 4.72GB
硬盘名称: /
硬盘容量: 2907.13GB
```

2.3.2 使用 OSHI 库获取系统信息

尽管使用 Java 标准库以及随 JDK 发布的一些库（如 OperatingSystemMXBean），可以获得系统信息，但这些系统信息并不全面。而且由于使用的 API 可能在不同平台存在差异，所以代码不一定在所有的 OS 平台上都是有效的。为了尽可能让代码跨平台，以及获取更多的信息，可以使用 OSHI 库。

OSHI（Open Source Health Indicator）库是一个在 Java 中使用的库，专注于提供跨平台的系统信息，不需要本地代理或本地代码。OSHI 库的目标是提供一个简单的方法来访问关键的系统和硬件信息，比如操作系统、物理和虚拟内存、CPU、磁盘、网络接口和传感器等。以下是一些主要特点和功能。

- 操作系统和硬件信息

（1）OSHI 库可以提供关于当前操作系统的详细信息，包括名称、版本、架构、文件系统以及正在运行的进程和线程。

（2）OSHI 库还能提供硬件信息，如 CPU 的详细数据（如型号、核心数、每个核心的利用率等）、物理和虚拟内存的统计数据（如总量、使用量、交换空间等）。

- 系统内存信息

（1）OSHI 库能够提供关于系统内存（RAM）的详细信息，包括总内存、可用内存、已用内存和虚拟内存（交换空间）的使用情况。

（2）OSHI 库还能提供有关物理内存的信息，如大小、时钟速率和物理内存的健康状况。

- 处理器（CPU）信息

OSHI库可以获取非常详细的处理器信息，包括处理器型号、家族、核心数、线程数、频率、负载和温度（如果可用）。

- 磁盘驱动器和文件系统

（1）OSHI 库能提供关于磁盘驱动器的信息，包括型号、序列号、大小以及读写统计。

（2）OSHI 库还可以提供文件系统的信息，包括文件系统类型、可用空间、总空间等。

- 网络接口

OSHI 库可以提供网络接口的详细信息，如 IP 地址、MAC 地址、数据包传输统计等。

- 电源和温度传感器

如果硬件和操作系统支持，OSHI 库还能获取电池状态、电源插拔状态和系统温度。

- 跨平台支持

另外，OSHI 库支持多种操作系统，包括 Windows、Linux、macOS 和 UNIX 系统。

- 不需要本地代码

OSHI 库的一个关键特点是它不依赖于本地代理或本地代码库。它通过 Java 本身的能力和一些跨平台兼容的方法来收集信息。

OSHI 库在系统监控和健康检查的应用程序中非常有用，尤其是在需要收集关于系统性能和资源利用率的详细信息时。它的易用性和跨平台的特性使其成为获取系统信息的理想选择。

读者可以通过下面的链接访问 OSHI 库在 github 上的源代码页面。

https://github.com/oshi/oshi

可以从 maven 仓库安装 OSHI 库，由于 OSHI 库有很多依赖，所以需要安装一些依赖，下面是 pom.xml 中与 OSHI 库相关的配置，输入这些配置后，IntelliJ IDEA 会自动下载 OSHI 相关的库。

```
<dependencies>
    <dependency>
        <groupId>com.github.oshi</groupId>
        <artifactId>oshi-core</artifactId>
        <version>6.4.7</version>
    </dependency>
    <dependency>
        <groupId>org.slf4j</groupId>
        <artifactId>slf4j-api</artifactId>
        <version>1.7.32</version> <!-- 使用最新版本 -->
    </dependency>
    <!-- Log4j SLF4J Binding -->
    <dependency>
        <groupId>org.apache.logging.log4j</groupId>
        <artifactId>log4j-slf4j-impl</artifactId>
        <version>2.14.1</version> <!-- 使用最新版本 -->
    </dependency>
```

```
        <!-- Log4j API and Core -->
        <dependency>
            <groupId>org.apache.logging.log4j</groupId>
            <artifactId>log4j-api</artifactId>
            <version>2.14.1</version> <!-- 使用最新版本 -->
        </dependency>
        <dependency>
            <groupId>org.apache.logging.log4j</groupId>
            <artifactId>log4j-core</artifactId>
            <version>2.14.1</version> <!-- 使用最新版本 -->
        </dependency>
    </dependencies>
```

下面的例子使用 OSHI 库获取了更详细的系统信息。

代码位置：src/main/java/system/info/OSHISystemInfo.java

```java
package system.info;
import oshi.SystemInfo;
import oshi.hardware.GlobalMemory;
import oshi.hardware.HWDiskStore;
import oshi.hardware.HWPartition;
import oshi.software.os.OSProcess;
import oshi.software.os.OperatingSystem;
import java.util.List;
public class OSHISystemInfo {
    public static void main(String[] args) {
        SystemInfo si = new SystemInfo();
        OperatingSystem os = si.getOperatingSystem();
        // 获取操作系统版本信息
        System.out.println("操作系统版本信息：" + os.toString());
        // 获取系统总内存、已经使用的内存、剩余内存
        GlobalMemory memory = si.getHardware().getMemory();
        long totalByte = memory.getTotal();
        long availableByte = memory.getAvailable();
        long usedByte = totalByte - availableByte;

        double totalGB = bytesToGigabytes(totalByte);
        double usedGB = bytesToGigabytes(usedByte);
        double freeGB = bytesToGigabytes(availableByte);

        System.out.println("系统总内存：" + totalGB + " GB");
        System.out.println("已经使用的内存：" + usedGB + " GB");
        System.out.println("剩余内存：" + freeGB + " GB");
        // 获取硬盘信息
        List<HWDiskStore> diskStores = si.getHardware().getDiskStores();
        for (HWDiskStore disk : diskStores) {
            long diskSize = 0;
            for (HWPartition partition : disk.getPartitions()) {
                diskSize += partition.getSize();
                System.out.println("硬盘名称：" + partition.getIdentification());
```

```
                System.out.println("分区容量: " + bytesToGigabytes(partition.
getSize()) + " GB");
            }
            System.out.println("硬盘总容量: " + bytesToGigabytes(diskSize) + " GB");
        }
        // 获取所有进程信息
        for (OSProcess process : os.getProcesses()) {
            System.out.println("进程 ID: " + process.getProcessID());
            System.out.println("进程名: " + process.getName());
            System.out.println("用户: " + process.getUser());
        }
    }
    private static double bytesToGigabytes(long bytes) {
        return bytes / (double)(1024 * 1024 * 1024);
    }
}
```

运行程序，会在终端输出如下内容（部分）：

```
操作系统版本信息: Apple macOS 11.7.6 (Big Sur) build 20G1231
系统总内存: 32.0 GB
已经使用的内存: 16.67049789428711 GB
剩余内存: 15.32950210571289 GB
硬盘名称: disk0s2
分区容量: 112.80464935302734 GB
硬盘名称: disk0s1
分区容量: 0.1953125 GB
硬盘总容量: 112.99996185302734 GB
硬盘名称: disk1s1
... ...
```

Java 标准库与 OSHI 库输出的系统信息可能会有差异，例如，已使用内存和可用内存。当遇到不同库报告不同的系统内存使用情况时，这种差异通常是由于不同的方法和内存度量标准导致的。让我们来看看提到的这两种情况。

- 使用 Java 标准库获取的内存信息

（1）Java 标准库通过 getTotalPhysicalMemorySize() 获取总物理内存大小，然后用 getFreePhysicalMemorySize() 获取空闲的物理内存大小。

（2）已使用的内存计算公式为：总内存 – 空闲内存。

（3）这种方法通常会计算出目前被操作系统及其所有进程占用的内存量。

- 使用 OSHI 库获取的内存信息

（1）OSHI 库是一个用于系统监控的开源库，提供了更多详细和特定的系统信息。

（2）OSHI 库通过 getTotal() 获取总内存，而 getAvailable() 则用于获取当前可用的内存。

（3）已使用的内存同样是通过计算总内存减去可用内存得到的。

（4）OSHI 库可能会考虑更多底层和特定于系统的因素来计算内存使用情况，例如缓存、缓冲区等。

那么两种情况为什么会有差异呢？

（1）度量标准不同：OSHI 库可能在计算内存使用时考虑了更多细节，比如缓存和缓冲区，而 Java 标准库可能仅考虑了基本的已用内存量。

（2）实时数据差异：内存使用情况是实时变化的，不同的库在不同时间点获取的数据可能有所不同。

（3）内存管理机制：操作系统的内存管理机制复杂，不同的工具可能采用不同的方式来解释和度量内存使用情况。

那么哪种方法更准确呢？

（1）这取决于需求。如果需要更通用的信息，Java 标准库提供的数据可能就足够了。如果需要更详细的、可能包括系统特有信息的数据，那么 OSHI 库可能是更好的选择。

（2）通常，没有一个绝对"正确"的数字，因为内存使用情况的度量取决于许多因素和上下文。这两种方法提供的数据都有其适用场景和准确度。

2.4 显示系统窗口

本节会介绍如何在 macOS、Windows 和 Linux 中使用 Java 显示系统窗口，例如，设置窗口、终端等。其实这 3 个操作系统显示系统窗口的方式尽管有一定的差异，但本质都是一样的。通过系统命令可以显示这些窗口，而要想用 Java 显示系统窗口，就需要用 Java 调用这些命令。

2.4.1 显示 macOS 中的系统窗口

在 macOS 中可以使用 open 命令显示系统窗口，也可以启动任何应用程序。例如，使用下面的命令可以显示"系统偏好设置"窗口。

```
open -a "System Preferences"
```

执行这行命令，会弹出如图 2-6 所示的"系统偏好设置"窗口。

open 命令中的 -a 参数是指显示指定应用程序并将文件传递给它。例如想使用 TextEdit 显示文件，则可以使用如下命令。

```
open -a TextEdit file.txt
```

如果只想显示 TextEdit，可以不传入任何文件，命令行如下：

```
open -a TextEdit
```

图 2-6 “系统偏好设置”窗口

这里的 TextEdit 是已经安装的 macOS 应用，在“应用程序”列表中可以查看，但要注意，在“应用程序”列表中显示的都是中文名称，而不是 macOS 应用程序实际的名称，读者可以在 /applications 目录（老版本 macOS 是 /System/Applications 目录）中查看对应的应用程序名称，所有的 macOS 应用程序都是扩展名为 app 的目录，不过可以直接使用 open 命令运行这些应用程序，例如，使用下面的命令可以运行“有道云笔记”。

```
open -a YoudaoNote
```

如果想显示“系统偏好设置”中某个子窗口，如“辅助功能”窗口，可以使用下面的命令（这里不需要加参数 -a）。

```
open /System/Library/PreferencePanes/UniversalAccessPref.prefPane
```

其中 UniversalAccessPref.prefPane 是“辅助功能”对应的应用程序目录，可以用 open 命令直接显示，效果如图 2-7 所示。

在 /System/Library/PreferencePanes 目录包含了“系统偏好设置”中所有子窗口的应用程序目录，读者可以进入该目录，使用 ls 命令查看目录中的内容，如图 2-8 所示。读者可以使用 open 命令运行其中的任何一个 prefPane 文件。

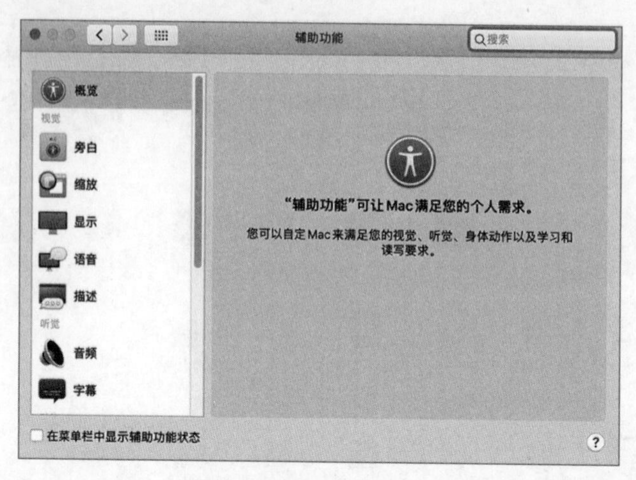

图 2-7 “辅助功能”窗口

```
(base) lining@lining PreferencePanes % ls
Accounts.prefPane                    FamilySharingPrefPane.prefPane    SharingPref.prefPane
Appearance.prefPane                  FibreChannel.prefPane             Sidecar.prefPane
AppleIDPrefPane.prefPane             InternetAccounts.prefPane         SoftwareUpdate.prefPane
Bluetooth.prefPane                   Keyboard.prefPane                 Sound.prefPane
ClassroomSettings.prefPane           Localization.prefPane             Speech.prefPane
DateAndTime.prefPane                 Mouse.prefPane                    Spotlight.prefPane
DesktopScreenEffectsPref.prefPane    Network.prefPane                  StartupDisk.prefPane
DigiHubDiscs.prefPane                Notifications.prefPane            TimeMachine.prefPane
Displays.prefPane                    PrintAndFax.prefPane              TouchID.prefPane
Dock.prefPane                        PrintAndScan.prefPane             Trackpad.prefPane
EnergySaver.prefPane                 Profiles.prefPane                 UniversalAccessPref.prefPane
Expose.prefPane                      ScreenTime.prefPane               Wallet.prefPane
Extensions.prefPane                  Security.prefPane
(base) lining@lining PreferencePanes %
```

图 2-8 “系统偏好设置”中所有子窗口的应用程序目录

在 Java 中可以通过执行 open 命令显示 macOS 下的任何应用程序，例如，下面的代码使用 Java 显示了“鼠标设置”窗口。

```
Runtime.getRuntime().exec(new String[]{"open",
        "/System/Library/PreferencePanes/Mouse.prefPane"});
```

下面的例子在终端建立了一个菜单，读者输入菜单序号，并按 Enter 键，系统会根据用户的选择显示“系统偏好设置”中对应的子窗口。

代码位置：src/main/java/system/window/SystemPreferencesOpener.java

```
package system.window;
import java.io.IOException;
import java.util.Scanner;
import java.util.Arrays;
import java.util.List;
public class SystemPreferencesOpener {
    // 显示选中的系统偏好窗口
    public static void openPreferencePane(String prefPane) {
```

```
        try {
            Runtime.getRuntime().exec(new String[]{"open", "/System/Library/
PreferencePanes/" + prefPane + ".prefPane"});
        } catch (IOException e) {
            e.printStackTrace();
        }
    }
    public static void main(String[] args) {
        // 定义系统偏好窗口对应的应用程序目录名
        List<String> preferencePanes = Arrays.asList(
                "DateAndTime",
                "Displays",
                "Extensions",
                "General",
                "iCloud",
                "Keyboard",
                "LanguageRegion",
                "Mouse",
                "Network",
                "Notifications",
                "ParentalControls",
                "PrintersScanners",
                "Profiles",
                "SecurityPrivacy",
                "Sharing"
        );
        Scanner scanner = new Scanner(System.in);
        // 创建可以循环输入序号的终端程序，输入 q 退出程序
        while (true) {
            System.out.println("请选择一个设置窗口：");
            for (int i = 0; i < preferencePanes.size(); i++) {
                System.out.println((i + 1) + ". " + preferencePanes.get(i));
            }
            System.out.println("输入 q 退出程序");
            System.out.println("请输入选项：");
            String choice = scanner.nextLine();
            if ("q".equals(choice)) {
                break;
            }
            try {
                int option = Integer.parseInt(choice);
                if (option < 1 || option > preferencePanes.size()) {
                    throw new NumberFormatException();
                }
                // 执行打开系统偏好窗口的命令
                openPreferencePane(preferencePanes.get(option - 1));
            } catch (NumberFormatException e) {
                System.out.println("无效的选项，请重新输入");
            }
        }
        scanner.close();
```

```
        }
}
```

执行代码，会输出如图 2-9 所示的选择菜单。

读者可以输入 1 到 15 的数字，例如，输入 9，按 Enter 键，就会弹出如图 2-10 所示的“网络”窗口。

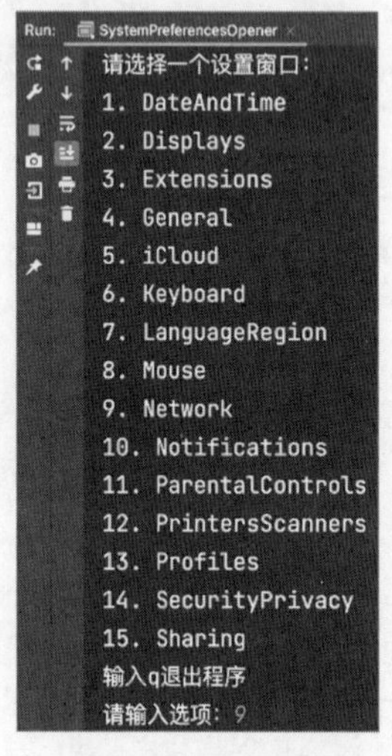

图 2-9 选择菜单

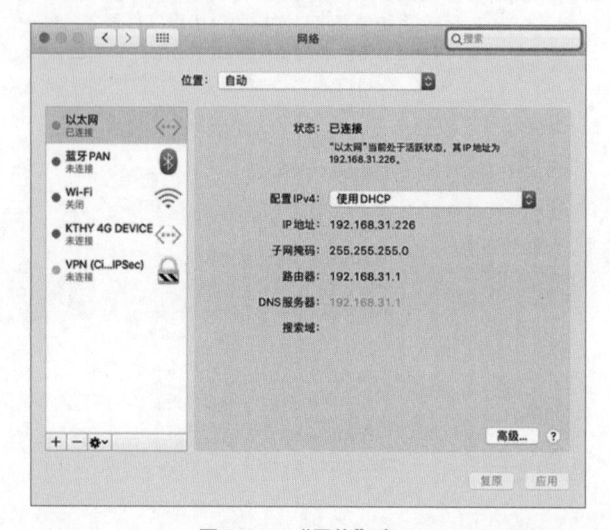

图 2-10 “网络”窗口

2.4.2 显示 Windows 中的系统窗口

在 Windows 中通过 control 命令可以显示控制面板的相关窗口，例如，“个性化设置”窗口、“鼠标设置”窗口等。下面的命令显示一些常用的设置窗口。

（1）control：显示控制面板。

（2）control admintools：显示管理工具。

（3）control desktop：显示个性化设置。

（4）control keyboard：显示键盘设置。

（5）control mouse：显示鼠标设置。

（6）control printers：显示打印机设置。

（7）control userpasswords2：显示用户账户设置。

在 Windows 终端中输入上面的命令，就会立刻显示对应的窗口，例如，执行 **control mouse** 命令，会弹出如图 2-11 所示的“鼠标属性”窗口。

control 命令还可以显示更多的设置窗口，例如，下面的命令显示 Windows 设置窗口。

```
control /name Microsoft.System
```

执行这行代码，会显示如图 2-12 所示的窗口。

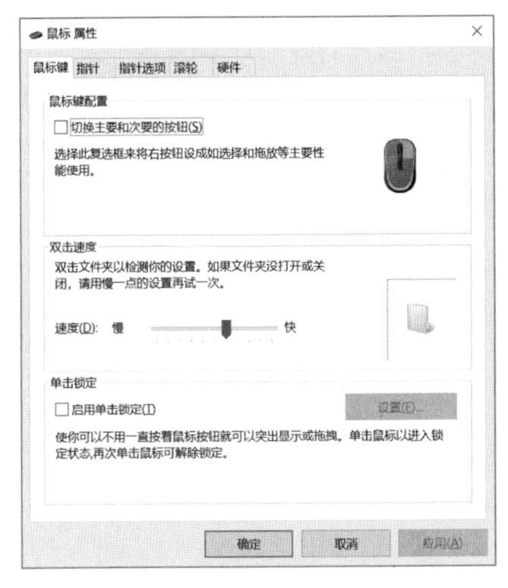

图 2-11　“鼠标属性”窗口

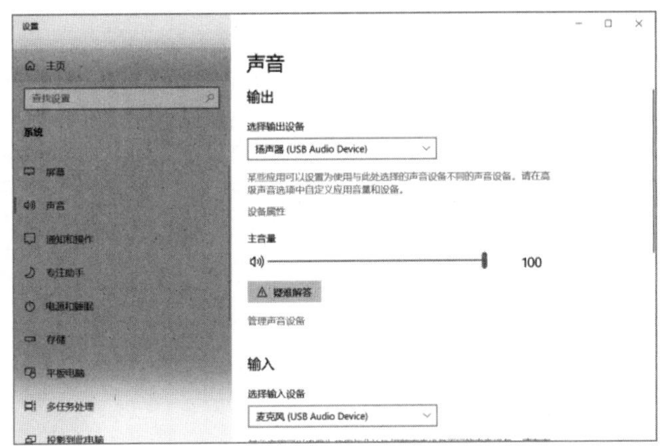

图 2-12　Windows 设置窗口

使用 Java 执行这些命令，就可以通过 Java 显示这些窗口，如下面的代码同样可以显示 Windows 设置窗口。

```
Runtime.getRuntime().exec("control /name Microsoft.System");
```

比 control 命令更强大的是 start 命令，该命令用于启动一个新的进程并显示指定的文件或应用程序。例如，下面的命令同样可以显示 Windows 设置窗口。

```
start ms-settings:
```

下面的命令用于显示“背景设置”窗口。

```
start ms-settings:personalization-background
```

下面的例子在终端建立一个菜单，读者输入菜单序号，并按 Enter 键，系统会根据用户的选择显示“Windows 设置”窗口中对应的子窗口。

代码位置：src/main/java/system/window/WindowsSettingsOpener.java

```
package system.window;
import java.io.IOException;
import java.util.Scanner;
public class WindowsSettingsOpener {
    public static void main(String[] args) {
        Scanner scanner = new Scanner(System.in);
        while (true) {
            // 在终端中显示菜单，并接收用户的选择
```

```
            System.out.println("请选择一个子窗口：\n1. 个性化\n2. 背景\n3. 锁屏\n4. 
任务栏\n5. 通知和动作\n6. 电源和睡眠\n7. 存储\n输入q退出。");
            String windowNum = scanner.nextLine();
            if (windowNum.matches("\\d+") && Integer.parseInt(windowNum) >= 1 
&& Integer.parseInt(windowNum) <= 7) {
                // 根据用户的选择显示对应的窗口
                openWindow(windowNum);
            } else if (windowNum.equalsIgnoreCase("q")) {
                // 如果用户输入q，则退出程序
                break;
            } else {
                System.out.println("输入无效。");
            }
        }
        scanner.close();
    }
    private static void openWindow(String windowNum) {
        String command = null;
        switch (windowNum) {
            case "1":
                command = "ms-settings:personalization";
                break;
            case "2":
                command = "ms-settings:personalization-background";
                break;
            case "3":
                command = "ms-settings:lockscreen";
                break;
            case "4":
                command = "ms-settings:taskbar";
                break;
            case "5":
                command = "ms-settings:notifications";
                break;
            case "6":
                command = "ms-settings:powersleep";
                break;
            case "7":
                command = "ms-settings:storagesense";
                break;
        }
        if (command != null) {
            try {
                // 执行打开Windows设置面板的命令
                Runtime.getRuntime().exec("cmd /c start " + command);
            } catch (IOException e) {
                e.printStackTrace();
            }
        }
    }
}
```

运行程序，会看到如图 2-13 所示的菜单。

输入一个菜单项序号，如 6，按 Enter 键，会显示如图 2-14 所示的“电源和睡眠”设置窗口。

图 2-13　选择菜单

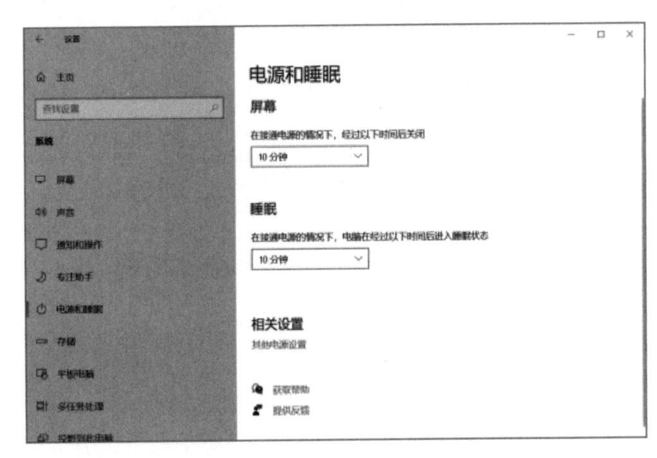

图 2-14　“电源和睡眠”设置窗口

2.4.3　显示 Linux 中的系统窗口

显示 Linux 中的系统窗口需要执行相关的命令，如果要用 Java 来显示这些窗口，就要在 Java 中执行这些命令。

下面的例子实现了一个终端程序，显示一个菜单，每一个菜单项对应“设置”窗口的一项，当用户输入菜单项序号，并按 Enter 键时，就会显示“设置”窗口，并切换到对应的设置页面。

代码位置：src/main/java/system/window/LinuxSettingsOpener.java

```
package system.window;

import java.io.IOException;
import java.util.Scanner;
public class LinuxSettingsOpener {
    public static void main(String[] args) {
        Scanner scanner = new Scanner(System.in);
        while (true) {
            // 显示菜单
            System.out.println("1. 背景 ");
            System.out.println("2. 蓝牙 ");
            System.out.println("3. 网络 ");
            System.out.println("4. 电源 ");
            System.out.println("5. 声音 ");
            System.out.println("6. 显示 ");
            System.out.println("7. 日期和时间 ");
```

```
            System.out.println("q. 退出");
            // 获取用户输入
            System.out.println("请输入菜单项序号：");
            String choice = scanner.nextLine();
            // 根据用户输入执行相应的操作
            switch (choice) {
                case "1":
                    openGnomeControlCenter("background");
                    break;
                case "2":
                    openGnomeControlCenter("bluetooth");
                    break;
                case "3":
                    openGnomeControlCenter("network");
                    break;
                case "4":
                    openGnomeControlCenter("power");
                    break;
                case "5":
                    openGnomeControlCenter("sound");
                    break;
                case "6":
                    openGnomeControlCenter("display");
                    break;
                case "7":
                    openGnomeControlCenter("datetime");
                    break;
                case "q":
                    System.exit(0);                         // 正常退出程序
                default:
                    System.out.println("输入无效。");
                    break;
            }
        }
    }
    private static void openGnomeControlCenter(String setting) {
        ProcessBuilder processBuilder = new ProcessBuilder("gnome-control-
center", setting);
        try {
            processBuilder.start();                         // 启动对应的控制中心面板
        } catch (IOException e) {
            e.printStackTrace();
        }
    }
}
```

运行程序，会显示一个菜单，如图 2-15 所示。

输入一个菜单项的序号，如 5，按 Enter 键后，就会显示与之对应的设置页面，如图 2-16 所示。

图 2-15　选择菜单

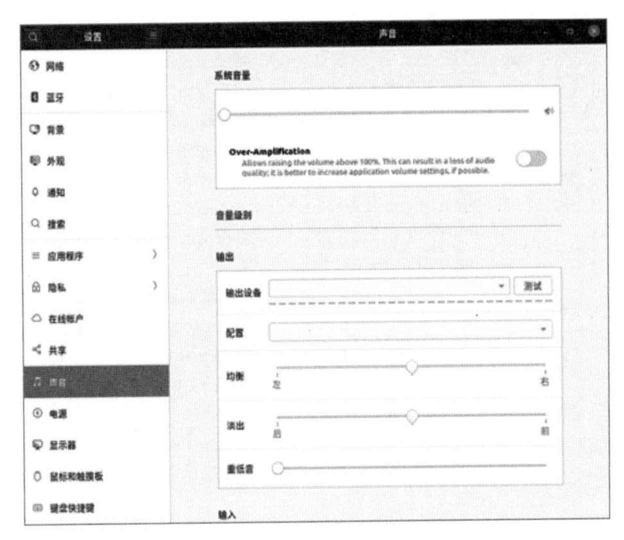

图 2-16　“声音”设置页面

2.5　打开文件夹

在很多场景，需要用程序控制打开操作系统的文件夹。例如，在某个软件系统中有一个缓存目录，可以提供一个按钮直接打开缓存目录，这样用户就可以直接定位到这个目录了。回收站是一类特殊的目录，所以也可以使用打开普通目录的方式打开回收站目录。

2.5.1　打开 macOS 文件夹与回收站

使用 **open** 命令可以打开文件夹，例如，使用下面的命令可以打开 ~/Documents 文件夹。

```
open ~/Documents
```

使用下面的命令可以打开回收站文件夹。

```
open ~/.Trash
```

用 Java 完成同样的操作，可以使用下面的代码。

代码位置： src/main/java/open/folder/OpenMacFolder.java

```
package open.folder;
import java.awt.Desktop;
import java.io.File;
import java.io.IOException;
public class OpenWindowsFolder {
    public static void main(String[] args) {
```

```
        // 打开 ~/Documents 文件夹
        openDirectory(System.getProperty("user.home") + "/Documents");
        // 打开回收站文件夹
        openDirectory(System.getProperty("user.home") + "/.Trash");
    }
    private static void openDirectory(String path) {
        if (Desktop.isDesktopSupported()) {
            Desktop desktop = Desktop.getDesktop();
            try {
                File directoryToOpen = new File(path);
                desktop.open(directoryToOpen);          // 使用桌面类打开文件夹
            } catch (IllegalArgumentException | IOException e) {
                e.printStackTrace();
            }
        } else {
            System.out.println("桌面功能不支持在当前平台上运行。");
        }
    }
}
```

2.5.2 打开 Windows 文件夹与回收站

使用 start 命令可以打开文件夹，例如，使用下面的命令可以打开 D:\test 目录。

```
start d:\test
```

使用下面的命令可以打开回收站文件夹。

```
start shell:RecycleBinFolder
```

用 Java 完成同样的操作，可以使用下面的代码。

代码位置：src/main/java/open/folder/OpenWindowsFolder.java

```
package open.folder;
import java.io.IOException;
public class OpenWindowsFolder {
    public static void main(String[] args) {
        // 打开 D:\test 目录
        openDirectory("d:\\test");
        // 打开回收站目录
        openDirectory("shell:RecycleBinFolder");
    }
    private static void openDirectory(String directory) {
        try {
            // 使用 Runtime.exec 方法执行 Windows 命令打开目录
            Runtime.getRuntime().exec("cmd /c start " + directory);
        } catch (IOException e) {
            // 打印异常信息
```

```
            e.printStackTrace();
        }
    }
}
```

2.5.3 打开 Linux 文件夹与回收站

使用 xdg-open 命令可以打开文件夹，例如，使用下面的命令可以打开“~/ 文档”目录。

```
start d:\test
```

使用下面的命令可以打开回收站文件夹。

```
gio open trash://
```

用 Java 完成同样的操作，可以使用下面的代码。

代码位置：src/main/java/open/folder/OpenLinuxFolder.java

```
package open.folder;
import java.io.IOException;
public class OpenLinuxFolder{
    public static void main(String[] args) {
        // 打开 "~/ 文档 " 目录
        openDirectory("xdg-open", System.getProperty("user.home") + "/ 文档 ");
        // 打开回收站目录
        openDirectory("gio", "open", "trash://");
    }
    private static void openDirectory(String... command) {
        try {
            // 使用 ProcessBuilder 类来启动系统命令
            new ProcessBuilder(command).start();
        } catch (IOException e) {
            // 如果发生错误，打印错误信息
            e.printStackTrace();
        }
    }
}
```

2.6 小结

经过对本章的学习，相信读者又解锁了很多新技能。原来 Java 还可以这样玩。本章使用 Java 操控 OS 主要用了三招：第三方库（如 OHSI、JNA 等）、读写系统文件和执行系统命令。通过这三板斧，使用 Java 几乎可以控制 OS 的一切。本章的内容只是抛砖引玉，读者可以利用这三招发掘出更多操控 OS 的方法。

第 3 章

GUI 工具包——JavaFX

本章介绍了 JavaFX 的核心使用方法，JavaFX 是基于 Java 的一种用于构建富客户端应用程序的库。JavaFX 提供了一套丰富的界面组件、布局、动画和图形渲染工具，使开发者能够创建跨平台的桌面应用程序。本章旨在通过实例和详细说明，使读者能够理解和掌握 JavaFX 的基础知识和核心概念。首先，本章讨论了 Java 支持的各种 GUI 框架，包括 Swing、AWT、Apache Pivot 等，并对比了它们与 JavaFX 的优缺点。接着，本章详细介绍了如何安装和配置 JavaFX 环境，涵盖了从下载 JavaFX SDK 到配置项目的所有必要步骤。本章还详细介绍了 JavaFX 的关键概念，如 Stage、Scene、Layout 等，并通过实例演示了这些概念的应用。此外，本章还介绍了 JavaFX 中常用的组件和布局，如 ListView、ComboBox、TableView、TreeView 等，以及如何使用它们创建复杂和响应式的用户界面。接下来，本章还涵盖了如何在 JavaFX 应用程序中创建和使用菜单和对话框，包括标准 JavaFX 对话框和 ControlsFX 提供的扩展对话框。最后，本章通过两个项目（自由绘画和旋转图像），让读者了解如何用 JavaFX 实现一个完整的项目。

3.1 Java 支持哪些 GUI 框架

Java 支持多种 GUI（Graphical User Interface 图形用户界面）框架，每个框架都有其独特的特点和适用场景。以下是一些流行的 Java GUI 框架，以及它们的优缺点。

❑ Swing

优点：Swing 是 JDK 的一部分，因此无须额外安装；Swing 提供丰富的组件和可定制性；跨平台兼容性好。

缺点：界面不如原生应用程序流畅，对初学者来说可能有点复杂。

Jetbrains 开发的 IDE 家族（如 IntelliJ IDEA、PyCharm、WebStorm 等）就使用了 Swing 进行开发，只不过进行了高度定制，让界面效果尽可能接近现代 UI 风格。

❑ JavaFX

优点：现代的 GUI 框架，支持 CSS 和 FXML，易于设计美观的界面；良好的图形和动画支持；更好的高分辨率支持。

缺点：自 Java 11 起不再包含在 JDK 中，需要单独安装；相比 Swing，生态系统较小。

- AWT (Abstract Window Toolkit)

优点：Java 的原生 GUI 工具包，简单易用；直接使用操作系统的 GUI 控件，因此具有原生外观。

缺点：组件较少，功能有限；界面和性能在不同平台上可能有差异。

- Apache Pivot

优点：结合了 Web 技术的特点，支持 XML 和 JSON 来描述界面；具有较好的性能。

缺点：社区和生态系统相对较小；学习曲线相对陡峭。

- SWT (Standard Widget Toolkit)

优点：由 Eclipse 开发，提供丰富的组件；使用操作系统的原生控件，性能较好。

缺点：与操作系统绑定较紧密，跨平台兼容性不如 Swing 或 JavaFX。

这些 GUI 框架都可以开发出非常强大的 GUI 程序，只不过难易程度不同。如果读者刚开始学习用 Java 开发 GUI 应用，推荐使用 JavaFX，因为 JavaFX 是目前 Java 推荐的 GUI 框架，社区非常活跃，而且 JavaFX 支持非常多的现代 GUI 个性，还支持 Java 与 JavaScript 混合开发。而 Swing 尽管不需要独立安装，也可以开发出像 PyCharm 一样强大的 IDE,但需要更多的定制。也就是说,使用 Swing 与 JavaFX 开发效果相同的复杂 GUI 应用，Swing 需要更多的代码，难度更大。所以本章将以 JavaFX 为主介绍如何用 Java 开发 GUI 应用。

3.2 安装和配置 JavaFX

JavaFX[①]并不是 JDK 的一部分，所以使用 JavaFX 之前首先要下载 JavaFX 安装包。进入下面的页面：

```
https://gluonhq.com/products/javafx
```

进入下载页面，在页面的中部会看到如图 3-1 所示的内容。

下载页面有两个 JavaFX 类型：SDK 和 jmods。其中 SDK 是用于开发的，如果读者是以开发为目的，可以下载 SDK 版本。jmods 包含了一些工具，用于生成可以运行 JavaFX 的最小环境，如果读者要发布使用 JavaFX 的应用，可以下载这个版本。

JavaFX 目前最新的版本号是 21.0.1，这是一个长期维护的版本。推荐读者下载这个版本。下载文件是一个压缩包，解压后，通常会有 lib 目录，Windows 版还会有一个 bin 目录。由于 JavaFX 调用了很多操作系统的 API，所以不同操作系统版本的 JavaFX 是不能通用的。

---

① JavaFX 目前由 OpenJFX 项目维护。OpenJFX 是 JavaFX 的开源版本，自 Oracle 将 JavaFX 开源以来，它已经成为 JavaFX 开发和维护的主要平台。OpenJFX 项目的主页是 https://openjfx.io。下载 JavaFX 时，会导航到 3.2 节给出的下载页面。

对于 macOS 平台的 JavaFX 压缩包中，所有的核心文件都在 lib 目录中。在 lib 目录中主要的文件类型是 jar 和 dylib，其中 jar 是 Java 要引用的 SDK 文件，dylib 是 macOS 下的动态链接库（类似 Windows 中的 dll），如图 3-2 所示。

Downloads

JavaFX version: 21.0.1 [LTS]　Operating System: [any]　Architecture: [any]　Type: [any]

Include older versions

Supported Platforms

OS	Version	Architecture	Type	Download
Linux	21.0.1	x64	SDK	Download [SHA256]
Linux	21.0.1	x64	jmods	Download [SHA256]
macOS	21.0.1	aarch64	SDK	Download [SHA256]
macOS	21.0.1	aarch64	jmods	Download [SHA256]
macOS	21.0.1	x64	SDK	Download [SHA256]
macOS	21.0.1	x64	jmods	Download [SHA256]
Windows	21.0.1	x64	SDK	Download [SHA256]
Windows	21.0.1	x64	jmods	Download [SHA256]
Javadoc	21.0.1		Javadoc	Download [SHA256]

图 3-1　JavaFX 下载页面

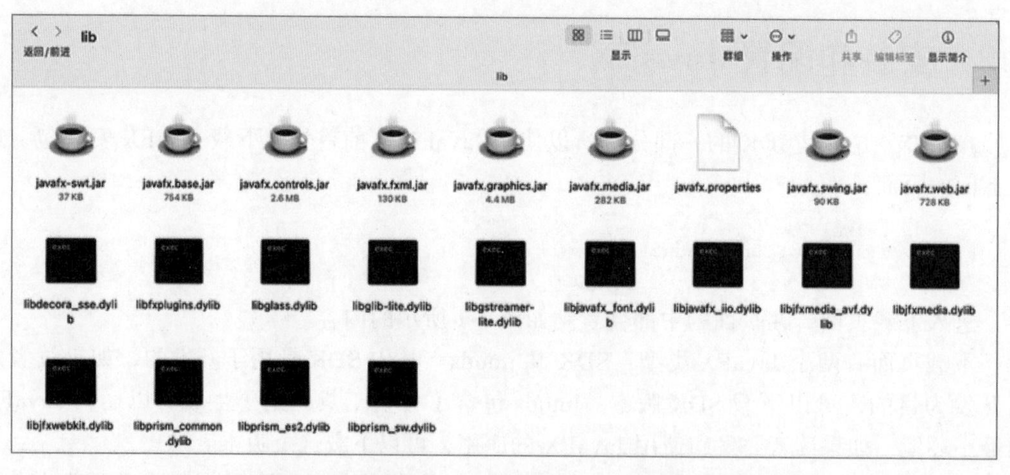

图 3-2　JavaFX for macOS 压缩包 lib 目录中的文件

Windows 平台的 JavaFX 压缩包中，lib 目录只有 jar 文件，而 bin 目录中有很多 dll 文件。

不管是哪一个操作系统平台，都将这些 jar 文件和 dylib 文件（或 Windows 的 dll 文件，Linux 是 .so 文件）统统复制到 Java 工程中。目录名可以随便起，后面需要通过 VM Options 指定这些路径。

复制完 JavaFX 压缩包中的文件后，在 IntelliJ IDEA 中引用所有的 jar 文件。到这里，就可以编写基于 JavaFX 的 GUI 程序了，不过要注意，只是编写，还不能运行。原因主要有如下两个：

（1）确认 JDK 版本。

（2）设置 VM Options。

不同版本的 JavaFX，是有最低 JDK 版本限制的。例如，JavaFX 21 要求 JDK 至少是 17，当然版本越高越好。如果 JDK 版本过低，运行基于 JavaFX 的程序时，就会输出类似下面的错误：

```
java: 无法访问 javafx.application.Application
  错误的类文件 : /System/Volumes/Data/MyStudio
  /marvellous_java_src/libs/javafx/javafx.graphics.jar!/javafx/application/
Application.class
  类文件具有错误的版本 61.0, 应为 58.0
 请删除该文件或确保该文件位于正确的类路径子目录中。
```

这个错误表明 JavaFX 的相关类是针对 JDK17（类的版本号是 61.0）编译的，而当前的 JDK 版本是 14（类的版本号是 58.0）。所以要么升级 JDK，要么降级 JavaFX。这里推荐升级 JDK，否则无法体验最新的 JavaFX 特性。

目前最新的 JDK 版本是 21，读者可以从下面的页面下载 JDK 21 或 JDK 17。

```
https://www.oracle.com/java/technologies/downloads
```

对于普通的第三方 Java 库，引用 jar 文件后就可以运行了，但 JavaFX 比较特殊，还需要使用 VM Options 添加一些选项。现在打开 Run/Debug Configurations 对话框，选择左侧 Application 中要运行的程序，在右侧单击 Modify Options，在弹出菜单中单击 Add VM Options 菜单项，就会在 Run/Debug Configurations 对话框中出现 VM Options 输入文本框，如图 3-3 所示。

要输入的 VM Options 格式如下：

```
--module-path jar_path --add-modules javafx.controls,javafx.fxml,
javafx.graphics -Djava.library.path=lib_path
```

这里面需要设置如下三个参数。

（1）--module-path：jar 文件的路径。

（2）--add_modules：程序中要使用的 JavaFX 模块。

（3）-Djava.library.path：动态链接库（.dylib、.dll、.so）文件所在的路径。

读者在设置 VM Options 时，应将 --module-path 和 -Djava.library.path 指定的路径（jar_path 和 lib_path）修改成自己机器上对应的路径。

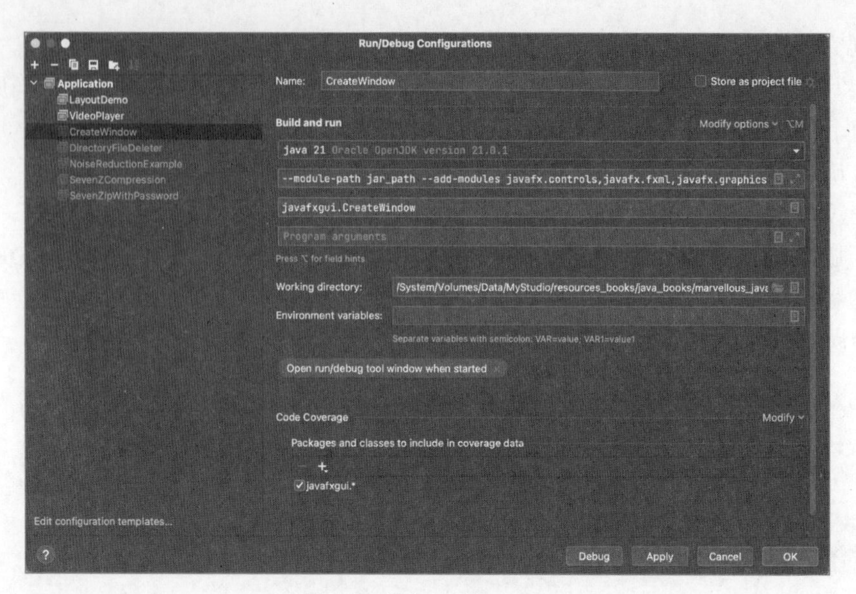

图 3-3　输入 VM Options

3.3　创建窗口

JavaFX 是一个用于构建富客户端应用程序的库。它提供了一套丰富的界面组件、布局、动画和图形渲染工具，允许开发者创建跨平台的桌面应用程序。在这一节我们会使用 JavaFX 创建一个红色背景的窗口，通过这个程序，读者可以了解使用 JavaFX 创建一个窗口程序的必要步骤。

在编写 JavaFX 程序的过程中，会涉及一些关键的类和方法，下面是对这些类和方法的详细描述。

（1）Application 类：JavaFX 应用程序的主要入口点。每个 JavaFX 应用程序必须继承自 Application 类。Application 类提供了 launch() 方法来启动应用程序。这个方法初始化应用程序的环境并调用 start() 方法。

（2）start(Stage primaryStage) 方法：该方法 Application 类的抽象方法，必须覆盖，用于设置 Stage（舞台），并开始 JavaFX 应用程序的用户界面。primaryStage 参数是主要的顶级容器。它代表了一个窗口，在这个窗口中，应用程序的所有可视部分都会显示。

（3）Stage 类：Stage 在 JavaFX 中代表一个窗口。它是一个独立的顶级容器，拥有自己的标题栏、边框，以及系统特定的窗口装饰。Stage 是所有 JavaFX 应用程序用户界面元素的外层容器。可以将一个或多个 Scene（场景）添加到 Stage 中，但在任何时候只能显示一个 Scene。

（4）StackPane 类：一种布局容器，可以在上面堆叠子节点（组件）。在 StackPane 中，子节点会按照它们被添加的顺序堆叠起来，最新添加的节点在顶部。

（5）Scene 类：代表 JavaFX 中的一个场景，它是容纳所有内容的容器，如 UI 组件（如按钮、文本框等）。可以设置 Scene（场景）的大小和背景色。场景需要一个根节点（如本例中的 StackPane），所有其他元素都是这个根节点的子节点。

编写 JavaFX 应用程序的基本步骤如下。

（1）应用启动：使用 launch(args) 启动 JavaFX 应用程序。

（2）创建 Stage（舞台）：start() 方法被调用，并接收一个 Stage 对象（在这里是 primaryStage）。

（3）设置根节点：创建 StackPane 对象作为根节点。

（4）创建 Scene（场景）：使用根节点创建 Scene 对象，并设置大小和背景色。

（5）配置 Stage：设置 Stage 的标题、图标，并将 Scene 放置到 Stage 上。

（6）显示 Stage：调用 primaryStage.show() 显示窗口。

下面的例子使用 JavaFX 创建一个红色背景的窗口，并设置了窗口尺寸、窗口图标和窗口标题。

代码位置：src/main/java/javafxgui/CreateWindow.java

```
package javafxgui;
import javafx.application.Application;
import javafx.scene.Scene;
import javafx.scene.layout.StackPane;
import javafx.scene.paint.Color;
import javafx.stage.Stage;
import javafx.scene.image.Image;

// 主类继承自 Application 类
public class CreateWindow extends Application {

    @Override
    public void start(Stage primaryStage) {
        // 创建一个栈面板作为根节点
        StackPane root = new StackPane();
        // 设置场景，包括根节点和窗口尺寸
        Scene scene = new Scene(root, 300, 220, Color.RED);   // 设置背景为红色
        // 设置窗口标题
        primaryStage.setTitle("窗口标题");
        // 设置窗口图标
        primaryStage.getIcons().add(new Image("file:images/tray.png"));
        // 设置窗口的场景
        primaryStage.setScene(scene);
        // 显示窗口
        primaryStage.show();
    }
    public static void main(String[] args) {
        launch(args);
    }
}
```

运行程序，会看到如图 3-4 所示的窗口。我们可以看到，图标显示在了窗口标题的左侧。这是 macOS 版的窗口，Windows 版和 Linux 版的窗口的效果略有差异，但总体的效果是一样的。

注意：运行程序之前，要将图标文件的路径修改为本机存在的路径。而且要像 3.2 节介绍的那样设置 VM Options。本章所有的程序都需要设置 VM Options。

图 3-4 红色背景的窗口（macOS 版）

3.4 布局

JavaFX 提供了一系列的布局容器，用于以一定的逻辑组织界面组件（如按钮、文本框等）。布局容器负责自动计算其子节点的大小和位置，从而响应界面大小的变化或内容的变动。下面是 JavaFX 中常用的几种布局容器。

（1）HBox：布局容器将其子节点水平排列，即从左到右。可以设置节点之间的间距，以及内边距（padding）。

（2）VBox：布局容器将其子节点垂直排列，即从上到下。同样可以设置节点间的间距和内边距，还可以设置子节点的最大宽度或高度，以允许子节点水平或垂直填满容器。

（3）GridPane：布局容器将子节点放置在网格中的行和列。这种布局非常适合创建复杂的基于网格的布局。可以独立设置每行和每列的间距，还可以控制单元格中子节点的对齐方式。

（4）BorderPane：布局容器将界面分为上、下、左、右、中五个区域。每个区域可以只放置一个节点，适用于整体布局的外围结构。

（5）StackPane：布局容器将所有子节点堆叠在一起，放置于容器的中心。通常用于重叠控件，如将文本放在图像上。

（6）TilePane：一个以网格形式排列相同大小块的布局容器。适合创建均匀的网格布局，类似 GridPane，但不需要为每个节点指定行列位置。

（7）FlowPane：按顺序排列子节点，当一行或一列满时，会自动换行或换列。类似 HTML 中的文本流，适用于创建不规则布局。

下面是一些布局的主要用法。

- HBox：水平布局

```
HBox hbox = new HBox(10);
hbox.setPadding(new Insets(20, 0, 0, 0));
```

❑ VBox：垂直布局

```
VBox vbox = new VBox(10);
Button btn = new Button(String.valueOf(i));
btn.setMaxWidth(Double.MAX_VALUE);
VBox.setMargin(btn, new Insets(0, 20, 0, 20));
```

❑ GridPane：网格布局

```
GridPane grid = new GridPane();
grid.setHgap(20);
grid.setVgap(20);
```

❑ GridPane（模拟表单布局）

```
GridPane form = new GridPane();
form.add(new Label("Name:"), 0, 0);
form.add(new Button("Button"), 1, 0);
```

使用这些布局容器，可以创建各种复杂和响应式的用户界面。每种布局都有自己的特性和适用场景，了解它们如何工作以及如何相互配合是进行 JavaFX 界面设计的关键。

下面的例子完整地演示了 HBox、VBox、GridPane 等布局的使用方法。

代码位置：src/main/java/javafxgui/LayoutDemo.java

```
package javafxgui;

import javafx.application.Application;
import javafx.geometry.Insets;
import javafx.geometry.Pos;
import javafx.scene.Scene;
import javafx.scene.control.Button;
import javafx.scene.control.Label;
import javafx.scene.layout.HBox;
import javafx.scene.layout.VBox;
import javafx.scene.layout.GridPane;
import javafx.stage.Stage;

public class LayoutDemo extends Application {
    @Override
    public void start(Stage primaryStage) {
        // 水平布局
        HBox hbox = new HBox(10);                              // 按钮之间的间距
        hbox.setAlignment(Pos.CENTER);
        hbox.setPadding(new Insets(20, 0, 0, 0));              // 上边缘距离
        for (int i = 0; i < 5; i++) {
            hbox.getChildren().add(new Button(String.valueOf(i)));
        }
```

```
        // 垂直布局
        VBox vbox = new VBox(10);
        vbox.setAlignment(Pos.CENTER);
        for (int i = 0; i < 3; i++) {
            Button btn = new Button(String.valueOf(i));
            btn.setMaxWidth(Double.MAX_VALUE);          // 按钮水平充满整个窗口
            VBox.setMargin(btn, new Insets(0, 20, 0, 20));
            // 设置按钮两端到窗口边缘的距离
            vbox.getChildren().add(btn);
        }
        // 网格布局
        GridPane grid = new GridPane();
        grid.setAlignment(Pos.CENTER);
        grid.setHgap(20);                               // 设置水平间距
        grid.setVgap(20);                               // 设置垂直间距
        for (int i = 0; i < 3; i++) {
            for (int j = 0; j < 3; j++) {
                Button button = new Button(String.valueOf(3 * i + j + 1));
                button.setPrefWidth(60);                // 按钮拉长一点
                grid.add(button, j, i);
            }
        }
        // 表单布局
        GridPane form = new GridPane();
        form.setAlignment(Pos.CENTER);
        form.setHgap(10);
        form.setVgap(10);
        String[] labels = {"Name:", "Age:", "Job:"};
        for (int i = 0; i < labels.length; i++) {
            Label label = new Label(labels[i]);
            Button btn = new Button("Button");
            form.add(label, 0, i);                      // 添加标签
            form.add(btn, 1, i);                        // 添加按钮
        }
        // 整体垂直布局
        VBox vbox_all = new VBox(10);
        vbox_all.getChildren().addAll(hbox, vbox, grid, form);

        // 创建场景
        Scene scene = new Scene(vbox_all, 400, 400);
        primaryStage.setTitle(" 布局演示 ");
        primaryStage.setScene(scene);
        primaryStage.show();
    }
    public static void main(String[] args) {
        launch(args);
    }
}
```

运行程序，会看到如图 3-5 所示的布局效果。

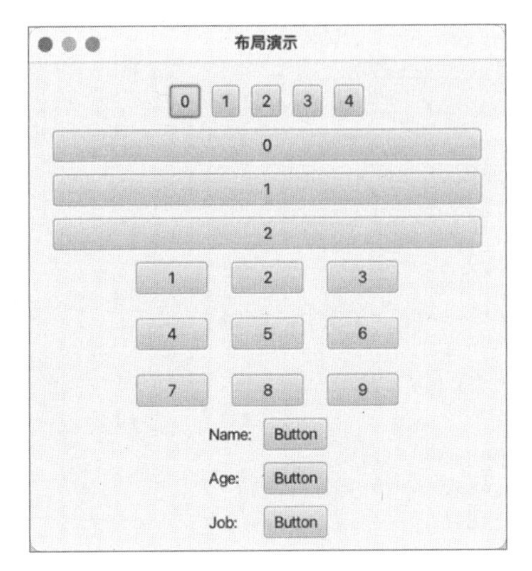

图 3-5　布局演示

3.5　常用组件

JavaFX 提供了一些常用的组件，如按钮、标签、文本框等，方便用户进行界面设计和交互。PyQt 6 中常用的组件包括 CheckBox（复选框）、Button（按钮）、Slider（滑竿）、ProgressBar（进度条）、DatePicker（选择日期）、TimePicker（选择时间）、Label（标签）、TextField（行文本编辑器）、TextArea（文本编辑器）等。

大多数组件的使用方法类似，每一个组件都对应一个类，类名就是前面给出的组件英文名，如 TextField、Button 等。创建一个组件，其实就是创建这个组件类的对象，然后将组件放到窗口上，例如，下面的代码创建一个 Button 组件。

```
Button button = new Button("显示");
```

创建完组件后，将 button 添加进 HBox、VBox 然后等布局，再将这些布局与 Scene 绑定，就可以将 button 显示在窗口上了。

下面的例子会用 JavaFX 创建一个窗口，并放置 3 个 TextField 组件、3 个 Label 组件和 3 个 CheckBox 组件。根据这些 CheckBox 组件是否选中，控制前面的 TextField 组件是否可以输入文本，如果不能输入文本，将 TextField 组件的背景色设置为红色，否则恢复为白色。在组件的下方还会放置一个按钮组件和一个 Label 组件，单击按钮组件，会将这 3 个 TextField 组件中输入的内容组合成一个字符串，显示在最下方的 Label 组件中。

代码位置：src/main/java/javafxgui/Components.java

```
package javafxgui;
```

```
import javafx.application.Application;
import javafx.geometry.Insets;
import javafx.scene.Scene;
import javafx.scene.control.*;
import javafx.scene.layout.HBox;
import javafx.scene.layout.Priority;
import javafx.scene.layout.VBox;
import javafx.stage.Stage;
public class Components extends Application {
    @Override
    public void start(Stage primaryStage) {
        // 创建垂直布局
        VBox vBox = new VBox();
        vBox.setPadding(new Insets(10));
        vBox.setSpacing(10);

        // 创建姓名行
        HBox nameHBox = new HBox();
        Label nameLabel = new Label("姓名");
        TextField nameTextField = new TextField();
        CheckBox nameCheckBox = new CheckBox();
        nameCheckBox.setSelected(true);                          // 默认选中
        nameHBox.getChildren().addAll(nameLabel, nameTextField, nameCheckBox);
        nameHBox.setSpacing(10);

        // 创建年龄行
        HBox ageHBox = new HBox();
        Label ageLabel = new Label("年龄");
        TextField ageTextField = new TextField();
        CheckBox ageCheckBox = new CheckBox();
        ageCheckBox.setSelected(true);                           // 默认选中
        ageHBox.getChildren().addAll(ageLabel, ageTextField, ageCheckBox);
        ageHBox.setSpacing(10);

        // 创建收入行
        HBox incomeHBox = new HBox();
        Label incomeLabel = new Label("收入");
        TextField incomeTextField = new TextField();
        CheckBox incomeCheckBox = new CheckBox();
        incomeCheckBox.setSelected(true);                        // 默认选中
        incomeHBox.getChildren().addAll(incomeLabel, incomeTextField,
incomeCheckBox);
        incomeHBox.setSpacing(10);

        // 设置复选框的事件处理
        nameCheckBox.setOnAction(event -> {
            nameTextField.setDisable(!nameCheckBox.isSelected());
```

```
            nameTextField.setStyle("-fx-background-color: " +
(nameCheckBox.isSelected() ? "white" : "red") + ";");
        });
        ageCheckBox.setOnAction(event -> {
            ageTextField.setDisable(!ageCheckBox.isSelected());
            ageTextField.setStyle("-fx-background-color: " +
(ageCheckBox.isSelected() ? "white" : "red") + ";");
        });
        incomeCheckBox.setOnAction(event -> {
            incomeTextField.setDisable(!incomeCheckBox.isSelected());
            incomeTextField.setStyle("-fx-background-color: " +
(incomeCheckBox.isSelected() ? "white" : "red") + ";");
        });
        // 创建按钮和显示标签
        Button displayButton = new Button("显示");
        Label displayLabel = new Label();
        // 让 displayLabel 水平充满整个屏幕
        displayButton.setMaxWidth(Double.MAX_VALUE);
        VBox.setVgrow(displayButton, Priority.ALWAYS);

        // 按钮单击事件处理
        displayButton.setOnAction(event -> {
            String name = nameTextField.getText();
            String age = ageTextField.getText();
            String income = incomeTextField.getText();
            String displayText = String.format("姓名: %s, 年龄: %s, 收入: %s",
name, age, income);
            displayLabel.setText(displayText);
        });
        // 添加所有元素到垂直布局中
        vBox.getChildren().addAll(nameHBox, ageHBox, incomeHBox, displayButton,
displayLabel);
        // 设置场景和舞台
        Scene scene = new Scene(vBox, 300, 220);
        primaryStage.setTitle("组件演示");
        primaryStage.setScene(scene);
        primaryStage.show();
    }
    public static void main(String[] args) {
        launch(args);
    }
}
```

运行代码，显示如图 3-6 所示的窗口，在 TextField 组件中输入一些文本，单击“显示”按钮，会在按钮下方显示在这 3 个 TextField 组件中输入的内容，如图 3-6 所示。

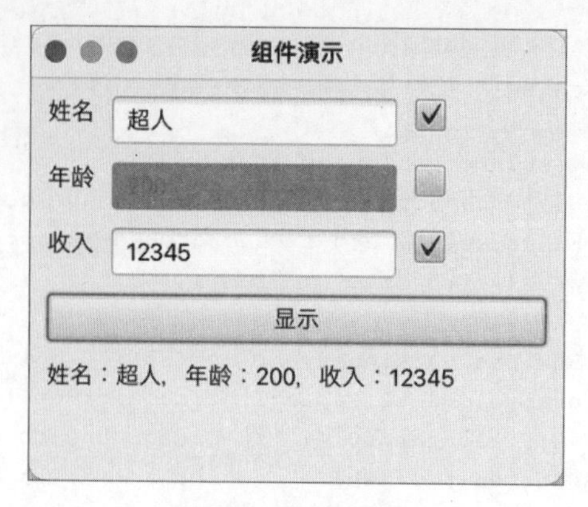

图 3-6 常用组件

3.6 列表组件

JavaFX 中的列表组件是 ListView。这个组件用于展示项（item）的列表，并允许用户选择一个或多个 item。ListView 组件可以显示静态列表，也可以绑定到动态数据源。它是一个泛型类，可以指定列表中项的类型。

ListView 组件的主要属性和方法如下。

（1）items：一个 ObservableList 对象，包含了所有列表项，可以动态修改。

（2）getSelectionModel()：返回列表的 SelectionModel，用于处理列表项的选择问题。

（3）getItems().add()：向 items 列表添加新项。

（4）getItems().remove()：从 items 列表中移除项。

下面的例子演示了如何使用 ListView 组件添加和删除 item。

```
// 创建一个 ObservableList 来存储列表项
ObservableList<String> listItems = FXCollections.observableArrayList(
    "列表项 1", "列表项 2", "列表项 3"
);
// 创建 ListView 并设置其项目
ListView<String> listView = new ListView<>(listItems);
// 添加一个新的列表项
listItems.add("新列表项");
// 删除第一个列表项
if (!listItems.isEmpty()) {
    listItems.remove(0);                          // 从列表中删除第一个项目
}
```

下面的例子在窗口上放置 1 个 ListView 组件和 2 个按钮（“添加”按钮和“删除”按钮），ListView 组件默认添加 6 个列表项。单击“添加”按钮，会向 ListView 组件中添加 1 个新的列表项，如果某一个列表项被选中，单击“删除”按钮，会删除当前的列表项。

代码位置：src/main/java/javafxgui/ListDemo.java

```
package javafxgui;
import javafx.application.Application;
import javafx.geometry.Insets;
import javafx.geometry.Pos;
import javafx.scene.Scene;
import javafx.scene.control.Button;
import javafx.scene.control.ListView;
import javafx.scene.layout.VBox;
import javafx.scene.layout.HBox;
import javafx.stage.Stage;
public class ListDemo extends Application {
    private ListView<String> listView;  // 创建 ListView 组件，用于显示列表项
    @Override
    public void start(Stage primaryStage) {
        listView = new ListView<>();                       // 初始化 ListView
        for (int i = 0; i < 6; i++) {                      // 默认添加 6 个列表项
            listView.getItems().add(" 列表项 " + i);
        }
        Button addButton = new Button(" 添加 ");             // 创建添加按钮
        addButton.setOnAction(event -> addItem());         // 设置添加按钮的事件处理

        Button delButton = new Button(" 删除 ");             // 创建删除按钮
        delButton.setOnAction(event -> delItem());         // 设置删除按钮的事件处理
        HBox hBox = new HBox(10, addButton, delButton); // 创建水平布局，放置两个按钮
       VBox vBox = new VBox(10, listView, hBox);
       // 创建垂直布局，将 ListView 和 HBox 放入其中
        hBox.setAlignment(Pos.CENTER);
        vBox.setPadding(new Insets(10, 10, 10, 10));
        Scene scene = new Scene(vBox, 360, 320);          // 创建场景，设置其大小
        primaryStage.setTitle("ListView 示例 ");            // 设置舞台标题
        primaryStage.setScene(scene);                      // 将场景放置在舞台上
        primaryStage.show();                               // 显示舞台
    }
    private void addItem() {
        listView.getItems().add(" 新列表项 ");               // 添加新的列表项
    }
    private void delItem() {
    int selectedIndex = listView.getSelectionModel().getSelectedIndex();
    // 获取当前选中的列表项索引
        if (selectedIndex != -1) {                         // 如果有选中的项
            listView.getItems().remove(selectedIndex); // 删除选中的列表项
        }
    }
    public static void main(String[] args) {
        launch(args);                                      // 启动应用程序
    }
}
```

运行程序，会显示如图 3-7 所示的窗口，读者可以单击“添加”按钮和“删除”按钮来体验 ListView 组件插入和删除列表项的效果。

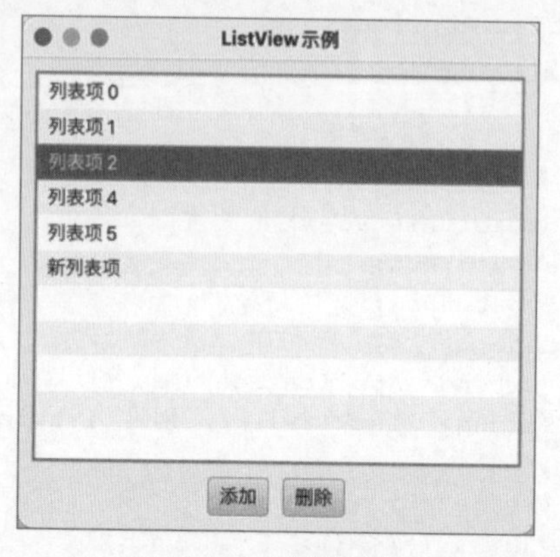

图 3-7　列表组件

3.7　下拉列表组件

ComboBox 是一个下拉列表组件，它允许用户从预定义的选项中选择一个值。使用 ComboBox.getItems().add() 方法可以向该组件中添加列表项，代码如下：

```
comboBox.getItems().add("New Item")
```

如果想删除当前选定的选项，可以使用 ComboBox.getItems().remove 方法，代码如下：

```
comboBox.getItems().remove(selectedIndex)
```

其中 selectedIndex 是当前选中 item 的索引。

下面的例子在窗口上放置一个 ComboBox 组件，里面默认添加 3 个 item。ComboBox 组件下方水平放置两个组件（“添加”按钮和“删除”按钮），单击“添加”按钮，向 ComboBox 组件添加一个列表项。单击“删除”按钮，会删除当前选中的列表项。

代码位置： src/main/java/javafxgui/ComboBoxDemo.java

```
package javafxgui;

import javafx.application.Application;
import javafx.geometry.Insets;
import javafx.geometry.Pos;
import javafx.scene.Scene;
import javafx.scene.control.Button;
import javafx.scene.control.ComboBox;
```

```
import javafx.scene.layout.VBox;
import javafx.scene.layout.HBox;
import javafx.stage.Stage;
public class ComboBoxDemo extends Application {
    @Override
    public void start(Stage primaryStage) {
        // 创建一个垂直布局 VBox，并设置间距和填充
        VBox vbox = new VBox(10);                              // 间距为 10
        vbox.setPadding(new Insets(10, 20, 10, 20)); // 上下间距为 10，左右间距为 20
        // 创建 ComboBox 对象
        ComboBox<String> comboBox = new ComboBox<>();
        comboBox.getItems().addAll("Item 1", "Item 2", "Item 3");
        comboBox.setMaxWidth(Double.MAX_VALUE);     // ComboBox 水平充满可用空间
        // 创建水平布局 HBox，并设置间距和填充
        HBox hbox = new HBox(10);
        hbox.setPadding(new Insets(10, 0, 10, 0)); // 上下间距为 10，左右间距为 0
        hbox.setAlignment(Pos.CENTER);                         // 使按钮水平居中
        // 创建添加按钮
        Button addButton = new Button("添加");
        addButton.setOnAction(e -> comboBox.getItems().add("New Item"));
        // 创建删除按钮
        Button removeButton = new Button("删除");
        removeButton.setOnAction(e -> {
            int selectedIndex = comboBox.getSelectionModel().getSelectedIndex();
            if (selectedIndex >= 0) {
                comboBox.getItems().remove(selectedIndex);
            }
        });
        // 将按钮添加到水平布局中
        hbox.getChildren().addAll(addButton, removeButton);
        // 将 ComboBox 和 HBox 添加到垂直布局中
        vbox.getChildren().addAll(comboBox, hbox);
        // 创建场景，设置场景的根节点为 vbox，并设置场景大小
        Scene scene = new Scene(vbox, 350, 100);
        // 设置舞台的标题
        primaryStage.setTitle("ComboBox 示例");
        // 将场景添加到舞台中
        primaryStage.setScene(scene);
        // 显示舞台
        primaryStage.show();
    }
    public static void main(String[] args) {
        launch(args);                                          // 启动应用程序
    }
}
```

运行程序，单击“添加”按钮会添加新的列表项，单击“删除”按钮会删除当前列表项，如图 3-8 所示。

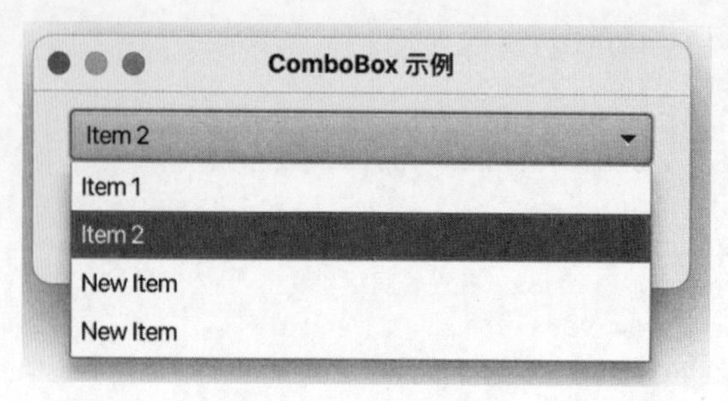

图 3-8 下拉列表组件

3.8 表格组件

JavaFX 的 TableView 是一个可以用来显示表格式数据的强大组件。它可以显示行和列，并允许用户对列进行排序和编辑。TableView 组件使用 ObservableList 来管理和显示数据。

TableView 的数据管理主要依赖以下几个类和接口。

（1）TableView<T>：这是主要的表格视图类，T 是表中每一行数据的类型。

（2）TableColumn<S, T>：用来定义表格的单个列，S 是行的数据类型，T 是列中单元格数据的类型。

（3）ObservableList<T>：一个列表，用来存储表中的所有行数据，通常是特定的模型对象列表。

（4）PropertyValueFactory<S, T>：用来从模型中提取出单元格值的工厂类。

以下是一个简单的表格配置的代码片段：

```
// 创建表格视图对象
TableView<Person> tableView = new TableView<>();
// 设置表格的列
TableColumn<Person, String> firstNameColumn = new TableColumn<>("First Name");
firstNameColumn.setCellValueFactory(new PropertyValueFactory<>("firstName"));

TableColumn<Person, String> lastNameColumn = new TableColumn<>("Last Name");
lastNameColumn.setCellValueFactory(new PropertyValueFactory<>("lastName"));

TableColumn<Person, Integer> ageColumn = new TableColumn<>("Age");
ageColumn.setCellValueFactory(new PropertyValueFactory<>("age"));

// 添加列到表格视图
tableView.getColumns().add(firstNameColumn);
tableView.getColumns().add(lastNameColumn);
tableView.getColumns().add(ageColumn);

// 创建和设置表格中的数据
ObservableList<Person> data = FXCollections.observableArrayList(
```

```
        new Person("John", "Doe", 30),
        new Person("Jane", "Doe", 25)
    );
    tableView.setItems(data);
```

在这个例子中，我们定义了一个 Person 类，它有 3 个属性：firstName、lastName 和 age。我们为每个属性创建了一个 TableColumn。PropertyValueFactory 被用来告诉每个列如何从 Person 对象中获取其值。

对于复杂的表格数据，可能需要自定义 cellValueFactory 来满足特定的要求，或者使用 CellFactory 来自定义单元格的显示方式。

注意： 在实际使用时，需要确保 Person 类有相应的属性和 JavaFX 属性绑定（如 SimpleStringProperty 和 SimpleIntegerProperty）以便于 TableView 可以监听属性值的变化。此外，为了使 PropertyValueFactory 正常工作，Person 类中的每个属性名需要有相应的 getter 方法，或者使用 JavaFX 属性模型。

在 JavaFX 中，模型类通常会有属性绑定的功能，这允许 TableView 在属性值发生变化时自动更新。为了使这种绑定工作，模型类的属性需要被封装在 JavaFX 的属性类中，比如 SimpleStringProperty 和 SimpleIntegerProperty。这些属性类实现了 ObservableValue 接口，这样 TableView 可以监听其值的改变。

下面是一个简单的模型类的例子，演示了如何设置这些属性：

```
import javafx.beans.property.SimpleIntegerProperty;
import javafx.beans.property.SimpleStringProperty;
import javafx.beans.property.StringProperty;

public class Person {
    private final SimpleStringProperty firstName;
    private final SimpleStringProperty lastName;
    private final SimpleIntegerProperty age;

    public Person(String fName, String lName, int age) {
        this.firstName = new SimpleStringProperty(fName);
        this.lastName = new SimpleStringProperty(lName);
        this.age = new SimpleIntegerProperty(age);
    }

    // FirstName 属性的 getter 和 setter
    public String getFirstName() {
        return firstName.get();
    }
    public void setFirstName(String fName) {
        firstName.set(fName);
    }
    public StringProperty firstNameProperty() {
```

```
        return firstName;
    }

    // LastName 属性的 getter 和 setter
    public String getLastName() {
        return lastName.get();
    }
    public void setLastName(String lName) {
        lastName.set(lName);
    }
    public StringProperty lastNameProperty() {
        return lastName;
    }
    // Age 属性的 getter 和 setter
    public int getAge() {
        return age.get();
    }
    public void setAge(int age) {
        this.age.set(age);
    }
    public SimpleIntegerProperty ageProperty() {
        return age;
    }
}
```

在这个 Person 类中有三个属性：firstName，lastName 和 age。每个属性都有相应的 get 和 set 方法，以及一个返回属性本身的方法（例如 firstNameProperty()）。这些属性方法非常重要，因为 PropertyValueFactory 将使用这些方法来访问属性的值。

现在，在 TableView 中设置列的 cellValueFactory 时，就可以使用这些属性名称来引用：

```
TableColumn<Person, String> firstNameCol = new TableColumn<>("First Name");
firstNameCol.setCellValueFactory(new PropertyValueFactory<>("firstName"));

TableColumn<Person, String> lastNameCol = new TableColumn<>("Last Name");
lastNameCol.setCellValueFactory(new PropertyValueFactory<>("lastName"));

TableColumn<Person, Number> ageCol = new TableColumn<>("Age");
ageCol.setCellValueFactory(new PropertyValueFactory<>("age"));
```

在上面的代码中，PropertyValueFactory 将会调用 Person 对象上的 firstNameProperty()、lastNameProperty() 和 ageProperty() 方法来获取 ObservableValue，这样 TableView 就可以监听这些值的变化了。这就是为什么模型类需要有一个与属性名称匹配的方法。如果这些方法不存在，PropertyValueFactory 将无法找到正确的属性来监听，因此将无法显示正确的数据或响应数据的变化。

下面的例子在窗口上放置 1 个 TableView 组件和 2 个按钮（“添加”和“删除”）。TableView 组件默认添加 2 行 4 列，单击“添加”按钮，会添加一个新表格行，单击“删除”

按钮，会删除当前选中的表格行。

代码位置：src/main/java/javafxgui/TableDemo.java

```
package javafxgui;

import javafx.application.Application;
import javafx.beans.property.ReadOnlyStringWrapper;
import javafx.beans.property.SimpleStringProperty;
import javafx.collections.FXCollections;
import javafx.collections.ObservableList;
import javafx.scene.Scene;
import javafx.scene.control.Button;
import javafx.scene.control.TableColumn;
import javafx.scene.control.TableView;
import javafx.scene.layout.HBox;
import javafx.scene.layout.VBox;
import javafx.stage.Stage;
// 数据模型类
class RowData {
    private final SimpleStringProperty[] data;
    public RowData(String... data) {
        this.data = new SimpleStringProperty[data.length];
        for (int i = 0; i < data.length; i++) {
            this.data[i] = new SimpleStringProperty(data[i]);
        }
    }
    public String getData(int index) {
        return data[index].get();
    }
    public void setData(int index, String value) {
        data[index].set(value);
    }
}
public class TableDemo extends Application {
    private final TableView<RowData> table = new TableView<>();
    private final ObservableList<RowData> data = FXCollections.
observableArrayList();
    @Override
    public void start(Stage stage) {
        // 设置窗口标题
        stage.setTitle("JavaFX TableView 示例");
        // 初始化表格列
        for (int i = 0; i < 4; i++) {
            TableColumn<RowData, String> column = new TableColumn<>
("header" + (i + 1));
            final int colIndex = i;
            column.setCellValueFactory(cellData -> new ReadOnlyStringWrapper
(cellData.getValue().getData(colIndex)));
            table.getColumns().add(column);
        }
```

```
        // 添加初始数据
        data.add(new RowData("0", "1", "2", "3"));
        data.add(new RowData("1", "2", "3", "4"));
        table.setItems(data);
        // 添加和删除按钮
        Button addButton = new Button("添加");
        addButton.setOnAction(e -> addRow());
        Button deleteButton = new Button("删除");
        deleteButton.setOnAction(e -> deleteRow());
        // 创建布局
        HBox buttonLayout = new HBox(10, addButton, deleteButton);
        VBox tableLayout = new VBox(10, table, buttonLayout);
        // 设置场景
        Scene scene = new Scene(tableLayout, 500, 300);
        stage.setScene(scene);
        stage.show();
    }
    // 添加行方法
    private void addRow() {
        int rowCount = data.size();
        data.add(new RowData(String.valueOf(rowCount), String.valueOf
(rowCount + 1), String.valueOf(rowCount + 2), String.valueOf(rowCount + 3)));
    }
    // 删除行方法
    private void deleteRow() {
        int selectedIndex = table.getSelectionModel().getSelectedIndex();
        if (selectedIndex != -1) {
            data.remove(selectedIndex);
        }
    }
    public static void main(String[] args) {
        launch(args);
    }
}
```

运行程序，会看到如图 3-9 所示的界面，单击“添加”按钮会添加新的行，单击“删除”按钮，会删除当前行。

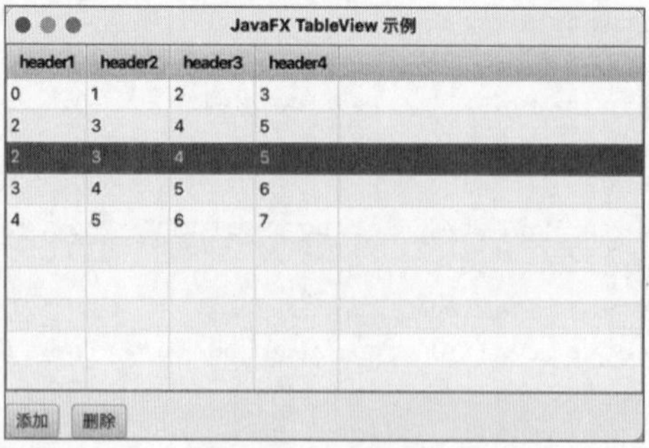

图 3-9　表格组件

3.9　树状组件

JavaFX 的 TreeView 组件用于显示层次化的数据结构，类似于文件系统的目录和文件结构。它由节点组成，每个节点都是 TreeItem 对象。

TreeView 管理树状数据的关键组件和概念如下所示。

（1）TreeView<T>：表示树状组件本身，T 是树中节点所持有的数据类型。

（2）TreeItem<T>：表示树中的一个节点，同样地，T 是节点持有的数据类型。每个 TreeItem 可以有子项，从而形成树的结构。

（3）ObservableList<TreeItem<T>>：每个 TreeItem 都有一个子项列表，这是一个 ObservableList，用于存储子节点。

下面的代码演示了如何创建一棵树并添加节点：

```
// 创建根节点
TreeItem<String> rootNode = new TreeItem<>("Root");
// 创建树视图，并设置根节点
TreeView<String> treeView = new TreeView<>(rootNode);
// 添加子节点
TreeItem<String> branchItem = new TreeItem<>("Branch");
rootNode.getChildren().add(branchItem);
// 添加叶子节点到 branchItem
TreeItem<String> leafItem = new TreeItem<>("Leaf");
branchItem.getChildren().add(leafItem);
// 设置根节点展开
rootNode.setExpanded(true);
```

在这个例子中，我们创建了一个根节点 rootNode，并将它设置为 TreeView 的根。接着，我们创建了一个分支节点 branchItem，并将其添加为根节点的子项。然后，我们创建了一个叶子节点 leafItem，并将其添加为分支节点的子项。

TreeItem 的 getChildren() 方法返回的 ObservableList 使得我们可动态地添加或删除节点，TreeView 将会自动更新显示。setExpanded() 方法用于控制节点是否展开显示其子项。

如果需要响应节点的选择变化，可以通过 TreeView 的 getSelectionModel() 方法获取到 SelectionModel，并添加监听器。

TreeItem 还可以包含任意的 Java 对象作为其值，这使得可以存储额外的数据或对象引用在树节点中。例如：

```
// 假设有一个用户类
public class User {
    private String name;
    // 其他属性和方法 ...
}

// 创建包含 User 对象的 TreeItem
```

```
User user = new User("John Doe");
TreeItem<User> userItem = new TreeItem<>(user);

// 现在 TreeView 类型是 TreeView<User>
TreeView<User> userTreeView = new TreeView<>(userItem);
```

使用 TreeView 时，也可以自定义节点的显示方式，通过设置 cellFactory 属性为树状组件添加编辑功能、拖放功能等高级特性。

下面的例子在窗口上放置一个 TreeView 组件和两个按钮（Add 按钮和 Delete 按钮）。TreeView 组件默认会随机添加一些节点。单击 Add 按钮，会在当前选中的节点添加一个子节点，单击 Delete 按钮，会删除当前选中的节点。

代码位置：src/main/java/javafxgui/TreeDemo.java

```
package javafxgui;

import javafx.application.Application;
import javafx.scene.Scene;
import javafx.scene.control.*;
import javafx.scene.layout.HBox;
import javafx.stage.Stage;

public class TreeDemo extends Application {
    @Override
    public void start(Stage primaryStage) {
        // 创建树视图对象
        TreeView<String> treeView = new TreeView<>();
        // 设置树视图的根节点
        TreeItem<String> rootItem = new TreeItem<>("Root");
        rootItem.setExpanded(true);                    // 默认展开根节点
        treeView.setRoot(rootItem);
        // 不显示根节点，只显示子节点
        treeView.setShowRoot(false);

        // 随机添加一些节点
        for (int i = 0; i < 10; i++) {
            TreeItem<String> item = new TreeItem<>("Node " + i);
            rootItem.getChildren().add(item);
        }
        // 创建添加按钮
        Button addButton = new Button("Add");
        addButton.setOnAction(e -> addNode(treeView));
        // 创建删除按钮
        Button deleteButton = new Button("Delete");
        deleteButton.setOnAction(e -> deleteNode(treeView));
        // 创建水平布局，并添加树视图和按钮
        HBox hBox = new HBox(10, treeView, addButton, deleteButton);
```

```
        // 创建场景，设置场景的根节点为 HBox
        Scene scene = new Scene(hBox, 400, 300);
        primaryStage.setTitle("TreeView 示例 ");
        primaryStage.setScene(scene);
        primaryStage.show();
    }
    // 添加节点的方法
    private void addNode(TreeView<String> treeView) {
        TreeItem<String> selectedItem = treeView.getSelectionModel().
getSelectedItem();
        if (selectedItem != null) {
            // 创建新的子节点
            TreeItem<String> newItem = new TreeItem<>("New Node");
            selectedItem.getChildren().add(newItem);
            selectedItem.setExpanded(true);
        }
    }
    // 删除节点的方法
    private void deleteNode(TreeView<String> treeView) {
        TreeItem<String> selectedItem = treeView.getSelectionModel().
getSelectedItem();
        if (selectedItem != null && selectedItem.getParent() != null) {
            // 删除选中的节点
            selectedItem.getParent().getChildren().remove(selectedItem);
        }
    }
    public static void main(String[] args) {
        launch(args);
    }
}
```

运行程序，会显示如图 3-10 所示的窗口，单击 Add 按钮，会在当前选中的节点下面添加子节点，单击 Delete 按钮，会将选中的节点（包括该节点的子节点）删除。

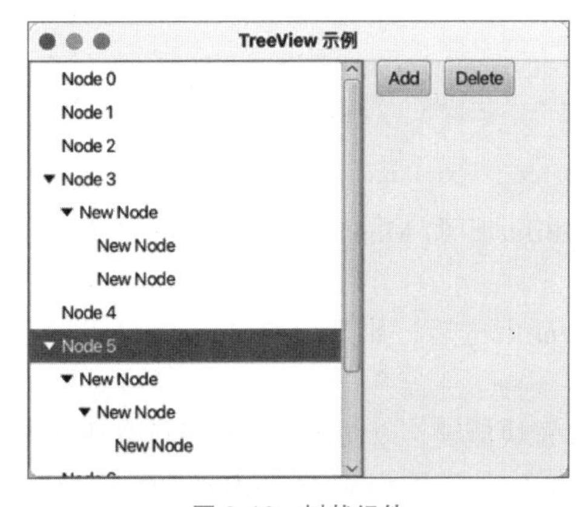

图 3-10　树状组件

3.10 菜单

在 JavaFX 中创建菜单涉及几个核心组件：MenuBar、Menu、MenuItem 和 ContextMenu。以下是这些组件的详细介绍。

❑ MenuBar

MenuBar 是容纳一组菜单的组件，通常位于应用程序窗口的顶部。在 macOS 系统中，如果指定使用系统菜单栏，它会显示在屏幕顶部的全局菜单栏上。MenuBar 主要用来包含应用程序的顶级菜单项。

```
MenuBar menuBar = new MenuBar();
```

❑ Menu

Menu 是一个可以包含多个 MenuItem 或者子 Menu 的组件。Menu 可以是 MenuBar 的直接子项，也可以是另一个 Menu 的子项，这样就可以创建子菜单。

```
Menu fileMenu = new Menu("文件");
menuBar.getMenus().add(fileMenu);                    // 将文件菜单添加到菜单栏
```

❑ MenuItem

MenuItem 是菜单中的一个选项，用户可以通过单击来触发一个操作。Menu 可以包含多个 MenuItem，代表不同的命令或操作。

```
MenuItem openItem = new MenuItem("打开");
fileMenu.getItems().add(openItem);                   // 将打开项添加到文件菜单
```

❑ ContextMenu

ContextMenu 是一个弹出式菜单，可以在用户右击特定组件时显示。它通常用于提供与上下文相关的操作选项。

```
ContextMenu contextMenu = new ContextMenu();
contextMenu.getItems().add(new MenuItem("编辑"));
```

使用菜单要了解如下几点。

（1）当用户单击 MenuBar 中的 Menu 时，该 Menu 下的 MenuItem 列表会显示出来，供用户选择。

（2）用户单击某个 MenuItem 后，如果为该 MenuItem 设置了事件处理器，相应的事件处理逻辑会被执行。

（3）MenuItem 可以设置快捷键，以便用户可通过键盘快捷方式直接触发。

（4）ContextMenu 通常不直接添加到 MenuBar，而是与特定的组件（如 TextField、ListView）关联，并在适当的上下文中显示。

下面的代码创建了基本的菜单和菜单项，并添加了事件处理功能。

```
// 创建菜单栏
MenuBar menuBar = new MenuBar();
// 创建菜单
Menu fileMenu = new Menu("文件");
// 创建菜单项
MenuItem newItem = new MenuItem("新建");
newItem.setOnAction(event -> {
    // 新建操作的代码
});
// 将菜单项添加到菜单
fileMenu.getItems().add(newItem);
// 将菜单添加到菜单栏
menuBar.getMenus().add(fileMenu);
```

下面的例子在窗口中放置一个标签组件。然后创建一个主菜单，菜单项包括：新建、打开、保存、分隔线、退出。单击“退出”菜单项，会退出程序，单击其他菜单项，会将菜单项文本显示在标签组件中。右击标签组件，会显示与主菜单一模一样的弹出菜单。菜单项的动作也完全相同。

代码位置：src/main/java/javafxgui/MenuDemo.java

```
package javafxgui;

import javafx.application.Application;
import javafx.scene.Scene;
import javafx.scene.control.*;
import javafx.scene.layout.VBox;
import javafx.stage.Stage;

public class MenuDemo extends Application {
    private Label statusLabel;                          // 用于显示菜单操作状态的标签
    @Override
    public void start(Stage primaryStage) {
        // 创建 VBox 作为主布局
        VBox root = new VBox();
        // 创建菜单栏
        MenuBar menuBar = new MenuBar();
      //  menuBar.setUseSystemMenuBar(true);            // 使用系统菜单栏
        Menu fileMenu = new Menu("文件");
        addMenuItems(fileMenu);
        menuBar.getMenus().add(fileMenu);
        // 创建标签并为其设置上下文菜单
        statusLabel = new Label("这是一个标签");
        statusLabel.setContextMenu(createContextMenu());
        // 将菜单栏和标签添加到布局中
        root.getChildren().addAll(menuBar, statusLabel);
```

```
        // 创建场景并设置舞台
        Scene scene = new Scene(root, 300, 250);
        primaryStage.setTitle("MenuDemo");
        primaryStage.setScene(scene);
        primaryStage.show();
    }
    // 为 Menu 添加菜单项
    private void addMenuItems(Menu menu) {
        MenuItem newMenuItem = new MenuItem("新建");
        newMenuItem.setOnAction(e -> statusLabel.setText("新建菜单项被单击"));
        MenuItem openMenuItem = new MenuItem("打开");
        openMenuItem.setOnAction(e -> statusLabel.setText("打开菜单项被单击"));
        MenuItem saveMenuItem = new MenuItem("保存");
        saveMenuItem.setOnAction(e -> statusLabel.setText("保存菜单项被单击"));
        MenuItem exitMenuItem = new MenuItem("退出");
        exitMenuItem.setOnAction(e -> System.exit(0));

        menu.getItems().addAll(newMenuItem, openMenuItem, saveMenuItem,
new SeparatorMenuItem(), exitMenuItem);
    }
    // 为 ContextMenu 添加菜单项
    private void addMenuItems(ContextMenu contextMenu) {
        MenuItem newMenuItem = new MenuItem("新建");
        newMenuItem.setOnAction(e -> statusLabel.setText("新建菜单项被单击"));
        MenuItem openMenuItem = new MenuItem("打开");
        openMenuItem.setOnAction(e -> statusLabel.setText("打开菜单项被单击"));
        MenuItem saveMenuItem = new MenuItem("保存");
        saveMenuItem.setOnAction(e -> statusLabel.setText("保存菜单项被单击"));
        MenuItem exitMenuItem = new MenuItem("退出");
        exitMenuItem.setOnAction(e -> System.exit(0));
        contextMenu.getItems().addAll(newMenuItem, openMenuItem, saveMenuItem,
new SeparatorMenuItem(), exitMenuItem);
    }
    // 创建并返回上下文菜单
    private ContextMenu createContextMenu() {
        ContextMenu contextMenu = new ContextMenu();
        addMenuItems(contextMenu);
        return contextMenu;
    }
    public static void main(String[] args) {
        launch(args);
    }
}
```

运行程序，会显示一个窗口，窗口中有一个“文件”菜单，单击“文件”菜单，会显示其中的菜单项，如图 3-11 所示。

在标签组件上右击，会显示如图 3-12 所示的弹出菜单。

在 Windows 和 Linux 环境下的菜单效果差不多，但在 macOS 下，默认仍然将菜单放在窗口上，这并不符合 macOS 的规范。macOS 会将窗口菜单放在 macOS 桌面顶部的菜单条，而不是窗口的顶部。为了实现这个效果，需要使用 MeanBar.setUseSystemMenuBar() 方法允许窗口使用系统的菜单条，代码如下：

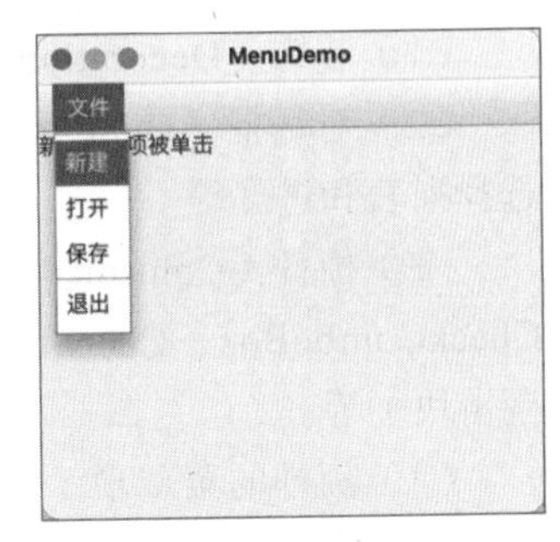

图 3-11　“文件”菜单

```
menuBar.setUseSystemMenuBar(true);
```

加上这行代码后，在 macOS 系统下，菜单就会显示在 macOS 桌面顶部的菜单条，如图 3-13 所示。

图 3-12　弹出菜单

图 3-13　主菜单和弹出菜单

3.11　对话框

JavaFX 支持多种对话框（controlsfx）类型，例如 Alert（消息对话框）、TextInputDialog（文本输入对话框）、FileChooser（打开文件对话框）、ColorPicker（颜色选择器）[①]、PrinterJob.showPrintDialog（打印对话框）等。尽管 JavaFX 支持很多常用的对话框，但也并不是所有类型的对话框都支持，例如，进度对话框并不包含在 JavaFX 标准库中。为了使用这些 JavaFX 不支持的对话框，最直接的方法就是自己写，不过比较麻烦。另一个方法就是使用第三方库中的对话框，这里推荐 ControlsFX。

ControlsFX 是一个开源项目，为 JavaFX 提供了一系列额外的组件和装饰功能，这些功能在 JavaFX 标准库中不可用或者实现起来较为复杂。以下是 ControlsFX 库的一些主要功能。

① ColorPicker 本身并不是对话框，只是用于选择颜色的组件。但可以将 ColorPicker 作为对话框的内容，这样就有了颜色选择对话框。

（1）装饰（Decorations）：提供了装饰组件的功能，比如图标、验证和自定义装饰。

（2）对话框（Dialogs）：提供了更丰富的对话框控件，例如信息、警告、确认、输入、选择列表和进度等。

（3）控件（Controls）：包含了许多额外的控件，如自动完成文本框、分页、面包屑条、CheckComboBox、CheckListView、PopOver（类似于弹出窗口的控件）、RangeSlider（范围滑块）等。

（4）额外的菜单项（Extra Menu Items）：提供了一些额外的菜单项类型，例如可用于检查的菜单项。

（5）图状组件（Graphic Components）：包括图状组件，如分级加载视图（TreeView 的改进版）、丰富的文本支持等。

（6）通知（Notifications）：支持在屏幕的四个角显示通知。

（7）Property Sheet：用于编辑属性的表格，支持多种数据类型。

（8）Validation：提供了输入验证功能。

（9）SpreadsheetView：类似 Excel 的电子表格视图。

ControlsFX 为开发者提供了一种快速、易于使用的方式来增强他们的 JavaFX 应用程序，使其拥有更加丰富和专业的用户界面。由于 ControlsFX 是社区驱动的项目，它还会不断地更新和增加新的控件和功能。

若要使用 ControlsFX 库，需要将其作为依赖项添加到 Java 项目中。对于 Maven 项目，可以在 pom.xml 文件中添加如下依赖：

```
<dependency>
    <groupId>org.controlsfx</groupId>
    <artifactId>controlsfx</artifactId>
    <version>11.2.0</version>
</dependency>
```

下面的例子使用 JavaFX 标准库以及 ControlsFX 库弹出了消息对话框、输入对话框、打开文件对话框、颜色选择对话框、打印对话框和进度条对话框，在对话框关闭后，如果单击的是“确定”或类似的按钮，会将对话框的选择结果显示在对应的标签组件中。

代码位置：src/main/java/javafxgui/DialogsDemo.java

```
package javafxgui;

import javafx.application.Application;
import javafx.concurrent.Task;
import javafx.event.ActionEvent;
import javafx.geometry.Insets;
import javafx.geometry.Pos;
import javafx.print.PrinterJob;
import javafx.scene.Scene;
import javafx.scene.control.*;
```

```
import javafx.scene.layout.VBox;
import javafx.scene.paint.Color;
import javafx.stage.FileChooser;
import javafx.stage.Stage;
import org.controlsfx.dialog.ProgressDialog;
import java.io.File;

public class DialogDemo extends Application {

    private Label label1 = new Label();
    private Label label2 = new Label();
    private Label label3 = new Label();
    private Label label4 = new Label();
    private Label label5 = new Label();
    private Label label6 = new Label();

    @Override
    public void start(Stage primaryStage) {
        VBox vbox = new VBox(10);

        Button btn1 = new Button("消息对话框");
        btn1.setOnAction(e -> showMessageBox());
        Button btn2 = new Button("输入对话框");
        btn2.setOnAction(e -> showInputDialog());
        Button btn3 = new Button("文件打开对话框");
        btn3.setOnAction(e -> showOpenFileDialog(primaryStage));
        Button btn4 = new Button("颜色选择对话框");
        btn4.setOnAction(e -> showColorDialog());
        Button btn5 = new Button("打印对话框");
        btn5.setOnAction(e -> showPrintDialog(primaryStage));
        Button btn6 = new Button("进度条对话框");
        btn6.setOnAction(e -> showProgressDialog());
        // 使所有按钮水平充满屏幕
        vbox.setPadding(new Insets(10));
        vbox.setSpacing(10);
        vbox.setAlignment(Pos.CENTER);
        vbox.getChildren().addAll(
                btn1, label1,
                btn2, label2,
                btn3, label3,
                btn4, label4,
                btn5, label5,
                btn6, label6
        );
        Scene scene = new Scene(vbox, 300, 500);
        primaryStage.setTitle("DialogsDemo");
        primaryStage.setScene(scene);
        primaryStage.show();
    }
    private void showMessageBox() {
        Alert alert = new Alert(Alert.AlertType.CONFIRMATION,
```

```
"QMessageBox 按钮已被按下。", ButtonType.OK, ButtonType.CANCEL);
        alert.setHeaderText("你选择了 messagebox");
        alert.showAndWait().ifPresent(response -> {
            if (response == ButtonType.OK) {
                label1.setText("ok");
            } else {
                label1.setText("cancel");
            }
        });
    }
    private void showInputDialog() {
        TextInputDialog dialog = new TextInputDialog();
        dialog.setTitle("输入对话框");
        dialog.setHeaderText("请输入你的名字：");
        dialog.showAndWait().ifPresent(name -> label2.setText(name));
    }
    private void showOpenFileDialog(Stage primaryStage) {
        FileChooser fileChooser = new FileChooser();
        File file = fileChooser.showOpenDialog(primaryStage);
        if (file != null) {
            label3.setText(file.getAbsolutePath());
        }
    }
    private void showColorDialog() {
        // 创建 ColorPicker 实例
        ColorPicker colorPicker = new ColorPicker();

        // 创建对话框并将 ColorPicker 设置为内容
        Dialog<Color> colorDialog = new Dialog<>();
        colorDialog.setTitle("颜色对话框");
        colorDialog.getDialogPane().setContent(colorPicker);
        colorDialog.getDialogPane().getButtonTypes().addAll(ButtonType.OK,
ButtonType.CANCEL);

        // 定义 OK 按钮响应事件
        Button okButton = (Button) colorDialog.getDialogPane().
lookupButton(ButtonType.OK);
        okButton.addEventFilter(ActionEvent.ACTION, event -> {
            // 获取 ColorPicker 的值并将其应用为按钮背景色
            Color color = colorPicker.getValue();
            label4.setStyle("-fx-background-color: #" + toHexString(color) + ";");
            label4.setMaxWidth(200);
        });
        // 显示对话框并等待结果
        colorDialog.showAndWait();
    }
    // 辅助方法，将 Color 对象转换为十六进制颜色字符串
    private String toHexString(Color color) {
        return String.format("%02X%02X%02X",
```

```
                (int) (color.getRed() * 255),
                (int) (color.getGreen() * 255),
                (int) (color.getBlue() * 255));
    }
    private void showPrintDialog(Stage primaryStage) {
        // JavaFX 打印对话框示例
        PrinterJob job = PrinterJob.createPrinterJob();
        if (job != null && job.showPrintDialog(primaryStage)) {
            label5.setText("成功");
            // 此处应添加实际的打印逻辑
            job.endJob();
        }
    }

    private void showProgressDialog() {
        // ControlsFX 进度对话框示例
        Task<Void> task = new Task<Void>() {
            @Override
            protected Void call() throws Exception {
                updateMessage("正在复制文件...");
                updateProgress(0, 100);
                for (int i = 0; i < 100; i++) {
                    updateProgress(i + 1, 100);
                    Thread.sleep(50);
                }
                return null;
            }
        };
        ProgressDialog progressDialog = new ProgressDialog(task);
        progressDialog.setTitle("进度对话框");
        new Thread(task).start();
    }
    // 辅助方法，将 Color 对象转换为 RGB 字符串
    private String toRgbString(Color c) {
        return String.format("#%02X%02X%02X",
                (int) (c.getRed() * 255),
                (int) (c.getGreen() * 255),
                (int) (c.getBlue() * 255));
    }
    public static void main(String[] args) {
        launch(args);
    }
}
```

运行程序，会显示一个带按钮的窗口，如图 3-14 所示。

单击任何一个按钮，都会弹出对应的对话框，例如，单击“进度条对话框”，会弹出如图 3-15 所示的进度条对话框，当进度到 100% 时会自动关闭对话框。

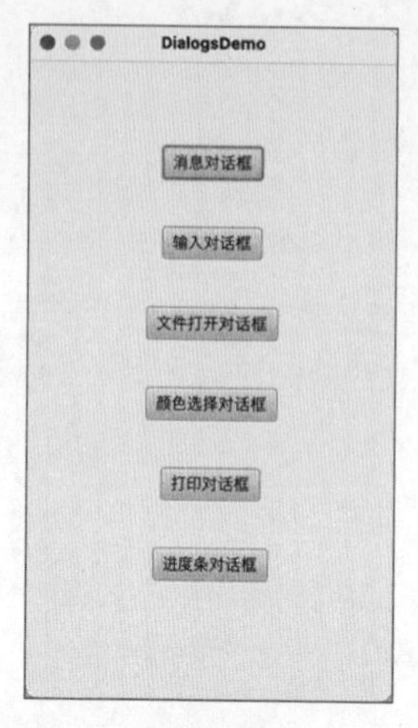

图 3-14 对话框演示

图 3-15 进度条对话框

对于一些对话框，如果 JavaFX 标准库、ControlsFX 库以及其他第三方库都没有，那就只能自己实现了。例如要实现字体选择对话框，首先要获取当前系统中安装的所有字体文件，使用下面的代码可以完成这个工作。而字体对话框的其他工作就非常简单了，从字体列表中选择一种字体，然后返回这种字体。也可以加更多的功能，例如，字体样式预览等。

代码位置：src/main/java/javafxgui/ListFonts.java

```
package javafxgui;
import java.awt.GraphicsEnvironment;
public class ListFonts {
    public static void main(String[] args) {
        GraphicsEnvironment ge = GraphicsEnvironment.
getLocalGraphicsEnvironment();
        String[] fontNames = ge.getAvailableFontFamilyNames();
        System.out.println("Available fonts on this system:");
        for (String fontName : fontNames) {
            System.out.println(fontName);
        }
    }
}
```

3.12 使用 CSS 设置 JavaFX 组件样式

在 3.7 节介绍的下拉列表组件中，默认情况下，当选中某一个 item 后，再次用鼠标划过其他的 item，那么上一次选中的 item 仍然处于选中状态，如图 3-16 所示。这与传统的下拉列表框的样式不同，为了与传统下拉列表框的表现一致，可以使用 CSS 设置下拉列表框的样式。

JavaFX 允许使用 CSS 设置 JavaFX 组件的样式。通常，需要将样式文件放到 src/main/resources 目录中，本例是 style.css，代码如下。

代码位置：src/main/resources/style.css

```
.list-cell {
    -fx-background-color: white;          /* 默认背景色为白色 */
    -fx-text-fill: black;                 /* 默认文本颜色为黑色 */
}
.list-cell:hover {
    -fx-background-color: #0096C9;        /* 悬停背景色为蓝色 */
    -fx-text-fill: white;                 /* 悬停文本颜色为白色 */
}
.list-cell:selected {
    -fx-background-color: white;          /* 选中项背景色为白色 */
    -fx-text-fill: black;                 /* 选中项文本颜色为黑色 */
}
.list-cell:selected:hover {
    -fx-background-color: #0096C9;        /* 选中项悬停背景色为蓝色 */
    -fx-text-fill: white;                 /* 选中项悬停文本颜色为白色 */
}
.list-cell:focused:selected, .list-cell:focused:selected:hover {
    -fx-background-color: white;          /* 聚焦选中项背景色为白色 */
    -fx-text-fill: black;                 /* 聚焦选中项文本颜色为黑色 */
}
```

然后使用下面的代码为下拉列表框组件设置样式文件即可。

```
scene.getStylesheets().add(getClass().getResource("/style.css").
toExternalForm());
```

修改样式后，当鼠标在 item 上划过后，只有被划过的 item 会显示蓝色背景，白色文字，其他的 item 会显示白色背景，黑色文字，效果如图 3-17 所示。

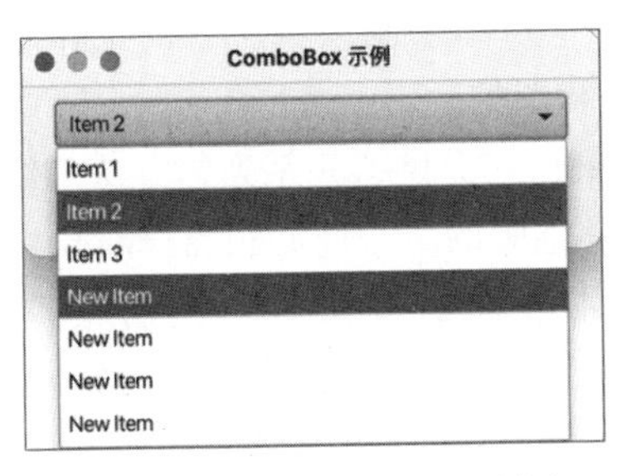

图 3-16　下拉列表框的默认样式

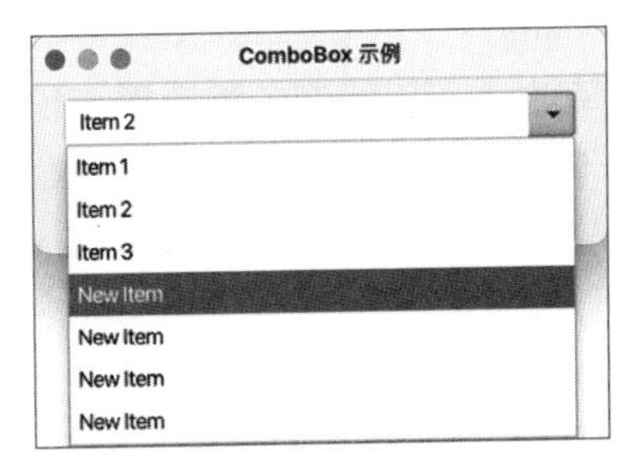

图 3-17　设置样式后的下拉列表框

在上面的 CSS 代码中，使用了一些特殊的属性前缀 -fx，这些是 JavaFX 样式属性的特殊标识符，表示它们是 JavaFX 的专有样式属性，不是标准的 Web CSS 属性。JavaFX 的 CSS 与 Web 的 CSS 在语法上是类似的，但是 JavaFX 的 CSS 拥有一些专有的属性，这些属性专门用于 JavaFX 应用程序中的 UI 控件。这意味着虽然两者在书写和结构上有相似之处，但它们的一些属性和值是不同的，因为它们是为不同的平台和用途设计的。

代码中的样式设置了 .list-cell 类的视觉效果，这个类通常用于 JavaFX 的 ListView 中的每个项。下面是 CSS 样式的详细描述。

（1）.list-cell：设置了所有 list cell 的默认背景色为白色，文本颜色为黑色。

（2）.list-cell:hover：当鼠标悬停在某个 list cell 上时，背景色变为蓝色（#0096C9），文本颜色变为白色。

（3）.list-cell:selected：设置了被选中的 list cell 的背景色为白色，文本颜色为黑色。

（4）.list-cell:selected:hover：当鼠标悬停在被选中的 list cell 上时，背景色变为蓝色，文本颜色为白色。

（5）.list-cell:focused:selected, .list-cell:focused:selected:hover：设置了当 list cell 被选中且获得焦点时（无论是否有鼠标悬停），背景色和文本颜色保持默认，也就是背景色为白色，文本颜色为黑色。

这样设置的目的是保持 UI 的一致性和可读性，尤其是在用户交互时。当用户移动鼠标或选择列表项时，他们可以清楚地看到他们的选择，因为背景色和文本颜色的对比会随着状态的变化而变化。

3.13 项目实战：自由绘画

这一节会用 JavaFX 做一个项目，这个项目的主要功能有如下 3 个。

（1）自由绘图：在窗口上通过鼠标绘制各种线条。

（2）设置画笔颜色：通过颜色选择对话框可以设置画笔颜色，从而绘制出各种颜色的线条。

（3）保存图像：将绘制的线条保存为 png 图像。

在 JavaFX 中，自由绘图通常涉及以下几个核心概念和组件。

（1）Canvas：一个 JavaFX 组件，它提供了一个空白的绘制区域。它的工作原理类似于画布，在上面可以绘制各种图形，包括线条、形状和图像。

（2）GraphicsContext：与 Canvas 绑定的一个对象，它提供了一系列的方法来绘制和填充颜色、渐变、图像等。可以认为 GraphicsContext 是绘画的笔，而 Canvas 是纸。

在 JavaFX 中，自由绘图的基本原理是捕捉用户的鼠标事件，然后根据这些事件在 Canvas 上进行绘制。当用户按下鼠标并移动时，可以监听 onMousePressed 和 onMouseDragged 事件，并使用 GraphicsContext 的 strokeLine 方法来绘制线条。

本项目完整的实现代码如下。

代码位置：src/main/java/javafxgui/Drawing.java

```
package javafxgui;
import javafx.application.Application;
```

```
import javafx.embed.swing.SwingFXUtils;
import javafx.geometry.Insets;
import javafx.geometry.Pos;
import javafx.scene.Scene;
import javafx.scene.canvas.Canvas;
import javafx.scene.canvas.GraphicsContext;
import javafx.scene.control.Button;
import javafx.scene.control.ColorPicker;
import javafx.scene.image.WritableImage;
import javafx.scene.layout.VBox;
import javafx.scene.paint.Color;
import javafx.stage.FileChooser;
import javafx.stage.Stage;
import javax.imageio.ImageIO;
import java.io.File;

public class Drawing extends Application {
    private GraphicsContext graphicsContext;
    private Color penColor = Color.BLACK;
    private double lastX = 0, lastY = 0;
    @Override
    public void start(Stage primaryStage) {
        // 设置窗口标题和大小
        primaryStage.setTitle("画图程序");
        VBox root = new VBox();
        root.setAlignment(Pos.CENTER);
        Scene scene = new Scene(root, 500, 550);

        // 创建画布并添加到布局中
        Canvas canvas = new Canvas(500, 450);
        graphicsContext = canvas.getGraphicsContext2D();
        graphicsContext.setFill(Color.WHITE);
        graphicsContext.fillRect(0, 0, canvas.getWidth(), canvas.getHeight());
        root.getChildren().add(canvas);

        // 画笔颜色选择器按钮
        ColorPicker colorPicker = new ColorPicker(penColor);
        colorPicker.setOnAction(event -> penColor = colorPicker.getValue());
        root.getChildren().add(colorPicker);
        VBox.setMargin(colorPicker, new Insets(20, 0, 0, 0)); // 上，右，下，左
        // 保存图像按钮
        Button saveButton = new Button("保存图像");

        saveButton.setOnAction(event -> saveImage(canvas));
        root.getChildren().add(saveButton);
        VBox.setMargin(saveButton, new Insets(10, 0, 0, 0));  // 上，右，下，左
        // 鼠标事件处理
        canvas.setOnMousePressed(e -> {
```

```
            lastX = e.getX();
            lastY = e.getY();
        });

        canvas.setOnMouseDragged(e -> {
            draw(lastX, lastY, e.getX(), e.getY());
            lastX = e.getX();
            lastY = e.getY();
        });
        canvas.setOnMouseReleased(e -> {
            draw(lastX, lastY, e.getX(), e.getY());
            lastX = 0;
            lastY = 0;
        });
        primaryStage.setScene(scene);
        primaryStage.show();
    }
    private void draw(double x1, double y1, double x2, double y2) {
        graphicsContext.setStroke(penColor);
        graphicsContext.setLineWidth(3);
        graphicsContext.strokeLine(x1, y1, x2, y2);
    }
    private void saveImage(Canvas canvas) {
        FileChooser fileChooser = new FileChooser();
        fileChooser.setTitle("保存图像");
        fileChooser.getExtensionFilters().add(new FileChooser.
ExtensionFilter("PNG Image (*.png)", "*.png"));
        File file = fileChooser.showSaveDialog(null);
        if (file != null) {
            try {
                WritableImage writableImage = new WritableImage((int)
canvas.getWidth(), (int) canvas.getHeight());
                canvas.snapshot(null, writableImage);
                ImageIO.write(SwingFXUtils.fromFXImage(writableImage, null),
"png", file);
            } catch (Exception e) {
                e.printStackTrace();
            }
        }
    }
    public static void main(String[] args) {
        launch(args);
    }
}
```

运行程序，可以在窗口的白色区域用鼠标自由绘制曲线了，并且可以通过颜色选择对话框设置画笔的颜色。绘制的效果如图 3-18 所示。单击“保存图像”按钮可以将当前绘制的图形保存为 png 文件。

图 3-18　自由绘画

3.14　项目实战：旋转图像

本节要实现的项目是旋转和保存图像。项目的主要功能是打开图像，逆时针和顺时针将图像旋转任意角度，并保存旋转后的效果。

在 JavaFX 中，图像旋转和保存涉及几个关键的类和概念。下面是对图像旋转原理的描述。

（1）ImageView：JavaFX 中用于显示图像的节点。它可以包含一个 Image 对象，并在场景图中渲染图像。

（2）Transform：在 JavaFX 中，图像的旋转通过应用变换来实现。这些变换可以是平移、缩放、旋转等。针对旋转，Rotate 类被用来定义旋转变换。它可以指定旋转的角度、旋转的中心点等信息。

（3）旋转操作：如果想旋转一个 ImageView 中的图像，可以直接对 ImageView 应用 Rotate 变换。这样，图像以及与之关联的任何变换都会显示在用户界面上。变换可以是累积的，即每次旋转操作都是相对于当前状态执行的。

下面是保存图像的步骤。

（1）调用 ImageView 的 snapshot 方法来生成一个快照，该快照反映了当前在屏幕上渲染的所有内容（包括旋转）。

（2）使用 SwingFXUtils.fromFXImage 方法将 WritableImage 转换为 BufferedImage。

（3）使用 ImageIO.write 方法将 BufferedImage 保存到文件系统中。

本项目完整的实现代码如下。

代码位置：src/main/java/javafxgui/RotateImage.java

```
package javafxgui;

import javafx.application.Application;
import javafx.geometry.Insets;
import javafx.geometry.Pos;
import javafx.scene.Scene;
import javafx.scene.control.*;
import javafx.scene.image.Image;
import javafx.scene.image.ImageView;
import javafx.scene.layout.VBox;
import javafx.stage.FileChooser;
import javafx.stage.Stage;
import javafx.embed.swing.SwingFXUtils;
import javax.imageio.ImageIO;
import java.awt.image.BufferedImage;
import java.io.File;

public class RotateImage extends Application {
    private ImageView imageView = new ImageView();
    private Image originalImage;

    @Override
    public void start(Stage primaryStage) {
        VBox vbox = new VBox(10);
        vbox.setAlignment(Pos.TOP_CENTER);
        vbox.setPadding(new Insets(20));

        // 创建按钮
        Button openButton = new Button("打开图像");
        Button rotateCCWButton = new Button("逆时针旋转");
        Button rotateCWButton = new Button("顺时针旋转");
        Button saveButton = new Button("保存图像");

        // 设置按钮的事件处理
        openButton.setOnAction(e -> openImage(primaryStage));
        rotateCCWButton.setOnAction(e -> rotateImageCCW(90)); // 示例逆时针旋转 90 度
        rotateCWButton.setOnAction(e -> rotateImageCW(90)); // 示例顺时针旋转 90 度
        saveButton.setOnAction(e -> saveImage(primaryStage));

        // 添加按钮和 ImageView 到布局容器
        vbox.getChildren().addAll(openButton, rotateCCWButton, rotateCWButton,
saveButton, imageView);

        // 设置场景和舞台
        primaryStage.setTitle("图像旋转工具");
        primaryStage.setScene(new Scene(vbox, 500, 600));
        primaryStage.show();
    }
    // 打开图像
```

```
    private void openImage(Stage stage) {
        FileChooser fileChooser = new FileChooser();
        fileChooser.setTitle(" 打开图像 ");
        File file = fileChooser.showOpenDialog(stage);
        if (file != null) {
            originalImage = new Image(file.toURI().toString());
            imageView.setImage(originalImage);
            imageView.setPreserveRatio(true);
            imageView.setFitHeight(300);          // 限制图像视图的高度，保持图像比例
        }
    }
    // 顺时针旋转图像
    private void rotateImageCW(double angle) {
        TextInputDialog dialog = new TextInputDialog("90");
        dialog.setTitle(" 输入旋转角度 ");
        dialog.setHeaderText(" 请输入旋转角度（度）:");
        dialog.setContentText(" 角度 :");
        dialog.showAndWait().ifPresent(input -> {
            double rotateAngle = Double.parseDouble(input);
            imageView.setRotate(imageView.getRotate() + rotateAngle);
        });
    }
    // 逆时针旋转图像
    private void rotateImageCCW(double angle) {
        TextInputDialog dialog = new TextInputDialog("90");
        dialog.setTitle(" 输入旋转角度 ");
        dialog.setHeaderText(" 请输入旋转角度（度）:");
        dialog.setContentText(" 角度 :");
        dialog.showAndWait().ifPresent(input -> {
            double rotateAngle = Double.parseDouble(input);
            imageView.setRotate(imageView.getRotate() - rotateAngle);
        });
    }
    // 保存图像
    private void saveImage(Stage stage) {
        if (imageView.getImage() != null) {
            FileChooser fileChooser = new FileChooser();
            fileChooser.setTitle(" 保存图像 ");
            File file = fileChooser.showSaveDialog(stage);
            if (file != null) {
                try {
                    // 保存图像
                    BufferedImage bufferedImage = SwingFXUtils.fromFXImage
(imageView.snapshot(null, null), null);
                    ImageIO.write(bufferedImage, "png", file);
                } catch (Exception e) {
                    e.printStackTrace();
                }
            }
        }
    }
```

```
    public static void main(String[] args) {
        launch(args);
    }
}
```

运行程序，单击“打开图像”按钮，打开一个图像文件，然后单击“顺时针旋转”按钮，会弹出一个角度输入对话框，输入 12，单击“确定”按钮，图像就会顺时针旋转 12°，如图 3-19 所示。

单击“保存图像”按钮，将旋转后的图像保存为 png 图像，保存后的效果如图 3-20 所示。

图 3-19 将图像顺时针旋转 12°

图 3-20 顺时针旋转 12° 后的图像

3.15 小结

本章为读者提供了 JavaFX 的全面介绍，从基础安装配置到构建完整的应用程序。通过对 JavaFX 的主要组件和概念的详细讲解，读者可以获得足够的知识来开始构建自己的 JavaFX 应用程序。各种布局容器和界面组件的讲解，可使读者能够创建美观且功能丰富的用户界面。通过实例和代码片段的引入，本章使复杂的概念变得更易理解。同时，本章还介绍了 ControlsFX 这个第三方库，它提供了 JavaFX 标准库中不可用或难以实现的附加功能，极大地丰富了 JavaFX 的能力。总的来说，本章为那些希望利用 JavaFX 开发现代桌面应用程序的开发者提供了一个坚实的基础。

第 4 章

有趣的 GUI 技术

有很多特殊的应用，在初学者看来，简直就和魔法一样。例如，在屏幕上显示一只青蛙，或是一只怪兽，用鼠标还可以来回拖动。在屏幕的任何位置绘制曲线，随时可以清除这些曲线，或者在系统托盘添加图标，弹出对话气泡等。这些看似很复杂，其实用 Java 实现起来相当简单，这是因为 Java 有大量的第三方库，通过这些库，只需要几行代码就可以搞定这些魔法。本章将通过大量完整的例子演示如何实现这些看似复杂，其实相当简单的功能。

4.1 特殊窗口

在很多场景中，往往需要实现各种特殊的窗口形态，例如，非矩形的窗口（称为异形窗口）、半透明窗口等，本节会介绍如何实现这些特殊窗口。

4.1.1 五角星窗口

窗口通常都是矩形的，但还有很多应用，尤其是游戏，窗口却是不规则的，例如，圆形、椭圆、三角形、五角星，甚至是一个怪兽的形态（如有些游戏程序），其实这些仍然是窗口，只不过通过掩模（mask）技术将某些部分变得透明，因此，用户看起来窗口就变成了不规则的图形。

在计算机科学中，掩模或位掩码（bitmask）是用于位运算的数据，特别是在位域中。使用掩模，可以在一个字节的多个位上设置开或关，或者在一次位运算中将开和关反转。

在计算机图形学中，掩模是用于隐藏或显示另一图像的部分的数字图像。掩模可以用于创建特殊效果或选择图像的区域进行编辑。掩模可以从零开始在图像编辑器中创建，也可以从现有图像生成。例如，可以使用一张照片作为掩模，来显示或隐藏另一张照片的部分，如果对窗口使用掩模，那么窗口就变成了异形窗口。

本节会使用 JavaFX 以及相关 API 实现五角星形态的异形窗口，也就是运行程序后，窗口会变成一个红色的五角星，其他部分是透明的，效果如图 4-1 所示。我们可以通过鼠标拖动五角星来移动窗口。

从图 4-1 所示的五角星可以看出，运行程序，只会显示一个五角星，其他部分是透明的，可以看到后面的 Java 文件目录。

要绘制五角星，需要先了解如何绘制五角星。绘制五角星需要使用如下几组数据。

（1）五角星的中心点坐标。

（2）五角星内切圆半径和外切圆半径。

（3）五角星 10 个顶点的坐标。

（4）五角星的旋转角度。

在这些数据中，（1）、（2）和（4）是直接指定的，而（3）可以通过计算获得。本节绘制的五角星是正五角星①，所以这里只讨论正五角星。五角星以及内切圆和外切圆如图 4-2 所示。其中 A 到 J 一共 10 个字母分别表示五角星的 10 个顶点。现在来计算顶点 B 的坐标，其他顶点坐标的计算方法类似。

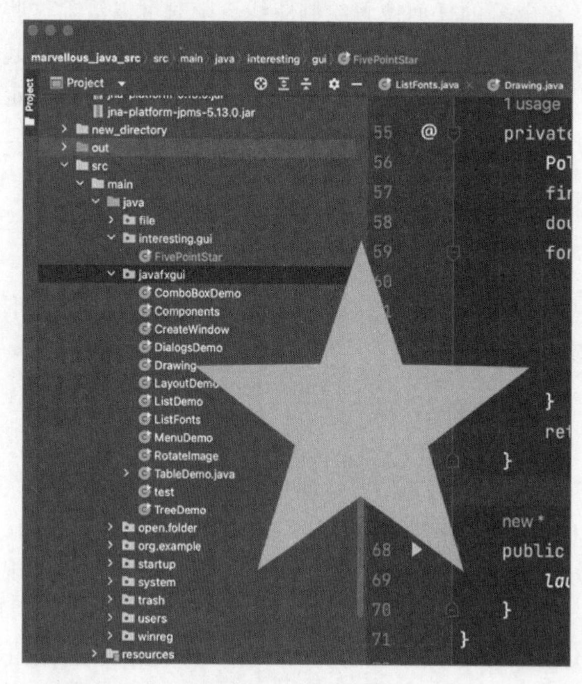

图 4-1　五角星窗口

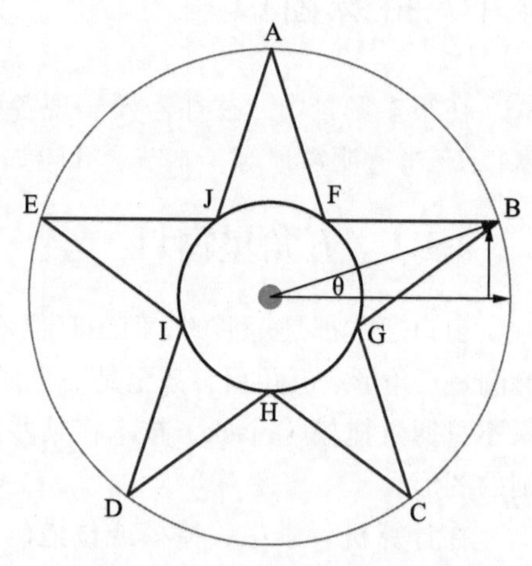

图 4-2　五角星顶点坐标的计算

假设五角星中心点的坐标是（centerX、centerY），顶点 B 与中心点的连线与水平线的夹角是 θ，外切圆半径是 r，那么 B 点的坐标如下：

```
(centerX + r * cos(θ), centerY - r * sin(θ))
```

① 正五角星是一个由五条对角线构成的正五边形的内接星形。它由 5 个相等的等腰三角形组成，每个三角形的顶点是正五边形的一个顶点。正五角星的每个内角是 36°，每个外角是 72°。正五角星有很多数学和几何性质，例如，它的对角线之比是黄金比。

按类似的方法计算完 10 个顶点的坐标，就可以绘制 10 个点的多边形，最终形成一个五角星。

下面描述如何用 JavaFX 来实现如图 4-1 所示的五角星异形窗口。

1. 创建主应用程序类

（1）继承自 Application 类，这是所有 JavaFX 应用程序的入口点。

（2）覆盖 start 方法，这是 JavaFX 生命周期的一部分，用于初始化和显示应用程序的主舞台（窗口）。

2. 设置舞台（Stage）样式

使用 StageStyle.TRANSPARENT 来创建一个没有边框且背景透明的窗口，使得我们能够实现异形效果。

3. 创建五角星形状（Polygon）

（1）createStar 方法负责生成五角星的顶点坐标，并创建一个 Polygon 对象。

（2）五角星通过交替连接外接圆和内接圆上的点来形成。外接圆上的点形成五角星的尖角，内接圆上的点位于尖角之间。

（3）createStar 方法接收中心点坐标、外接圆半径、内接圆半径和射线数量（五角星为 5）作为参数。

（4）使用两个半径交替绘制五角星的顶点，通过循环计算每个点的坐标。循环迭代次数是射线数量的两倍，因为每个射线上有一个外接圆上的点和一个内接圆上的点。

（5）每次迭代计算当前角度下的点坐标，使用三角函数（Math.cos 和 Math.sin）来计算。

（6）角度从正上方开始计算（即 Math.PI / 2.0），然后每次减去一个固定的角度步长（Math.PI / numRays），这样可以确保五角星的对称性。

4. 设置五角星的样式

使用 setFill 和 setStroke 方法为五角星设置填充颜色和描边颜色。

5. 创建布局并添加五角星

（1）创建一个 Pane，作为五角星的容器，并将其设置为场景（Scene）的根节点。

（2）将五角星形状添加到面板中，并通过 CSS 设置面板的背景为透明。

6. 实现五角星的拖动效果

（1）通过为根面板设置鼠标事件处理器来实现窗口的拖动效果。

（2）在鼠标按下事件（setOnMousePressed）中记录鼠标的初始位置和窗口的初始位置。

（3）在鼠标拖动事件（setOnMouseDragged）中，根据鼠标的当前位置和初始位置的差值更新窗口的新位置。这样，当用户按下并拖动鼠标时，五角星窗口就会跟随鼠标移动。

7. 显示舞台

（1）创建场景，设置其大小和填充色，并将其设置到舞台（Stage）上。

（2）调用 show 方法显示舞台。

在整个过程中，createStar 方法是绘制五角星的核心。通过精确计算并绘制五角星的顶点，我们创建了一个异形效果的五角星。通过处理鼠标事件，实现了窗口的拖动效果，使用户可以自由移动五角星窗口。这些技术的结合使得我们能够在 JavaFX 中实现一个有趣且互动性强的图形界面。

下面是实现五角星异形窗口的完整代码。

代码位置：src/main/java/interesting/gui/FivePointStar.java

```
package interesting.gui;

import javafx.application.Application;
import javafx.scene.Scene;
import javafx.scene.layout.Pane;
import javafx.scene.paint.Color;
import javafx.scene.shape.Polygon;
import javafx.stage.Stage;
import javafx.stage.StageStyle;

public class FivePointStar extends Application {
    private double xOffset = 0;
    private double yOffset = 0;
    @Override
    public void start(Stage primaryStage) {
        // 创建一个无边框的透明场景
        primaryStage.initStyle(StageStyle.TRANSPARENT);
        // 创建五角星形状，外接圆半径 200，内接圆半径 80
        Polygon star = createStar(300, 300, 200, 80, 5);
        // 填充颜色和描边
        star.setFill(Color.RED);
        star.setStroke(Color.RED);
        star.setStrokeWidth(2);
        // 创建根面板并添加五角星
        Pane root = new Pane();
        root.getChildren().add(star);
        root.setStyle("-fx-background-color: transparent;");  // 背景透明
        // 鼠标按下时记录位置
        root.setOnMousePressed(event -> {
            xOffset = event.getSceneX();
            yOffset = event.getSceneY();
        });
        // 鼠标拖动时更新窗口位置
        root.setOnMouseDragged(event -> {
            primaryStage.setX(event.getScreenX() - xOffset);
            primaryStage.setY(event.getScreenY() - yOffset);
        });
        // 注意：窗口尺寸需要调整以适应更大的五角星
        Scene scene = new Scene(root, 600, 600);
        scene.setFill(Color.TRANSPARENT);                    // 场景填充透明

        primaryStage.setScene(scene);
```

```
        primaryStage.show();
    }
    // 创建五角星的 createStar 方法
    private Polygon createStar(double centerX, double centerY, double
outerRadius, double innerRadius, int numRays) {
        Polygon star = new Polygon();
        final double angleStep = Math.PI / numRays;
        double startAngle = Math.PI / 2.0;
        for (int i = 0; i < numRays * 2; i++) {
            double angle = startAngle - i * angleStep;
            double radius = (i % 2 == 0) ? outerRadius : innerRadius;
            star.getPoints().add(centerX + Math.cos(angle) * radius);
            star.getPoints().add(centerY - Math.sin(angle) * radius);
        }
        return star;
    }
    public static void main(String[] args) {
        launch(args);
    }
}
```

4.1.2　异形美女机器人窗口

在 4.1.1 节使用 JavaFX 实现了一个异形窗口，但这个异形窗口只是绘制了一个简单的五角星，如果需要更复杂的异形窗口，就不能直接绘制了，需要使用透明 png 图像处理，例如，本节实现的如图 4-3 所示的异形美女机器人窗口，就是用了一个透明的 png 图像。

图 4-3　异形美女机器人窗口

实现这个程序主要需要如下 3 个技术点。

（1）装载图像：使用 Image 对象装载 png 图像，并在 ImageView 组件中显示。

（2）设置透明度：使用 primaryStage.initStyle(StageStyle.TRANSPARENT) 设置。

（3）拖动事件：使用 Pane.setOnMousePressed 和 Pane.setOnMouseDragged 处理。

这 3 个技术点的后 2 个与 4.1.1 节实现的异形五角星窗口完全相同，本节不再重复讲解。这里要着重讲解的是装载图像。图像路径可以使用相对路径，也可以使用绝对路径。如果使用相对路径，通常是相对于 main/resources 目录，如下面的代码从 main/resources/images 目录装载 robot.png 图像文件。

```
Image image = new Image(getClass().getResourceAsStream("/images/robot.png"));
```

如果要从绝对路径装载图像文件，需要使用 file:// 作为前缀，代码如下：

```
Image image = new Image("file:///temp/images/robot.png"));
```

这行代码从 /temp/images 目录装载 robot.png 图像文件。

实现异形美女机器人窗口的完整代码如下。

代码位置：src/main/java/interesting/gui/GirlWindow.java

```
package interesting.gui;

import javafx.application.Application;
import javafx.scene.Scene;
import javafx.scene.image.Image;
import javafx.scene.image.ImageView;
import javafx.scene.layout.Pane;
import javafx.stage.Stage;
import javafx.stage.StageStyle;

public class GirlWindow extends Application {
    private double xOffset = 0;
    private double yOffset = 0;

    @Override
    public void start(Stage primaryStage) throws Exception {
        // 加载透明 PNG 图像
        Image image = new Image(getClass().getResourceAsStream("/images/robot.png"));
        ImageView imageView = new ImageView(image);
        // 创建根面板并添加图像视图
        Pane root = new Pane(imageView);
        // 设置无边框和透明度
        primaryStage.initStyle(StageStyle.TRANSPARENT);         // 无边框
        // 创建场景并设置透明
        Scene scene = new Scene(root, image.getWidth(), image.getHeight());
```

```
        scene.setFill(null);                                        // 场景背景透明
        // 鼠标事件处理
        root.setOnMousePressed(event -> {
            xOffset = event.getSceneX();
            yOffset = event.getSceneY();
        });
        root.setOnMouseDragged(event -> {
            primaryStage.setX(event.getScreenX() - xOffset);
            primaryStage.setY(event.getScreenY() - yOffset);
        });
        // 显示舞台
        primaryStage.setScene(scene);
        primaryStage.show();
    }
    public static void main(String[] args) {
        launch(args);
    }
}
```

在运行程序前，要确保 main/resources/images 目录存在 robot.png 文件。

4.1.3　半透明窗口

在半透明窗口中，不管是窗口背景还是组件，都是半透明的，实现这种效果的总体思路与前两节实现的异形窗口类似，只是不仅需要设置 Stage 为透明类型，还需要使用 setOpacity 方法设置透明度，代码如下：

```
primaryStage.initStyle(StageStyle.TRANSPARENT);
// 设置透明度为 0.7，0 是完全透明，1 是不透明。如果设置为 0，就什么都看不见了
primaryStage.setOpacity(0.7);
```

半透明窗口的完整实现代码如下。

代码位置：src/main/java/interesting/gui/TranslucentWindow.java

```
package interesting.gui;

import javafx.application.Application;
import javafx.scene.Scene;
import javafx.scene.control.Button;
import javafx.scene.control.Label;
import javafx.scene.layout.Pane;
import javafx.stage.Stage;
import javafx.stage.StageStyle;

public class TranslucentWindow extends Application {
    private double xOffset = 0;
```

```
    private double yOffset = 0;
    @Override
    public void start(Stage primaryStage) {
        // 创建一个 Pane 来作为根节点
        Pane root = new Pane();

        // 创建一个标签，显示 "Hello, world!"
        Label label = new Label("Hello, world!");
        label.setLayoutX(150);
        label.setLayoutY(100);
        label.setStyle("-fx-font-size: 20px;");

        // 创建一个按钮，显示 "Close"
        Button button = new Button("Close");
        button.setLayoutX(150);
        button.setLayoutY(200);
        // 绑定按钮单击事件来关闭窗口
        button.setOnAction(event -> primaryStage.close());
        // 将标签和按钮添加到根节点
        root.getChildren().addAll(label, button);

        // 创建一个场景，设置透明度
        Scene scene = new Scene(root, 400, 300);
        scene.setFill(null);

        // 设置窗口无边框
        primaryStage.initStyle(StageStyle.TRANSPARENT);
        primaryStage.setOpacity(0.7);

        // 监听鼠标事件来实现窗口的拖动
        root.setOnMousePressed(event -> {
            xOffset = event.getSceneX();
            yOffset = event.getSceneY();
        });
        root.setOnMouseDragged(event -> {
            primaryStage.setX(event.getScreenX() - xOffset);
            primaryStage.setY(event.getScreenY() - yOffset);
        });
        primaryStage.setScene(scene);
        primaryStage.show();
    }
    public static void main(String[] args) {
        launch(args);
    }
}
```

运行程序，会显示一个如图 4-4 所示的半透明窗口。

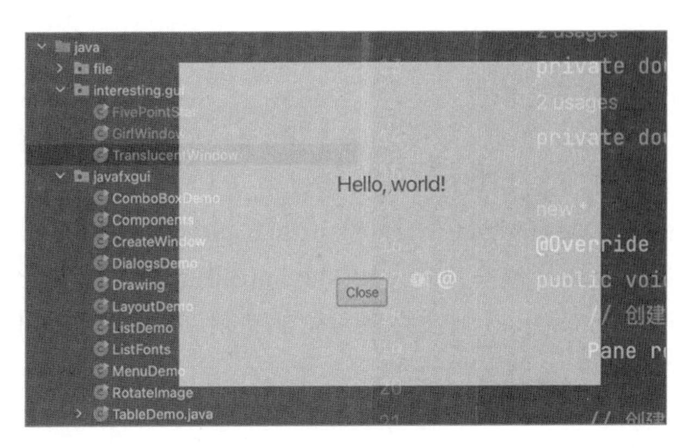

图 4-4　半透明窗口

4.2　在屏幕上绘制曲线

有很多画笔应用，可以在整个电脑屏幕上绘制曲线和各种图形，这种应用很适合在线教学或演示。例如，讲解代码的编写过程，可以一边展示代码，一边在代码上绘制曲线、手写一些文字等，效果如图 4-5 所示。

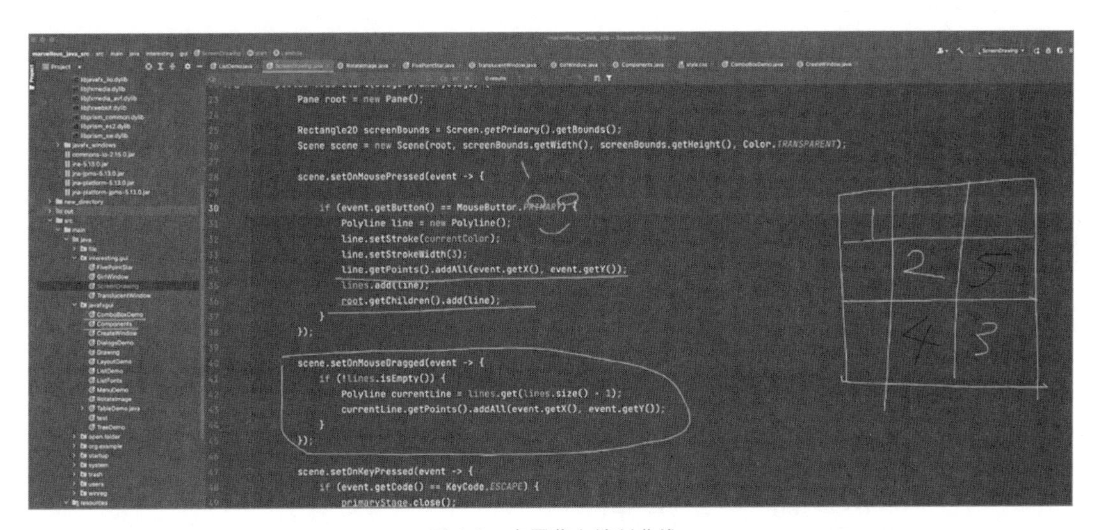

图 4-5　在屏幕上绘制曲线

实现这个功能的基本做法是让窗口充满整个屏幕，并且隐藏边框和标题栏，以及让窗口完全透明，这样就可以看到窗口后面的内容了。后面的事情就简单多了，例如，可以直接在窗口上绘制曲线，放置组件，绘制各种简单和复杂的图形，放置图像等。

本节会使用 JavaFX 实现这个程序，其中会涉及如下的主要技术点。

（1）Rectangle2D 对象：这个对象表示二维空间中的一个矩形区域，并提供了获取该区域宽度和高度的方法。当我们调用 Screen.getPrimary().getBounds() 时，它返回当前主屏幕的边界，即整个屏幕的宽度和高度。

（2）Color.TRANSPARENT：这是设置场景背景颜色为透明。这样做的目的是让窗口背后的内容（例如桌面或其他应用程序）能够透过来。

（3）primaryStage.setMaximized(true) 和 primaryStage.setFullScreen(true)：Stage 是 JavaFX 中的顶级容器，代表了整个窗口。setMaximized(true) 会最大化窗口，但保留任务栏和窗口装饰。setFullScreen(true) 则会将窗口切换到全屏模式，隐藏任务栏和窗口装饰，提供不受干扰的全屏体验。

使用这些技术，可以创建一个能够覆盖整个屏幕的画布，用户可以在上面进行绘制，类似于在一个透明的覆盖层上画图。这对于某些特定的应用程序很有用，比如屏幕标注工具或者绘图程序。通过设置为全屏并且背景透明，用户可以在任何他们想要的背景上进行绘制，同时也保留了其他程序的可见性。

在编写代码之前，先来明确一下这个例子的主要功能。

（1）绘制线条时，鼠标抬起再按下，会绘制新线条，旧的线条仍然保留，所以需要保存绘图历史。

（2）按 Ctrl + C 组合键，清除所有的绘图。

（3）按 Esc 键退出程序。

（4）按数字 1、2、3，分别设置线条颜色为黑色、红色和蓝色，默认是黑色。

（5）设置线条颜色时，以前绘制的线条的颜色不变。

本例的完整代码如下。

代码位置：src/main/java/interesting/gui/ScreenDrawing.java

```
package interesting.gui;

import javafx.application.Application;
import javafx.geometry.Rectangle2D;
import javafx.scene.Scene;
import javafx.scene.input.*;
import javafx.scene.layout.Pane;
import javafx.scene.paint.Color;
import javafx.scene.shape.Polyline;
import javafx.stage.Screen;
import javafx.stage.Stage;
import javafx.stage.StageStyle;
import java.util.ArrayList;
import java.util.List;

public class ScreenDrawing extends Application {
    private List<Polyline> lines = new ArrayList<>();  // 存储所有线条的列表
    private Color currentColor = Color.BLACK;          // 当前线条的颜色，默认为黑色
    @Override
    public void start(Stage primaryStage) {
```

```
        Pane root = new Pane();
        Rectangle2D screenBounds = Screen.getPrimary().getBounds();
        Scene scene = new Scene(root, screenBounds.getWidth(), 
screenBounds.getHeight(), Color.TRANSPARENT);
        scene.setOnMousePressed(event -> {
            // 鼠标按下开始绘制
            if (event.getButton() == MouseButton.PRIMARY) {
                Polyline line = new Polyline();
                line.setStroke(currentColor);
                line.setStrokeWidth(3);
                line.getPoints().addAll(event.getX(), event.getY());
                // 保存绘制历史
                lines.add(line);
                // 将当前绘制的线条添加到root中
                root.getChildren().add(line);
            }
        });
        // 滑动鼠标绘制线条
        scene.setOnMouseDragged(event -> {
            if (!lines.isEmpty()) {
                Polyline currentLine = lines.get(lines.size() - 1);
                currentLine.getPoints().addAll(event.getX(), event.getY());
            }
        });
        // 捕获键盘按键
        scene.setOnKeyPressed(event -> {
            if (event.getCode() == KeyCode.ESCAPE) {
                primaryStage.close();
            } else if (event.getCode() == KeyCode.DIGIT1) {
                currentColor = Color.BLACK;
            } else if (event.getCode() == KeyCode.DIGIT2) {
                currentColor = Color.RED;
            } else if (event.getCode() == KeyCode.DIGIT3) {
                currentColor = Color.BLUE;
            } else if (new KeyCodeCombination(KeyCode.C, 
KeyCombination.CONTROL_DOWN).match(event)) {
                // 清除root中的所有线条
                root.getChildren().removeAll(lines);
                lines.clear();
            }
        });
        primaryStage.initStyle(StageStyle.TRANSPARENT);
        // 让窗口总是在顶端显示
        primaryStage.setAlwaysOnTop(true);
        primaryStage.setScene(scene);
        primaryStage.show();
    }
    public static void main(String[] args) {
        launch(args);
    }
}
```

运行程序，仍然可以看到当前桌面上已经运行的软件，这是因为 ScreenDrawing 的窗口是透明的，现在可以在屏幕上绘制任意曲线了。

4.3 控制状态栏

状态栏一直是各类程序争夺的主阵地之一，有很多主流程序都会在状态栏添加一个或多个图标，以及图标弹出菜单、对话气泡等功能。本节将详细讲解如何用 Java 攻占这块主阵地。

4.3.1 在状态栏上添加图标

Windows、macOS 和 Linux 操作系统都有状态栏，而且都可以添加图标。只是添加图标的区域不同。**Windows** 称为通知区域，也就是任务栏右侧的部分，它包含了一些常用的图标和通知，比如电池、Wi-Fi、音量、时钟和日历等。**macOS** 称为菜单栏，在菜单栏左侧显示 macOS 菜单，右侧显示各种图标，单击这些图标，会有不同的动作。**Linux** 右上角显示图标的区域的称呼可能因不同的桌面环境而有所不同，但一般可以称为状态栏或者通知区域。为了统一，后面统称为状态栏。

通过 Java 的抽象窗口工具包（Abstract Window Toolkit，AWT）可以实现在系统托盘区域添加一个应用程序图标，并且提供一个单击图标时弹出的菜单。具体实现原理和步骤如下。

（1）检查系统托盘支持：程序首先检查操作系统是否支持系统托盘。这是通过调用 SystemTray.isSupported() 方法实现的。如果返回 True，则系统托盘可用。

（2）创建弹出菜单：接下来程序创建一个 PopupMenu 对象，这个对象将承载所有的菜单项。

（3）添加菜单项：通过实例化 MenuItem 对象并将它们添加到 PopupMenu 对象中，程序创建了可供用户选择的菜单项。

（4）添加事件监听器：对于每个菜单项，程序通过调用 addActionListener 方法并传入一个 ActionListener 来定义菜单项被单击时的行为。

（5）加载图标图片：使用 ImageIO.read() 方法加载一个图片文件，这个图片将作为系统托盘图标显示。

（6）创建托盘图标：程序创建了一个 TrayIcon 实例，将之前加载的图片、一个标题字符串（当鼠标悬停在图标上时显示），以及弹出菜单传递给它的构造函数。

（7）添加图标到系统托盘：最后，程序调用 SystemTray.getSystemTray().add(trayIcon) 方法，将创建的 TrayIcon 实例添加到系统托盘。

下面详细解释一下如何响应菜单事件：当用户单击某个菜单项时，与该菜单项关联的 ActionListener 的 actionPerformed 方法会被调用。在这个方法中，可以定义单击后的行为，例如打印一句话或者退出程序。

例如，以下代码片段显示了如何为 Exit 菜单项添加一个事件监听器，使程序在单击时退出：

```
MenuItem exitItem = new MenuItem("Exit");
exitItem.addActionListener(e -> System.exit(0));                    // 单击后退出程序
popup.add(exitItem);
```

下面的例子完整地演示了如何使用 AWT 相关 API 在状态栏添加一个菜单，以及相应菜单项的单击动作。本例的菜单中有两个菜单项：Hello 和“退出”。单击 Hello 菜单项会在终端输出 Hello 字符串，单击“退出”菜单项，会退出应用程序。

代码位置：src/main/java/interesting/gui/TrayDemo.java

```
package interesting.gui;
import java.awt.*;
import java.awt.event.ActionListener;
import java.net.URL;
import javax.imageio.ImageIO;
public class TrayDemo {
    public static void main(String[] args) throws Exception {
        // 确保系统托盘支持
        if (!SystemTray.isSupported()) {
            System.out.println("System tray is not supported!");
            return;
        }
        // 创建一个弹出菜单
        PopupMenu popup = new PopupMenu();

        // 创建菜单项
        MenuItem helloItem = new MenuItem("Hello");
        helloItem.addActionListener(e -> System.out.println("Hello, World!"));
        popup.add(helloItem);

        MenuItem exitItem = new MenuItem("退出");
        exitItem.addActionListener(e -> System.exit(0));
        popup.add(exitItem);

        // 从资源文件中加载图标图片
        URL imageURL = TrayDemo.class.getResource("/images/tray.png");
        Image image = ImageIO.read(imageURL);

        // 创建托盘图标并添加到系统托盘
        TrayIcon trayIcon = new TrayIcon(image, "Tray Demo", popup);
        trayIcon.setImageAutoSize(true);
        SystemTray.getSystemTray().add(trayIcon);
```

```
    }
}
```

运行程序，会看到在状态栏上显示一个绿色的图标，在图标上右击，会弹出一个菜单，图 4-6 是 Windows 的效果，图 4-7 是 macOS 的效果，图 4-8 是 Ubuntu Linux 的效果。

图 4-6 Windows 通知区域的图标和菜单

图 4-7 macOS 菜单栏的图标和菜单

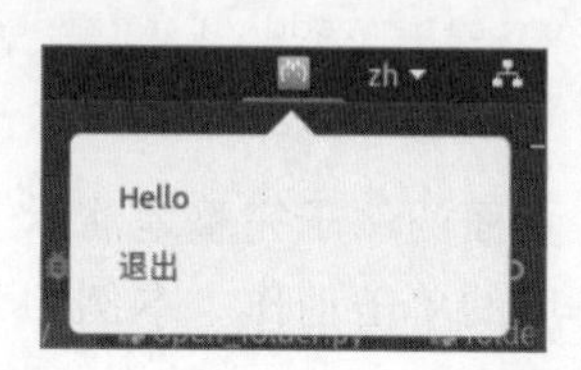

图 4-8 Ubuntu Linux 菜单栏的图标和菜单

4.3.2 添加 Windows 10 风格的 Toast 消息框

我们可以使用 AWT 的 SystemTray 类和 TrayIcon 类来实现在 Windows 10 中显示一个 toast 消息框的功能，下面是具体的实现步骤。

（1）系统托盘支持检查：SystemTray.isSupported() 方法用于检查当前平台是否支持系统托盘功能。如果不支持，程序会打印一条消息并退出。

（2）加载和设置托盘图标：通过 ImageIO.read() 方法从资源文件中加载图像。由于托盘图标通常需要是较小的尺寸，所以使用 getScaledInstance() 方法将图像缩放到合适的大小。

（3）添加托盘图标到系统托盘：创建 TrayIcon 对象，并将处理过的图像设置为其图标。然后获取系统托盘的实例并通过 add() 方法添加这个托盘图标。

（4）显示通知消息：调用 TrayIcon 对象的 displayMessage() 方法来显示一个 toast 消息。这个方法接收三个参数：通知的标题、文本内容和消息类型。在此例中，消息类型被设置为 MessageType.INFO，表示这是一个信息性质的通知。

（5）添加通知单击事件处理器：为 TrayIcon 对象添加一个 ActionListener，当用户单击通知时会触发这个监听器。在这个监听器中，使用 Desktop.getDesktop().browse() 方法打开一个网页。

（6）打开网页的方法：openWebPage 方法接收一个字符串形式的 URL，并使用 Desktop 类的 browse() 方法尝试在用户的默认浏览器中打开这个网址。

下面的例子使用 AWT 中相关 API 在 Windows 通知区域添加一个图标，以及在图标上方显示一个 Toast 消息框，单击消息框，会打开浏览器，并在浏览器中显示 webbrowser.open 函数打开的页面。如果在非 Windows 系统（如 macOS、Linux 等）上运行例子，通常不会显示这个 Toast 信息框。但在图标上右击，通常会使用浏览器打开指定的网页。

代码位置：src/main/java/interesting/gui/ToastDemo.java

```
package interesting.gui;
import javax.imageio.ImageIO;
import java.awt.*;
import java.awt.TrayIcon.MessageType;
import java.io.IOException;
import java.net.URI;
import java.net.URL;
public class ToastDemo {
    public static void main(String[] args) throws IOException, AWTException {
        // 检查系统托盘是否支持
        if (!SystemTray.isSupported()) {
            System.out.println("System tray not supported!");
            return;
        }
        // 转换并设置托盘图标
        URL imageURL = TrayDemo.class.getResource("/images/tray.png");
        Image image = ImageIO.read(imageURL);
        Image trayIconImage = image.getScaledInstance(16, 16, Image.SCALE_DEFAULT);
        TrayIcon trayIcon = new TrayIcon(trayIconImage);

        // 添加托盘图标到系统托盘
        SystemTray tray = SystemTray.getSystemTray();
        tray.add(trayIcon);

        // 显示通知消息，这里需要使用第三方库来实现单击通知后的回调
        // 以下代码只是为了展示如何显示通知，不包括回调功能
        trayIcon.displayMessage("软件更新", "UnityMarvel 已经更新到 2.01 版本，
单击下载", MessageType.INFO);

        // 使用 Java 的桌面 API 打开网页
        trayIcon.addActionListener(e -> openWebPage("https://www.unitymarvel.com"));
    }
    // 使用 Java 的桌面 API 打开网页的函数
    private static void openWebPage(String url) {
        try {
            Desktop.getDesktop().browse(URI.create(url));
        } catch (IOException e) {
            e.printStackTrace();
        }
    }
}
```

运行程序，会在 Windows 右下角状态栏中显示如图 4-9 所示的 Toast 消息框，单击消息框，会打开指定的网页。

图 4-9 Toast 消息框

4.4 小结

本章介绍的内容相当有意思，尽管这些功能对于大多数应用程序不是必需的，但如果自己的应用程序有这些功能，会显得更酷，更专业。尤其是在状态栏中添加图标，读者可以将一些常用的功能添加到图标菜单中，这样用户就可以很方便地使用这些功能了。

第 5 章 动　画

Java 有很多第三方库可以实现各种有趣的动画，如 gif 动画、数学动画、CSS 动画等。利用这些第三方库，可以设计出堪比专业动画制作软件的系统。本章会选一些非常流行的第三方库，用来展示 Java 到底有多强大。

5.1 属性动画

在很多场景下，窗口中的组件会完成很多动作，如移动、缩放、旋转等。这些动作本质上就是组件属性的变化。如果用数学的语言就是组件中属性在某一时刻的值是时间的函数。尽管我们可以自己设计算法来控制这些属性值的变化，但在 JavaFX 中并不用这么复杂。下面详细解释如何使用 JavaFX 通过属性动画实现 Button 的颜色渐变动画和形状动画，并且让这两种动画并行。

1. 颜色渐变动画

由于 Button 不是一个形状对象（如 Rectangle 或 Circle），我们不能直接使用 FillTransition 来改变它的背景色。相反，我们需要采用一个间接的方法。

首先，我们需要创建一个 ObjectProperty<Color> 对象 color。这个属性用于跟踪颜色的变化。

```
ObjectProperty<Color> color = new SimpleObjectProperty<>(Color.RED);
```

接下来，我们给这个颜色属性添加一个监听器。当颜色属性发生变化时，监听器会更新按钮的背景色样式。

```
color.addListener((obs, oldColor, newColor) -> {
        String rgb = String.format("#%02X%02X%02X",
                (int)(newColor.getRed() * 255),
                (int)(newColor.getGreen() * 255),
                (int)(newColor.getBlue() * 255));
        button.setStyle("-fx-background-color: " + rgb + ";");
});
```

我们可以使用 Timeline 创建颜色渐变动画。Timeline 允许我们定义一系列的 KeyFrame（关键帧），每个关键帧指定了特定时间点的目标值。在这种情况下，我们定义了从红色到绿色的渐变。通过 Timeline 控制 color 的变化，再利用 color 监听器，控制按钮背景色的变化，这就是实现按钮背景色渐变动画的基本原理。

```
Timeline colorTimeline = new Timeline(
        new KeyFrame(Duration.ZERO, new KeyValue(color, Color.RED)),
        new KeyFrame(Duration.seconds(3), new KeyValue(color, Color.GREEN))
);
```

2. 形状动画（尺寸变化）

尺寸变化动画是通过 ScaleTransition 实现的。这个类是为了方便地创建形状（或任何节点）的缩放动画而设计的。

首先需要创建一个 ScaleTransition 对象，指定动画的持续时间和目标对象（在这个案例中是按钮）。

```
ScaleTransition scaleTransition = new ScaleTransition(Duration.seconds(2), button);
```

接下来，需要设定动画的目标尺寸（在这里是原始尺寸的 3 倍）。

```
scaleTransition.setToX(3.0);                              // 宽度放大到 3 倍
scaleTransition.setToY(3.0);                              // 高度放大到 3 倍
```

3. 实现并行动画效果

为了同时运行这两个动画，需要使用 ParallelTransition 类。这个类允许将多个动画组合在一起，并且同时执行它们。

为了实现并行效果，需要创建一个 ParallelTransition 对象，并将之前创建的颜色渐变动画（colorTimeline）和形状动画（scaleTransition）添加到其中。

```
ParallelTransition parallelTransition = new ParallelTransition
(colorTimeline, scaleTransition);
```

当按钮被单击时，parallelTransition 的 play 方法被调用，这会同时启动两个动画。

```
button.setOnAction(e -> parallelTransition.play());
```

通过这种方式，我们可以实现通过 JavaFX 对 Button 进行颜色渐变和尺寸变化的动画效果，并使这两种动画并行执行。这种方法的关键是利用 Timeline 和属性监听机制来间接实现颜色动画，以及使用 ScaleTransition 来实现形状动画，最后通过 ParallelTransition 将它们结合在一起。

下面的代码完整地演示了如何使用 JavaFX 同时让按钮的背景色和尺寸发生变化，从而实现并行属性动画的效果。

代码位置：src/main/java/animations/PropertyAnimation.java

```
package animations;

import javafx.animation.KeyFrame;
import javafx.animation.KeyValue;
import javafx.animation.ScaleTransition;
import javafx.animation.Timeline;
import javafx.application.Application;
import javafx.scene.Scene;
import javafx.scene.control.Button;
import javafx.scene.layout.StackPane;
import javafx.stage.Stage;
import javafx.util.Duration;
import javafx.animation.ParallelTransition;
import javafx.beans.property.ObjectProperty;
import javafx.beans.property.SimpleObjectProperty;
import javafx.scene.paint.Color;

public class PropertyAnimation extends Application {
    @Override
    public void start(Stage primaryStage) {
        Button button = new Button("单击我");
        // 创建颜色属性
        ObjectProperty<Color> color = new SimpleObjectProperty<>(Color.RED);
        color.addListener((obs, oldColor, newColor) -> {
            String rgb = String.format("#%02X%02X%02X",
                    (int)(newColor.getRed() * 255),
                    (int)(newColor.getGreen() * 255),
                    (int)(newColor.getBlue() * 255));
            button.setStyle("-fx-background-color: " + rgb + ";");
        });
        // 创建颜色渐变动画
        Timeline colorTimeline = new Timeline(
                new KeyFrame(Duration.ZERO, new KeyValue(color, Color.RED)),
                new KeyFrame(Duration.seconds(3), new KeyValue(color, Color.GREEN))
        );
        // 创建尺寸变换动画
        ScaleTransition scaleTransition = new ScaleTransition
 (Duration.seconds(2), button);
        scaleTransition.setToX(3.0);
        scaleTransition.setToY(3.0);
        // 创建并行动画，同时执行颜色和尺寸变化
        ParallelTransition parallelTransition = new ParallelTransition
 (colorTimeline, scaleTransition);
        // 设置按钮单击事件
        button.setOnAction(e -> parallelTransition.play());
```

```
        // 设置场景和舞台
        StackPane root = new StackPane();
        root.getChildren().add(button);
        Scene scene = new Scene(root, 400, 400);
        primaryStage.setScene(scene);
        primaryStage.setTitle("属性动画");
        primaryStage.show();
    }

    public static void main(String[] args) {
        launch(args);
    }
}
```

运行程序，单击“单击我”按钮开始运行动画。

5.2 缓动动画

本节的例子演示如何使用JavaFX内置的动画和插值功能来实现缓动动画。这种动画在用户界面中常用于指示状态的改变，提供视觉反馈，或者简单地增加视觉吸引力。

1. 缓动动画(Easing Animation)

缓动动画是动画中常用的一种技术，它可以使动画的开始和（或）结束时速度变化更自然，而不是线性变化。这是通过使用缓动函数实现的。在JavaFX中，Interpolator类提供了多种缓动函数。

如果在代码中使用Interpolator.EASE_BOTH，这意味着动画在开始和结束时都会慢下来，中间部分速度较快，这样看起来更平滑。

2. Timeline和KeyFrame

Timeline是JavaFX中用于创建动画的核心类之一。它可以包含一个或多个KeyFrame对象，每个KeyFrame定义动画的一个特定时间点及其应达到的状态。

```
Timeline fadeIn = new Timeline(
    new KeyFrame(Duration.seconds(1),
        new KeyValue(buttonColor, Color.BLUE),
        new KeyValue(textColor, Color.WHITE))
);
```

在上述代码中，fadeIn动画在1秒钟内将按钮的背景色从当前颜色渐变到蓝色，并将文本颜色渐变到白色。类似地，fadeOut动画将颜色从蓝色和白色渐变回红色和黑色。

3. 颜色属性

我们使用了ObjectProperty<Color>来存储和监听背景色和文本色的变化。这样，当颜色属性变化时，监听器会触发并更新按钮的样式。

```
ObjectProperty<Color> buttonColor = new SimpleObjectProperty<>(Color.RED);
ObjectProperty<Color> textColor = new SimpleObjectProperty<>(Color.BLACK);
```

4. updateButtonStyle 方法

updateButtonStyle 方法负责将颜色对象转换为十六进制颜色字符串，并更新按钮的样式。这个方法确保颜色值正确地转换为十六进制字符串格式，这是设置 CSS 样式所必需的。

```
private void updateButtonStyle(Button button, Color bg, Color text) {
    String style = String.format("-fx-background-color: #%02X%02X%02X;
-fx-text-fill:
                                 #%02X%02X%02X;",
        (int)(bg.getRed() * 255), (int)(bg.getGreen() * 255), (int)
(bg.getBlue() * 255),
        (int)(text.getRed() * 255), (int)(text.getGreen() * 255), (int)
(text.getBlue() * 255));
    button.setStyle(style);
}
```

在这个方法中，我们将颜色对象的红、绿、蓝分量乘以 255（因为 JavaFX 的 `Color` 对象将这些值作为 0 到 1 之间的浮点数），然后强制转换为整数，以便使用 `String.format` 方法生成正确的 CSS 颜色值。

5. 整体效果

将所有这些元素结合起来，当鼠标进入按钮时，fadeIn 动画被触发，背景色和文本色通过渐变动画变成蓝色和白色。当鼠标离开按钮时，fadeOut 动画被触发，背景色和文本色通过渐变动画变回红色和黑色。

使用 JavaFX 内置的动画和插值功能，我们可以创建平滑的颜色过渡效果，不需要外部库支持。这种动画不仅可以提高用户体验，还能使界面更加动态地响应用户的交互。

下面的例子完整地演示了缓动动画的实现过程，当鼠标进入按钮和离开按钮后，按钮的背景色和文字演示都会发生渐变。

代码位置：src/main/java/animations/ColorAnimationButton.java

```
package animations;
import javafx.application.Application;
import javafx.beans.property.ObjectProperty;
import javafx.beans.property.SimpleObjectProperty;
import javafx.scene.Scene;
import javafx.scene.control.Button;
import javafx.scene.layout.StackPane;
import javafx.scene.paint.Color;
import javafx.stage.Stage;
import javafx.animation.Interpolator;
import javafx.animation.KeyFrame;
import javafx.animation.KeyValue;
import javafx.animation.Timeline;
```

```
import javafx.util.Duration;
public class ColorAnimationButton extends Application {
    @Override
    public void start(Stage primaryStage) {
        // 创建按钮背景色和文本颜色的属性，初始值分别为红色和黑色
        ObjectProperty<Color> buttonColor = new SimpleObjectProperty<>(Color.RED);
        ObjectProperty<Color> textColor = new SimpleObjectProperty<>(Color.BLACK);
        // 创建按钮，设置初始文字
        Button button = new Button("Hello");
        // 调用 updateButtonStyle 方法，根据当前颜色属性设置按钮样式
        updateButtonStyle(button, buttonColor.get(), textColor.get());

        // 为颜色属性添加监听器，当属性值发生变化时，更新按钮的样式
        buttonColor.addListener((obs, oldColor, newColor) ->
updateButtonStyle(button, newColor, textColor.get()));
        textColor.addListener((obs, oldColor, newColor) ->
updateButtonStyle(button, buttonColor.get(), newColor));
        // 创建鼠标进入时的颜色渐变动画，持续时间为 1 秒
        Timeline fadeIn = new Timeline(
                new KeyFrame(Duration.seconds(1),
                        new KeyValue(buttonColor, Color.BLUE, Interpolator.
EASE_BOTH),
                        new KeyValue(textColor, Color.WHITE, Interpolator.EASE_BOTH))
        );
        // 创建鼠标离开时的颜色渐变动画，持续时间为 1 秒
        Timeline fadeOut = new Timeline(
                new KeyFrame(Duration.seconds(1),
                        new KeyValue(buttonColor, Color.RED, Interpolator.EASE_BOTH),
                        new KeyValue(textColor, Color.BLACK, Interpolator.EASE_BOTH))
        );
        // 设置鼠标事件处理器，当鼠标进入和离开按钮时，播放对应的动画
        button.setOnMouseEntered(e -> fadeIn.playFromStart());
        button.setOnMouseExited(e -> fadeOut.playFromStart());
        // 使用 StackPane 作为根布局，并将按钮添加到布局中
        StackPane root = new StackPane(button);
        // 创建场景，设置其大小为 200×200 像素
        Scene scene = new Scene(root, 200, 200);
        // 设置舞台的场景
        primaryStage.setScene(scene);
        // 显示舞台
        primaryStage.show();
        // 将按钮的宽高属性绑定到场景的宽高属性，使按钮的大小始终填满整个场景
        button.prefWidthProperty().bind(scene.widthProperty());
        button.prefHeightProperty().bind(scene.heightProperty());
    }
    // updateButtonStyle 方法用于更新按钮的背景色和文本色
    private void updateButtonStyle(Button button, Color bg, Color text) {
        // 格式化颜色值为十六进制字符串，并设置按钮的样式
        String style = String.format("-fx-background-color: #%02X%02X%02X;
-fx-text-fill: #%02X%02X%02X;",
                (int)(bg.getRed() * 255), (int)(bg.getGreen() * 255), (int)
```

```
    (bg.getBlue() * 255),
                (int)(text.getRed() * 255), (int)(text.getGreen() * 255), (int)
    (text.getBlue() * 255));
            button.setStyle(style);
        }
        // main 方法，JavaFX 应用程序的启动点
        public static void main(String[] args) {
            launch(args);                                   // 启动 JavaFX 应用程序
        }
    }
```

运行程序，会显示如图 5-1 所示的窗口，将鼠标移入移出按钮，可看到按钮的背景色会逐渐改变。

图 5-1　缓动动画

5.3　正弦波动画 GIF

正弦波是一种数学动画。数学动画是指通过计算机程序生成的动画，用于展示数学概念、定理、公式等。Java 有第三方库可以生成数学动画。本节将结合这些第三方库，演示如何用 Java 生成正弦波动画 GIF。正弦波会从右向左不断移，每次移动 π/10 个长度。

JDK 的标准 API 并没有提供生成动画 GIF 以及相关的绘图 API，所以要实现生成正弦波动画 GIF 的功能，需要使用两个第三方库：AnimatedGifEncoder 和 JFreeChart。这两个库在 Java 中得到了广泛应用，它们在数据可视化和多媒体处理方面发挥着重要作用。

1. AnimatedGifEncoder

AnimatedGifEncoder 是一个用于生成 GIF 动画的 Java 类库。它提供了一个简单的 API 来创建 GIF 动画，允许用户设置循环次数、每帧持续时间和质量等参数。这个库通常不包含在标准的 Java 发行版中，因此需要单独下载和添加到项目中。

AnimatedGifEncoder 库的主要方法如下。

（1）start(String filepath)：初始化 GIF 编码器，准备将生成的 GIF 保存到指定的文件路径。

（2）setDelay(int ms)：设置两帧之间的延迟时间，以毫秒为单位。

（3）setRepeat(int repeat): 设置动画循环的次数。设置为 0 可使动画无限循环。

（4）addFrame(BufferedImage image)：添加一个帧到 GIF 动画中。帧是以 BufferedImage 的形式提供的。

（5）finish()：结束 GIF 动画的创建并关闭文件流。

用 AnimatedGifEncoder 生成 GIF 动画的步骤如下。

（1）初始化：通过创建 AnimatedGifEncoder 的实例开始。

（2）设置循环：使用 setRepeat 方法设置动画循环播放。传入的参数 0 表示动画将无限循环。

```
AnimatedGifEncoder encoder = new AnimatedGifEncoder();
encoder.setRepeat(0);
```

（3）开始写入：start 方法用于开始一个新的 GIF 动画文件的写入过程。参数是文件的路径。

（4）添加帧：addFram 方法用于添加一个帧到 GIF 动画中。在循环中对每个 BufferedImage 调用此方法来构建动画序列。

```
for (BufferedImage frame : frames) {
    encoder.addFrame(frame);
}
```

（5）设置延迟：setDelay 方法用于设置每一帧之间的延迟，以毫秒为单位。

（6）完成动画 **：finish 方法调用后，GIF 动画写入完成，文件关闭。

安装 AnimatedGifEncoder

如果使用 Maven 管理依赖，需要在 pom.xml 文件中添加如下依赖：

```
<dependency>
    <groupId>com.madgag</groupId>
    <artifactId>animated-gif-lib</artifactId>
    <version>1.4</version>
</dependency>
```

如果使用 Gradle 管理依赖，需要在 build.gradle 文件中添加如下依赖：

```
implementation 'com.madgag:animated-gif-lib:1.4'
```

2. JFreeChart

JFreeChart 是一个广泛使用的 Java 库，用于创建各种图表，如线图、柱状图和饼图等。JFreeChart 的主要功能和用法如下。

（1）ChartFactory：提供了多种静态方法来创建不同类型的图表。

（2）XYSeries 和 XYSeriesCollection：用于存储 X 和 Y 坐标系列的数据。

（3）createXYLineChart：创建一个XY线图，可以指定标题、X轴标签、Y轴标签和数据集。

（4）XYPlot：获取图表的一个实例，用于更细粒度的图表定制，如设置颜色、范围等。

（5）createBufferedImage：将图表转换为BufferedImage对象，这对于将图表保存为图像或将其添加到GIF动画中非常有用。

JFreeChart库的使用通常包括以下几个步骤。

（1）创建数据集：首先创建一个数据集，该数据集包含所有需要在图表中显示的数据。这里使用的是XYSeries和XYSeriesCollection。

```
XYSeries series = new XYSeries("正弦波");
XYSeriesCollection dataset = new XYSeriesCollection(series);
```

（2）创建图表：使用ChartFactory类的静态方法createXYLineChart创建图表。此方法需要图表的标题、X轴标签、Y轴标签和数据集。

（3）自定义图表：可以获取JFreeChart对象并进行自定义，例如设置颜色、文字和其他样式。在此示例中，主要设置了坐标轴的范围。

```
XYPlot plot = chart.getXYPlot();
plot.getDomainAxis().setRange(0, 2 * Math.PI);
plot.getRangeAxis().setRange(-1.1, 1.1);
```

（4）生成图像：最后，createBufferedImage方法用于将JFreeChart对象转换为BufferedImage对象，以便可以将其加入GIF动画中。

安装JFreeChart

如果使用Maven管理依赖，需要在pom.xml文件中添加如下依赖：

```
<dependency>
    <groupId>org.jfree</groupId>
    <artifactId>jfreechart</artifactId>
    <version>1.5.4</version>
</dependency>
```

如果使用Gradle管理依赖，需要在build.gradle文件中添加如下依赖：

```
implementation 'org.jfree:jfreechart:1.5.4'
```

3. 结合业务逻辑

在本节的例子中，首先使用JFreeChart生成一系列表示正弦波的图像。对于每一帧，计算出一组新的Y坐标来模拟正弦波的运动，然后通过JFreeChart绘制出对应的图像。之后，这些图像被转换为BufferedImage对象，并传递给AnimatedGifEncoder，它将这些单独的帧合成为一个动画GIF文件，此文件可以在Web浏览器中循环播放。这个过程体现了两个库之间的协同工作：JFreeChart 负责图像的生成，而 AnimatedGifEncoder 负责动画的创建和保存。

下面的例子完整地演示了如何使用这两个第三方库生成正弦波动画 GIF 文件的过程。

代码位置：src/main/java/animations/SineWave.java

```
package animations;

import java.awt.image.BufferedImage;
import java.util.ArrayList;
import org.jfree.chart.ChartFactory;
import org.jfree.chart.JFreeChart;
import org.jfree.chart.plot.XYPlot;
import org.jfree.data.xy.XYSeries;
import org.jfree.data.xy.XYSeriesCollection;
import com.madgag.gif.fmsware.AnimatedGifEncoder;

// 创建一个类来模拟 gif.frame 装饰器的功能
public class SineWave {
    public static void main(String[] args) {
        final int framesCount = 20;                                    // 动画帧数
        final double delta = Math.PI / 10;                             // 每次移动的距离
        ArrayList<BufferedImage> frames = new ArrayList<>();  // 存储动画帧的列表

        // 生成横坐标数据
        double[] x = new double[30];
        for (int i = 0; i < x.length; i++) {
            x[i] = i * 2 * Math.PI / (x.length - 1);
        }
        // 生成动画帧
        for (int i = 0; i < framesCount; i++) {
            double[] y = new double[x.length];
            for (int j = 0; j < x.length; j++) {
                y[j] = Math.sin(x[j] + delta * i);                     // 计算纵坐标
            }
            frames.add(createChartImage(x, y));                   // 创建图表并添加到帧列表
        }

        // 保存为 GIF 文件
        saveAsGif(frames, "sine_wave.gif");
    }
    // 使用 JFreeChart 创建图表，并返回其图像
    private static BufferedImage createChartImage(double[] x, double[] y) {
        XYSeries series = new XYSeries("正弦波");
        for (int i = 0; i < x.length; i++) {
            series.add(x[i], y[i]);
        }
        XYSeriesCollection dataset = new XYSeriesCollection(series);
        JFreeChart chart = ChartFactory.createXYLineChart(
                null, "X", "Y", dataset
```

```
        );
        XYPlot plot = chart.getXYPlot();
        plot.getDomainAxis().setRange(0, 2 * Math.PI);
        plot.getRangeAxis().setRange(-1.1, 1.1);

        return chart.createBufferedImage(500, 500);
    }
    // 保存动画帧为 GIF 文件
    private static void saveAsGif(ArrayList<BufferedImage> frames, String filepath) {
        AnimatedGifEncoder encoder = new AnimatedGifEncoder();
        encoder.setRepeat(0);                           // 设置循环播放，0 表示无限循环
        encoder.start(filepath);
        for (BufferedImage frame : frames) {
            encoder.setDelay(100);                      // 设置每帧的延迟时间为 100 毫秒
            encoder.addFrame(frame);
        }
        encoder.finish();
    }
}
```

运行程序，会生成一个 sine_wave.gif 文件，需要用支持动画 GIF 的软件打开该文件，如浏览器，效果如图 5-2 所示。

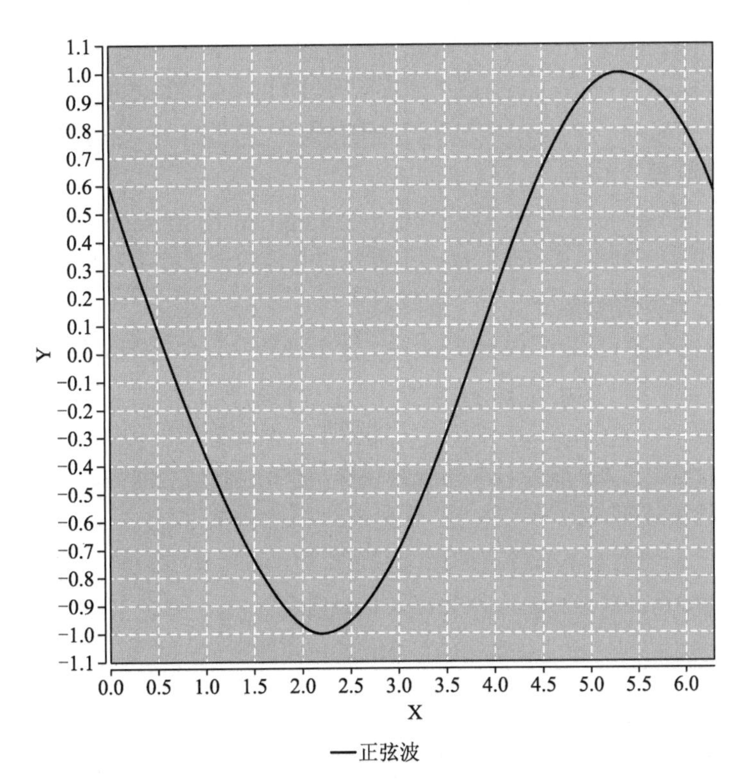

图 5-2　正弦波动画 GIF

5.4 动画 GIF

本节会介绍如何根据静态图像生成动画 GIF 文件，以及如何播放动画 GIF 文件。

5.4.1 使用静态图像生成动画 GIF 文件

在这一节会将目录中指定的图像文件转换为动画 GIF 文件。在这个例子中，会编写一个 createAnimatedGif 方法，用于创建动画 GIF 文件。要实现这个功能，需要如下几步。

1. 初始化 GIF 编码器

创建 AnimatedGifEncoder 对象。这个对象负责处理 GIF 的创建和帧的添加。

2. 设置 GIF 文件的输出路径和基本属性

（1）encoder.start(gifFile)：开始写入 GIF 到指定的文件路径。

（2）encoder.setDelay(delay)：设置每帧之间的延迟时间（毫秒）。这个延迟决定了动画的播放速度。

（3）encoder.setRepeat(-1)：设置动画循环播放的次数。-1 表示动画将无限循环。

3. 遍历并处理每个图像文件

对于 imageFiles 列表中的每个图像路径：

（1）使用 AnimatedGif.class.getResource(imageFile) 获取图像文件的 URL。

（2）使用 ImageIO.read(imageURL) 读取图像文件，将其转换成 BufferedImage 对象。BufferedImage 是 Java 中用于处理具有可访问图像数据缓冲区的图像的类。

4. 添加图像帧到 GIF 中

对于每个 BufferedImage 对象，调用 encoder.addFrame(bufferedImage) 将其作为一个独立的帧添加到 GIF 中。这一步骤重复执行，直到所有的图像文件都被处理。

5. 完成 GIF 文件的生成

调用 encoder.finish() 结束 GIF 编码过程。这个调用确保所有的数据都被写入文件，并且文件格式正确。

通过上述步骤，createAnimatedGif 方法能够从一系列静态图像创建出一个连续播放的动画 GIF 文件，每个图像文件成为 GIF 动画的一帧。通过控制帧之间的延迟时间和循环次数，可以调整动画的播放效果。

下面的代码完整地演示了将静态图像文件转换为动画 GIF 文件的过程。

代码位置：src/main/java/animations/AnimatedGif.java

```
package animations;

// 导入需要的类
import com.madgag.gif.fmsware.AnimatedGifEncoder;
import java.awt.image.BufferedImage;
import java.io.IOException;
```

```
import java.net.URL;
import java.util.ArrayList;
import javax.imageio.ImageIO;
// 定义一个类，用于生成动画 GIF
public class AnimatedGif {
    // 定义一个方法，用于将多个静态图像文件转换为一个动画 GIF 文件
    public static void createAnimatedGif(ArrayList<String> imageFiles,
String gifFile, int delay) throws IOException {
        // 创建一个 AnimatedGifEncoder 对象，用于生成动画 GIF
        AnimatedGifEncoder encoder = new AnimatedGifEncoder();
        // 设置动画 GIF 的输出文件
        encoder.start(gifFile);
        // 设置动画 GIF 的延迟时间，单位为毫秒
        encoder.setDelay(delay);
        // 设置动画 GIF 的重复次数，-1 表示无限循环
        encoder.setRepeat(-1);
        // 遍历静态图像文件数组
        for (String imageFile : imageFiles) {
            URL imageURL = AnimatedGif.class.getResource(imageFile);
            // 读取静态图像文件为 BufferedImage 对象
            BufferedImage bufferedImage  = ImageIO.read(imageURL);
            // 将 BufferedImage 对象添加到动画 GIF 中
            encoder.addFrame(bufferedImage);
        }
        // 结束动画 GIF 的生成
        encoder.finish();
    }
    // 定义一个主方法，用于测试
    public static void main(String[] args) throws IOException {
        ArrayList<String> imageFiles = new ArrayList<>();
        for(int i = 1; i < 10; i++) {
            imageFiles.add("/images/robot"+i+".jpg");
        }
        // 定义一个动画 GIF 文件的路径
        String gifFile = "animated.gif";
        // 定义一个延迟时间，单位为毫秒
        int delay = 500;
        // 调用 createAnimatedGif 方法，将静态图像文件转换为动画 GIF 文件
        createAnimatedGif(imageFiles, gifFile, delay);
        // 打印提示信息
        System.out.println(" 动画 GIF 文件已生成 ");
    }
}
```

在运行程序之前，要确保工程目录中的 src/main/resources/images 子目录中有 robot1.jpg 到 robot9.jpg 共 9 个 JPG 图像文件。程序运行后，会在当前目录生成一个 animated.gif 文件，需要在支持动画 GIF 的软件（如浏览器）中打开。本例设置的是每个静态图像停留 500 毫秒。读者可以设置不同的停留时间，以实现不同的动画效果。

5.4.2 播放动画 GIF

本节会使用 JavaFX 实现一个播放动画 GIF 文件的程序，该程序的关键是读取并播放动画 GIF 文件。JavaFX 可以使用 ImageView 组件来显示动画 GIF。

首先创建 Image 对象来加载 GIF 文件。Image 对象的构造函数接收 GIF 文件的路径作为参数。

接下来创建 ImageView 对象来显示 Image 对象。ImageView 是 JavaFX 中用于显示图像的组件。

为了让动画 GIF 与窗口适配，需要使用下面的代码保持图像的宽高比。

```
imageView.setPreserveRatio(true)          // 确保在调整大小时保持图像的宽高比
```

还需要使用 imageView.setFitWidth 和 imageView.setFitHeight 用于设置图像的初始宽度和高度，这里设为场景的宽度和高度。

下面是播放动画 GIF 的完整代码。

代码位置：src/main/java/animations/GifPlayer.java

```
package animations;
// 导入需要的类
import javafx.application.Application;
import javafx.beans.value.ChangeListener;
import javafx.beans.value.ObservableValue;
import javafx.scene.Scene;
import javafx.scene.image.Image;
import javafx.scene.image.ImageView;
import javafx.scene.layout.StackPane;
import javafx.stage.Stage;

// 定义一个类，继承 Application 类，用于创建 JavaFX 应用程序
public class GifPlayer extends Application {
    // 定义一个主方法，用于启动应用程序
    public static void main(String[] args) {
        launch(args);
    }
    // 重写 start 方法，用于设置舞台和场景
    @Override
    public void start(Stage primaryStage) throws Exception {
        // 创建一个 StackPane 对象，用于布局
        StackPane root = new StackPane();
        // 创建一个 Image 对象，用于加载 GIF 文件
        Image image = new Image("/images/animated.gif");
        // 创建一个 ImageView 对象，用于显示 Image 对象
        ImageView imageView = new ImageView(image);
        // 设置保持原始宽高比
        imageView.setPreserveRatio(true);
        // 将 ImageView 对象添加到 StackPane 对象中
```

```
        root.getChildren().add(imageView);
        // 创建一个 Scene 对象，用于显示 StackPane 对象
        Scene scene = new Scene(root, 500, 500);
        // 设置图像的宽度和高度为场景的宽度和高度
        imageView.setFitWidth(scene.getWidth());
        imageView.setFitHeight(scene.getHeight());
        // 给场景添加一个 ChangeListener，监听宽度和高度的变化
        scene.widthProperty().addListener(new ChangeListener<Number>() {
            @Override
            public void changed(ObservableValue<? extends Number> observable,
Number oldValue, Number newValue) {
                // 更新图像的宽度
                imageView.setFitWidth(newValue.doubleValue());
            }
        });
        scene.heightProperty().addListener(new ChangeListener<Number>() {
            @Override
            public void changed(ObservableValue<? extends Number> observable,
Number oldValue, Number newValue) {
                // 更新图像的高度
                imageView.setFitHeight(newValue.doubleValue());
            }
        });
        // 设置舞台的标题
        primaryStage.setTitle("Gif Player");
        // 设置舞台的场景
        primaryStage.setScene(scene);
        // 显示舞台
        primaryStage.show();
    }
}
```

在运行程序之前，要确保 src/main/resources/images 目录存在 animated.gif 文件，程序运行后，会显示如图 5-3 所示的窗口，里面播放着 GIF 动画。

图 5-3　播放动画 GIF 文件

5.5 模拟物理运动

在这一节我们使用 JavaFX、FFmpeg 以及其他第三方库模拟小球弹跳的物理运动，将捕获每一帧的画面，最后用这些画面生成 MP4 文件。

下面详细描述这个例子的实现步骤和原理。

1. 实现弹跳球模拟物理操作的动画

首先使用 JavaFX 创建一个简单的弹跳球动画。这是通过修改球的垂直位置和速度来模拟重力和弹性碰撞效应来实现的。

```
if (isMoving) {
    // 小球运动逻辑
    if (ball.getLayoutY() >= HEIGHT - RADIUS) {
        yVelocity *= -0.7;
        if (Math.abs(yVelocity) < 1) {
            isMoving = false;                                      // 动画结束
        }
    } else {
        yVelocity += GRAVITY;
    }
    ball.setLayoutY(ball.getLayoutY() + yVelocity);
}
```

在这段代码中，首先检查小球是否正在移动。如果是，我们检查小球是否接触到窗口的底部（通过比较它的 layoutY 属性和窗口高度）。如果小球触底，我们将其垂直速度 yVelocity 乘以 –0.7，这表示速度方向反转，并且速度大小减小（模拟能量损失）。如果速度降低到一定程度（绝对值小于 1），我们认为小球已停止移动。如果小球没有触底，我们增加其垂直速度（模拟重力作用）。

2. 捕获每一帧保存成图像

接下来我们在每次动画更新时捕获当前的帧，并将其保存为 PNG 格式的图像文件。

```
private void captureFrame(Pane pane) throws IOException {
    WritableImage image = pane.snapshot(null, null);
    BufferedImage bufferedImage = SwingFXUtils.fromFXImage(image, null);
    File outputFile = new File("frame_" + frameNumber++ + ".png");
    ImageIO.write(bufferedImage, "png", outputFile);
```

在这段代码中，pane.snapshot() 方法用于捕获 JavaFX 窗格的当前视觉状态，生成一个 WritableImage 对象。然后，我们使用 SwingFXUtils.fromFXImage() 方法将这个 WritableImage 转换为 BufferedImage 对象，这样我们就可以使用 ImageIO.write() 方法将其保存为 PNG 文件。

3. 调用 FFmpeg 生成 MP4 视频

动画结束后，我们调用 FFmpeg 命令将这些图像文件合成为一个 MP4 视频文件。

```
private void createVideo() throws IOException, InterruptedException {
    String cmd = "ffmpeg -framerate 30 -i frame_%d.png -c:v libx264 -r 30
-pix_fmt yuv420p out.mp4";
    Process process = Runtime.getRuntime().exec(cmd);
    process.waitFor();                                    // 等待视频创建过程完成
}
```

在这个方法中构建了一个 FFmpeg 命令字符串。这个命令告诉 FFmpeg 以每秒 30 帧的速率读取名为 frame_%d.png 的图像文件，并使用 libx264 编解码器以及 yuv420p 像素格式将它们编码成 MP4 视频。Runtime.getRuntime().exec(cmd) 用于在 Java 程序中执行这个命令，而 process.waitFor() 确保 Java 程序等待直到视频文件创建完成。

4. FFmpeg 命令行格式

```
ffmpeg -framerate 30 -i frame_%d.png -c:v libx264 -r 30 -pix_fmt yuv420p out.mp4
```

（1）-framerate 30：设置输入帧率为 30 帧 / 秒。

（2）-i frame_%d.png：指定输入文件的格式，这里使用 %d 作为通配符匹配所有编号的帧。

（3）-c:v libx264：使用 libx264 编解码器进行视频编码。

（4）-r 30：设置输出视频的帧率为 30 帧 / 秒。

（5）-pix_fmt yuv420p：设置像素格式。

（6）out.mp4：输出文件的名称。

下面的代码完整地演示了使用 JavaFX、FFmpeg 等技术模拟物理运动，并将物理运动过程转换为 MP4 文件的过程。

代码位置：src/main/java/animations/BouncingBallSimulation.java

```
package animations;
import javafx.animation.KeyFrame;
import javafx.animation.Timeline;
import javafx.application.Application;
import javafx.embed.swing.SwingFXUtils;
import javafx.scene.Scene;
import javafx.scene.image.WritableImage;
import javafx.scene.layout.Pane;
import javafx.scene.paint.Color;
import javafx.scene.shape.Circle;
import javafx.stage.Stage;
import javafx.util.Duration;
import javax.imageio.ImageIO;
import java.awt.image.BufferedImage;
import java.io.File;
import java.io.IOException;

public class BouncingBallSimulation extends Application {
```

```
// 设置窗口的宽度和高度
private static final int WIDTH = 400;
private static final int HEIGHT = 400;

// 设置小球的半径和重力加速度
private static final double RADIUS = 20;
private static final double GRAVITY = 0.2;

// 定义小球的垂直速度、运动状态和帧编号
private double yVelocity = 0;
private boolean isMoving = true;
private int frameNumber = 0;

// 重写 start 方法，这是 JavaFX 应用程序的入口
@Override
public void start(Stage primaryStage) throws IOException {
    Pane pane = new Pane();                              // 创建一个面板
    Circle ball = new Circle(RADIUS, Color.BLUE);  // 创建一个蓝色的圆形代表小球
    ball.relocate(WIDTH / 2 - RADIUS, 0);            // 将小球置于窗口顶部中央
    pane.getChildren().add(ball);                    // 将小球添加到面板中
    Scene scene = new Scene(pane, WIDTH, HEIGHT);
    // 创建一个场景，包含面板和设置的尺寸

    // 创建时间线动画，每个关键帧间隔 20 毫秒
    Timeline timeline = new Timeline(new KeyFrame(Duration.millis(20), e -> {
        if (isMoving) {
            // 小球运动逻辑
            if (ball.getLayoutY() >= HEIGHT - RADIUS) {
                yVelocity *= -0.7;                       // 反弹时减小速度
                if (Math.abs(yVelocity) < 1) {
                    isMoving = false;                    // 若速度太小，则停止运动
                }
            } else {
                yVelocity += GRAVITY;                    // 在空中时，增加下落速度
            }

            ball.setLayoutY(ball.getLayoutY() + yVelocity);
            // 更新小球的垂直位置

            // 尝试捕获当前帧并保存为图片
            try {
                captureFrame(pane);
            } catch (Exception ee) {
                // 异常处理
            }
        } else {
            // 动画结束，生成视频
            try {
                createVideo();
                primaryStage.close();                    // 关闭窗口
```

```
            } catch (IOException | InterruptedException ex) {
                ex.printStackTrace();                          // 打印异常信息
            }
        }
    }));
    timeline.setCycleCount(Timeline.INDEFINITE);    // 设置时间线无限循环
    timeline.play();                                // 开始播放动画

    primaryStage.setTitle("JavaFX Bouncing Ball Simulation"); // 设置窗口标题
    primaryStage.setScene(scene);                   // 将场景放置在舞台上
    primaryStage.show();                            // 显示窗口
}
// 方法：捕获当前帧并保存为 PNG 图片
private void captureFrame(Pane pane) throws IOException {
    WritableImage image = pane.snapshot(null, null);
    BufferedImage bufferedImage = SwingFXUtils.fromFXImage(image, null);
    File outputFile = new File("frame_" + frameNumber++ + ".png");
    ImageIO.write(bufferedImage, "png", outputFile);
}
// 方法：使用 FFmpeg 命令行工具将捕获的帧组合成视频
private void createVideo() throws IOException, InterruptedException {
    String cmd = "ffmpeg -framerate 30 -i frame_%d.png -c:v libx264 -r
30 -pix_fmt yuv420p out.mp4";
    Process process = Runtime.getRuntime().exec(cmd);
    process.waitFor();                              // 等待视频创建过程完成
}

// main 方法，Java 应用程序的入口
public static void main(String[] args) {
    launch(args);                                   // 启动 JavaFX 应用程序
}
}
```

运行程序，会弹出一个如图 5-4 所示的窗口，有一个蓝色的小球上下弹跳，同时，弹跳结束，会在当前目录生成一个 out.mp4 文件，这是最终生成的模拟小球弹跳的视频，大概 150 帧。

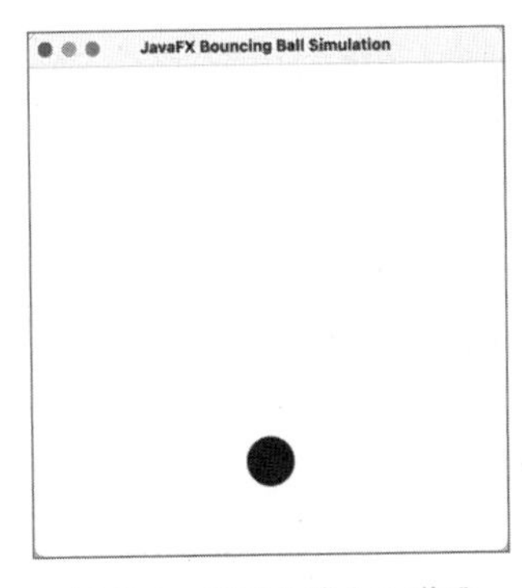

图 5-4　模拟小球上下弹跳

5.6 CSS 动画

JavaFX 支持使用 CSS（Cascading Style Sheet，层叠样式表）来定制其组件的样式，包括动画效果。CSS 动画是一种使用 CSS 来创建动画效果的技术。这种方式可以让元素在一段时间内逐渐改变其样式。在 JavaFX 中，这通常是通过改变场景图中节点的样式来实现的。

CSS 动画通常涉及以下几个关键概念。

（1）选择器：确定哪些 HTML 元素或 JavaFX 节点将被应用样式。

（2）属性：定义要动画化的具体样式属性，例如颜色、大小、位置等。

（3）持续时间：动画持续的时间。

（4）时间函数：控制动画的速度曲线，例如线性、缓入、缓出等。

（5）关键帧：定义动画过程中的中间步骤和样式。

在 JavaFX 中，CSS 用于定义和修改界面组件的样式，类似于在 Web 开发中的作用。JavaFX 的 CSS 支持是专门为 JavaFX 组件设计的，有一些特殊的属性前缀（如 -fx-），它们与标准的 Web CSS 略有不同。这些样式属性是 JavaFX 特有的，用于控制 JavaFX 组件的视觉表现。

下面的代码演示了如何在 JavaFX 中使用 CSS 动画让一个小球从窗口的左侧匀速移动到窗口的右侧。

代码位置：src/main/java/animations/CSSAnimation.java

```
package animations;
// 导入 JavaFX 相关的类
import javafx.application.Application;
import javafx.scene.Scene;
import javafx.scene.layout.Pane;
import javafx.scene.shape.Circle;
import javafx.stage.Stage;
import javafx.animation.Animation;
import javafx.animation.KeyFrame;
import javafx.animation.KeyValue;
import javafx.animation.Timeline;
import javafx.util.Duration;
public class CSSAnimation extends Application {
    // 重写 start 方法，设置舞台和场景
    @Override
    public void start(Stage primaryStage) {
        // 创建一个面板，作为根节点
        Pane root = new Pane();
        // 设置面板的宽度和高度
        root.setPrefSize(600, 400);
        // 创建一个场景，将面板添加到场景中
        Scene scene = new Scene(root);
```

```
        // 设置场景的标题
        scene.getStylesheets().add("/anim_style.css");
        // 创建一个圆形，作为动画的对象
        Circle circle = new Circle(50);
        // 设置圆形的初始位置
        circle.setLayoutX(50);
        circle.setLayoutY(200);
        // 设置圆形的样式，使用 CSS 选择器
        circle.getStyleClass().add("circle");
        // 将圆形添加到面板中
        root.getChildren().add(circle);
        // 创建一个时间轴，用于控制动画的进度
        Timeline timeline = new Timeline();
        // 设置时间轴的循环次数，无限循环
        timeline.setCycleCount(Animation.INDEFINITE);
        // 设置时间轴的自动反转，使动画来回播放
        timeline.setAutoReverse(true);
        // 创建一个关键帧，表示动画的结束状态
        KeyFrame keyFrame = new KeyFrame(
                // 设置关键帧的持续时间，5 秒
                Duration.seconds(5),
                // 设置关键帧的属性值，圆形的 X 坐标变为 500
                new KeyValue(circle.layoutXProperty(), 500)
        );
        // 将关键帧添加到时间轴中
        timeline.getKeyFrames().add(keyFrame);
        // 启动时间轴，开始动画
        timeline.play();
        // 将场景添加到舞台中
        primaryStage.setScene(scene);
        // 设置舞台的标题
        primaryStage.setTitle("CSS Animation");
        // 显示舞台
        primaryStage.show();
    }
    // 主方法，启动应用程序
    public static void main(String[] args) {
        launch(args);
    }
}
```

在这个程序中使用了 anim_style.css 文件，该文件通过 CSS 定义了 JavaFX 组件的样式。

代码位置：src/main/resources/anim_style.css

```
/* 定义 CSS 样式表，用于设置圆形的样式 */
.circle {
    /* 设置圆形的填充颜色，渐变色 */
    -fx-fill: linear-gradient(from 0% 0% to 100% 100%, red, yellow);
    /* 设置圆形的描边颜色，黑色 */
    -fx-stroke: black;
```

```
        /* 设置圆形的描边宽度，5 像素 */
        -fx-stroke-width: 5;
        /* 设置圆形的阴影效果，偏移量为 10，模糊半径为 20，颜色为灰色 */
        -fx-effect: dropshadow(gaussian, gray, 20, 0, 10, 10);
    }
    .root {
        -fx-background-color: black;
    }
```

运行程序，会显示如图 5-5 所示的窗口，在窗口中会呈现动画效果。

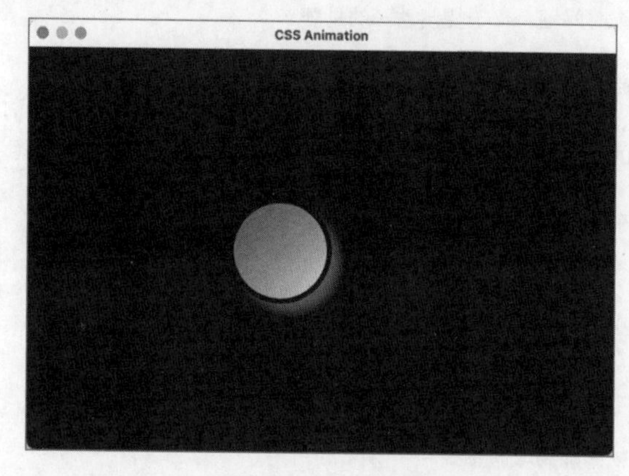

图 5-5 CSS 动画

5.7 小结

可能读者读完本章的内容后，感觉信息量有点大。其实，本章的内容也只不过介绍了 Java 动画功能的沧海一粟。本章涉及了多个第三方库，如 JavaFX、madgag 等，这些库中的任何一个的功能都非常强大，甚至可以写一本非常厚的书，所以这些库的功能，根本不可能通过一章介绍完。本章的目的只是抛砖引玉，当我们了解到 Java 到底有多强大时，就会不知不觉想拥有这种力量，如果读者有这些想法，那么互联网会成为你最好的老师。

第 6 章 音　频

Java 给人的印象一直是用于做企业级的软件，以及 Web 应用或桌面应用。其实，Java 远比我们想象的强大。利用其强大的第三方库，可以实现很多以前只有 C++ 才能实现的软件，例如，音频就是 Java 比较擅长的领域之一。不光是非常基础的播放音频和录制音频，还可以进行任意音频格式的转换，以及音频编辑。相信通过对本章的学习，读者会对 Java 有一个全新的认识。

6.1　音乐播放器

在这一节会使用 JavaFX 实现一个音频播放器，包含“打开”“播放”“停止”“暂停 / 继续”功能，还可以显示播放进度，效果如图 6-1 所示。

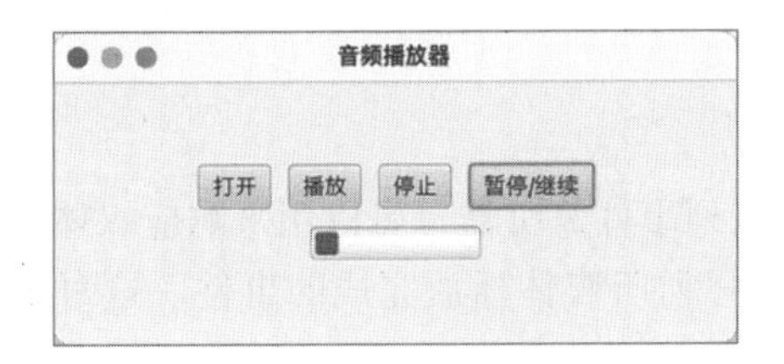

图 6-1　音频播放器

在 JavaFX 中，播放音频需要使用 MediaPlayer 类。 MediaPlayer 是 JavaFX 中用于媒体播放的核心类，它提供了播放音频和视频文件的功能。以下是对 MediaPlayer 类及其用法的详细说明。

1. 创建和初始化 MediaPlayer

（1）创建 Media 对象。

首先，需要创建一个 Media 对象，它表示要播放的媒体文件。Media 对象通过文件的 URL 初始化，这个 URL 可以是本地文件路径或网络地址。

```
Media media = new Media(file.toURI().toString());
```

（2）创建 MediaPlayer。

使用 Media 对象创建 MediaPlayer 实例。MediaPlayer 负责控制媒体的播放，包括播放、暂停、停止、跳转等。

```
MediaPlayer mediaPlayer = new MediaPlayer(media);
```

2. 控制媒体播放

MediaPlayer 提供了多种方法来控制媒体播放。

（1）播放和暂停。

play()：开始播放媒体。

pause()：暂停播放。

```
mediaPlayer.play();
mediaPlayer.pause();
```

（2）停止播放。

stop()：停止播放并将播放头重置到起始位置。

```
mediaPlayer.stop();
```

3. 跳转到特定时间

可以使用 seek(Duration) 方法跳转到媒体的特定时间点。

```
mediaPlayer.seek(Duration.seconds(10));                    // 跳转到第 10 秒
```

4. 监听播放事件

MediaPlayer 提供了一系列事件监听器，可以用来响应媒体播放中的各种事件。

（1）播放结束监听：可以监听媒体播放完成的事件。这对于视频播放结束后执行某些操作很有用。

```
mediaPlayer.setOnEndOfMedia(() -> {
    // 媒体播放结束时执行的代码
});
```

（2）播放进度监听：通过监听 currentTimeProperty() 的变化，可以实时获取媒体的当前播放时间。

```
mediaPlayer.currentTimeProperty().addListener((observable, oldTime, newTime) -> {
   // 当播放时间变化时执行的代码
});
```

5. 音频播放进度

MediaPlayer 提供了方法来获取和设置媒体的播放进度。getCurrentTime() 方法返回当前播放时间，而 getTotalDuration() 方法返回媒体的总时长。这些信息通常用于更新进度条或显示播放时间。

```
Duration currentTime = mediaPlayer.getCurrentTime();
Duration totalDuration = mediaPlayer.getTotalDuration();
```

6. 处理媒体错误

使用 MediaPlayer 时可能会遇到各种错误，如文件格式不支持、文件损坏或网络问题。MediaPlayer 提供了 setOnError() 方法来设置一个当出现错误时执行的处理器。

```
mediaPlayer.setOnError(() -> {
    // 当播放出现错误时执行的代码
});
```

MediaPlayer 是一个功能强大的类，用于在 JavaFX 应用程序中播放音频和视频。它提供了丰富的控制方法和事件监听功能，使得开发者可以方便地实现各种媒体播放相关的功能。正确地使用和管理 MediaPlayer 对于创建流畅、响应良好的媒体播放应用至关重要。

下面的代码演示了如何用 JavaFX 实现本节描述的音频播放器。

代码位置：src/main/java/audio/AudioPlayer.java

```
package audio;
import javafx.application.Application;
import javafx.geometry.Pos;
import javafx.scene.Scene;
import javafx.scene.control.Button;
import javafx.scene.control.ProgressBar;
import javafx.scene.layout.HBox;
import javafx.scene.layout.VBox;
import javafx.scene.media.Media;
import javafx.scene.media.MediaPlayer;
import javafx.stage.FileChooser;
import javafx.stage.Stage;
import javafx.util.Duration;
import java.io.File;

public class AudioPlayer extends Application {
    private MediaPlayer mediaPlayer;
    @Override
    public void start(Stage primaryStage) {
        // 创建按钮和进度条
        Button openButton = new Button("打开");
        Button playButton = new Button("播放");
        Button stopButton = new Button("停止");
```

```
        Button pauseButton = new Button("暂停/继续");
        ProgressBar progressBar = new ProgressBar(0);
        // 为打开按钮设置事件处理器
        openButton.setOnAction(e -> {
            FileChooser fileChooser = new FileChooser();
            File file = fileChooser.showOpenDialog(primaryStage);
            if (file != null) {
                Media media = new Media(file.toURI().toString());
                if (mediaPlayer != null) {
                    mediaPlayer.stop();
                }
                mediaPlayer = new MediaPlayer(media);
                mediaPlayer.currentTimeProperty().addListener(ov ->
updateProgress(progressBar, mediaPlayer));
            }
        });
        // 设置播放、停止、暂停/继续按钮的事件处理器
        playButton.setOnAction(e -> {
            if (mediaPlayer != null) {
                mediaPlayer.play();
            }
        });
        stopButton.setOnAction(e -> {
            if (mediaPlayer != null) {
                mediaPlayer.stop();
            }
        });
        pauseButton.setOnAction(e -> {
            if (mediaPlayer != null) {
                if (mediaPlayer.getStatus() == MediaPlayer.Status.PLAYING) {
                    mediaPlayer.pause();
                } else {
                    mediaPlayer.play();
                }
            }
        });
        // 按钮布局
        HBox buttonBox = new HBox(10, openButton, playButton,
stopButton, pauseButton);
        buttonBox.setAlignment(Pos.CENTER);
        // 主布局
        VBox root = new VBox(10, buttonBox, progressBar);
        root.setAlignment(Pos.CENTER);
        // 设置场景和舞台
        primaryStage.setScene(new Scene(root, 400, 150));
        primaryStage.setTitle("音频播放器");
        primaryStage.show();
    }
    // 更新进度条的方法
    private void updateProgress(ProgressBar progressBar, MediaPlayer mediaPlayer) {
        Duration currentTime = mediaPlayer.getCurrentTime();
```

```
            Duration totalDuration = mediaPlayer.getTotalDuration();
            progressBar.setProgress(currentTime.toMillis() / totalDuration.toMillis());
        }

        public static void main(String[] args) {
            launch(args);
        }
    }
```

注意，由于播放音频需要使用 JavaFX 的媒体功能，所以在运行程序之前，需要在 VM 选项中添加媒体库（javafx.media）的引用，具体如下：

```
--add-modules javafx.controls,javafx.fxml,javafx.graphics,javafx.media
```

6.2　录音机

Java 提供了 javax.sound.sampled 包，用于处理音频，包括录音。这个包允许控制音频数据的录制、播放、文件读写和格式转换等。下面将详细解释如何使用 Java 进行录音，包括 AudioFormat、麦克风控制和 AudioInputStream 的使用。

1. AudioFormat（设置音频格式）

AudioFormat 类定义了音频数据的特定格式。要录制音频，首先需要创建一个 AudioFormat 对象以确定录制的音频格式。

（1）Sample Rate（采样率）：表示每秒采样的次数。常见的采样率有 44100 Hz、22050 Hz、16000 Hz 等。

（2）Sample Size（采样大小）：每个样本的位数，通常是 16 位或 8 位。

（3）Channels（通道数）：1 表示单声道，2 表示立体声。

（4）Signed/Unsigned（有符号 / 无符号）：通常音频数据是有符号的。

（5）Big/Little Endian（字节序）：数据的字节顺序。

```
AudioFormat format = new AudioFormat(16000.0f, 16, 2, true, true);
```

2. TargetDataLine（控制麦克风）

TargetDataLine 是一个可以从中读取音频数据的数据行。它通常用来从麦克风捕获音频。

（1）获取并打开麦克风。使用 AudioSystem.getTargetDataLine 方法获取一个与指定 AudioFormat 兼容的 TargetDataLine。通过调用 open() 方法打开它。

```
TargetDataLine microphone = AudioSystem.getTargetDataLine(format);
microphone.open(format);
```

（2）开始和停止录音调用。start() 方法用于开始捕获音频。stop() 方法用于停止捕获音频。

```
microphone.start();
microphone.stop();
```

3. AudioInputStream（处理音频流）

AudioInputStream 类代表音频输入流。它可以用于从 TargetDataLine 读取音频数据。创建 AudioInputStream 时，将 TargetDataLine 作为源传递给它。可以从 AudioInputStream 读取数据并将其写入文件。

```
AudioInputStream audioInputStream = new AudioInputStream(microphone);
```

4. 录制音频并保存到文件

录制音频通常包括从 TargetDataLine 读取数据并将其写入文件。这可以通过 AudioSystem.write 方法实现，该方法将 AudioInputStream 转换为指定的音频文件格式。

```
File outputFile = new File("recording.wav");
AudioSystem.write(audioInputStream, AudioFileFormat.Type.WAVE, outputFile);
```

基于以上描述，一个完整的录音过程大致如下。

（1）创建 AudioFormat 来指定录音格式。

（2）获取并打开 TargetDataLine（麦克风）。

（3）创建 AudioInputStream 来从麦克风读取数据。

（4）开始录音，并将数据从 AudioInputStream 写入文件。

（5）录音完成后，关闭麦克风并释放资源。

下面的代码实现了一个完整的录音机程序，支持“录音”“暂停 / 继续”和“停止”功能。单击“停止”按钮，会停止录音，并且在当前目录保存当前的路由文件，文件名由当前时间（年月日时分秒）组成，如 20231204142042.wav。

代码位置：src/main/java/audio/AudioRecorder.java

```
package audio;

import javafx.application.Application;
import javafx.geometry.Pos;
import javafx.scene.Scene;
import javafx.scene.control.Button;
import javafx.scene.layout.HBox;
import javafx.scene.layout.VBox;
import javafx.stage.Stage;
import javax.sound.sampled.*;
import java.io.File;
```

```
import java.io.IOException;
import java.text.SimpleDateFormat;
import java.util.Date;

public class AudioRecorder extends Application {
    private AudioFormat format;
    private TargetDataLine microphone;
    private AudioInputStream audioInputStream;
    private File audioFile;
    private Thread recordingThread;
    private boolean isRecording = false;
    private boolean isPaused = false;

    @Override
    public void start(Stage primaryStage) {
        // 设置录音格式
        format = new AudioFormat(16000.0f, 16, 2, true, true);
        // 创建按钮
        Button recordButton = new Button("录音");
        Button pauseButton = new Button("暂停/继续");
        Button stopButton = new Button("停止");
        // 录音按钮事件
        recordButton.setOnAction(event -> {
            if (!isRecording) {
                startRecording();
                isRecording = true;
            }
        });

        // 暂停/继续按钮事件
        pauseButton.setOnAction(event -> {
            if (isRecording) {
                if (!isPaused) {
                    pauseRecording();
                    isPaused = true;
                } else {
                    resumeRecording();
                    isPaused = false;
                }
            }
        });
        // 停止按钮事件
        stopButton.setOnAction(event -> {
            if (isRecording) {
                stopRecording();
                isRecording = false;
                isPaused = false;
            }
        });
        // 按钮布局
        HBox buttonBox = new HBox(10, recordButton, pauseButton, stopButton);
```

```
        buttonBox.setAlignment(Pos.CENTER);
        // 主布局
        VBox root = new VBox(10, buttonBox);
        root.setAlignment(Pos.CENTER);
        // 设置场景和舞台
        primaryStage.setScene(new Scene(root, 300, 100));
        primaryStage.setTitle("录音机");
        primaryStage.show();
    }
    private void startRecording() {
        try {
            microphone = AudioSystem.getTargetDataLine(format);
            microphone.open(format);
            microphone.start();
            audioInputStream = new AudioInputStream(microphone);
            recordingThread = new Thread(new Runnable() {
                @Override
                public void run() {
                    saveAudio();
                }
            });
            recordingThread.start();
        } catch (LineUnavailableException e) {
            e.printStackTrace();
        }
    }
    private void pauseRecording() {
        microphone.stop();
        microphone.flush();
    }
    private void resumeRecording() {
        microphone.start();
    }

    private void stopRecording() {
        microphone.stop();
        microphone.close();
        recordingThread.interrupt();
    }
    private void saveAudio() {
        try {
            String fileName = new SimpleDateFormat("yyyyMMddHHmmss").format
(new Date()) + ".wav";
            audioFile = new File( fileName);
            AudioSystem.write(audioInputStream, AudioFileFormat.Type.WAVE,
audioFile);
        } catch (IOException e) {
            e.printStackTrace();
        }
    }
    public static void main(String[] args) {
```

```
        launch(args);
    }
}
```

注意，由于录制音频需要使用 JavaFX 的媒体功能，这就需要在运行程序之前，在 VM 选项中添加媒体库（javafx.media）的引用，具体如下：

```
--add-modules javafx.controls,javafx.fxml,javafx.graphics,javafx.media
```

运行程序，会显示如图 6-2 所示的录音机界面，单击“录音”按钮开始录音，单击“停止”按钮停止录音，并在当前目录生成音频文件（WAV 文件）。

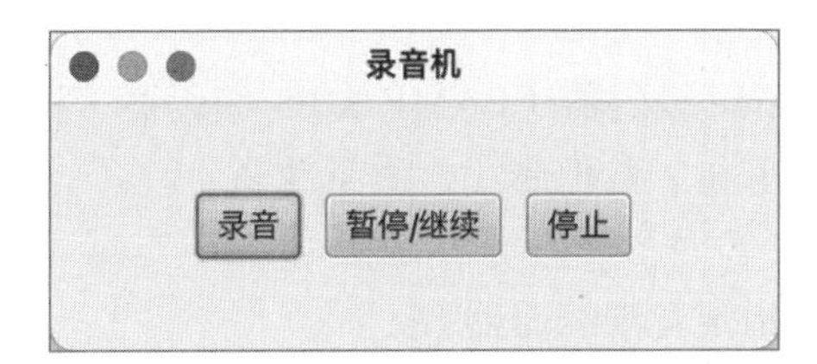

图 6-2　录音机

6.3　音频格式转换

本节的例子可以实现 MP3 和 WAV 音频格式之间的转换。最直接的方式是使用 FFmpeg 命令。可以使用 Java 的 Runtime 或 ProcessBuilder 类来执行 FFmpeg 命令，就好像 FFmpeg 是 Java 的一部分一样。

FFmpeg 是一个非常强大的多媒体处理工具，广泛用于处理视频和音频文件。它支持转换格式、编码、解码、混流、滤镜处理等多种功能。以下是 FFmpeg 命令的一些基本组成部分，以及用于音频格式转换的常见命令行参数。

1. 基本命令结构

FFmpeg 的基本命令结构如下：

```
ffmpeg [全局选项] {[输入文件选项] -i 输入文件} ... {[输出文件选项] 输出文件} ...
```

（1）FFmpeg：命令行调用 FFmpeg。

（2）[全局选项]：影响整个程序的选项，例如日志级别。

（3）[输入文件选项]：应用于输入文件的选项，例如指定解码器。

（4）-i 输入文件：指定输入文件的路径。

（5）[输出文件选项]：应用于输出文件的选项，例如指定编码器或调整比特率。

（6）输出文件：指定输出文件的路径。

2. 音频格式转换

对于音频格式转换，FFmpeg 主要依赖于指定输入文件和输出文件的格式。FFmpeg 会自动根据输出文件的扩展名推断所需的编码器。以下是一些常见的音频格式转换命令。

❑ MP3 转 WAV：

```
ffmpeg -i input.mp3 output.wav
```

在这个命令中，FFmpeg 会将一个 MP3 文件转换为 WAV 格式。

❑ WAV 转 MP3：

```
ffmpeg -i input.wav -codec:a libmp3lame -qscale:a 2 output.mp3
```

这个命令使用 libmp3lame 编码器（LAME MP3 编码器）将 WAV 文件转换为 MP3 格式。-qscale:a 2 设置了音频质量，数值越小，质量越高。

其他常用的音频参数如下。

（1）-codec:a 或 -c:a：指定音频编码器。

（2）-b:a：设置音频比特率，例如 -b:a 128k。

（3）-ar：设置音频采样率，例如 -ar 44100。

（4）-ac：设置音频通道数，例如 -ac 2 表示立体声。

注意事项如下。

（1）FFmpeg 支持的格式和编码器取决于它的构建配置。某些特定的编码器可能不包含在所有构建中。

（2）FFmpeg 命令非常灵活，可以通过组合不同的参数来实现复杂的处理逻辑。

（3）在处理大型文件或流媒体时，可能需要考虑更多的性能和资源管理参数。

在使用 FFmpeg 之前，需要先安装 FFmpeg，读者可以到 https://ffmpeg.org/download.html 下载 FFmpeg 的对应操作系统版本。然后在操作系统的终端中输入 FFmpeg，如果找到了命令，就表示成功安装了 FFmpeg。

下面的代码利用 FFmpeg 命令实现了一个 MP3 和 WAV 互转的命令行工具。

代码位置：src/main/java/audio/AudioConverter.java

```
package audio;
import java.io.IOException;
public class AudioConverter {
    public static void main(String[] args) {
        if (args.length != 2) {
            System.out.println("用法：java AudioConverter <输入文件.wav>
<输出文件.mp3>");
            System.exit(1);
        }
       String inputFilePath = args[0];
       String outputFilePath = args[1];
```

```
        // 构建用于将 WAV 转换为 MP3 的 FFmpeg 命令
        String[] command = {
                "ffmpeg",
                "-i", inputFilePath,
                "-codec:a", "libmp3lame",
                "-qscale:a", "2",
                outputFilePath
        };
        try {
            // 使用 ProcessBuilder 执行命令
            ProcessBuilder processBuilder = new ProcessBuilder(command);
            Process process = processBuilder.start();
            // 等待进程结束
            process.waitFor();
            System.out.println("转换完成：" + inputFilePath + " -> " +
outputFilePath);
        } catch (IOException | InterruptedException e) {
            e.printStackTrace();
        }
    }
}
```

在运行 AudioConverter 时，需要提供两个命令行参数，分别是输入音频文件名和输出音频文件名，如 audio.mp3 和 audio.wav。如果读者使用了 Intellij IDEA 或其他 Java IDE，可以添加这两个命令行参数。如果执行程序输出如下的信息，就说明音频格式转换成功。

```
转换完成：audio.mp3 -> audio.wav
```

注意：如果目标音频文件删除，需要先删除目标音频文件，才能正常转换。读者也可以在 Java 代码中添加删除目标音频文件的代码。

6.4　音频分析和处理

本节主要介绍如何使用 Java 以及第三方库实现音频分析和处理功能，主要包括音高检测、添加噪声、降低噪声等。

6.4.1　PCM 格式的 WAV 文件

音频分析通常需要 PCM（Pulse Code Modulation，脉冲编码调制）格式的 WAV 文件，原因在于 PCM 提供了一种直接且无损的方式来表示音频波形的样本值。这与 WAV 文件中可能使用的其他编码格式相比，使得音频处理和分析变得更简单和直接。下面详细解释一下使用 PCM 格式 WAV 文件进行音频分析的原因。

（1）直接性和无损质量：PCM 是一种线性的、未压缩的、无损的音频格式，它以原始的形式直接表示音频波形的样本值。这对于精确分析音频信号非常重要。

（2）简化处理：在 PCM 编码的音频文件中，每个样本值直接对应于音频信号在特定时间点的振幅。这简化了音频的处理，如音高检测、节拍分析等，因为不需要额外的解码步骤。

（3）广泛支持：大多数音频处理和分析库都支持 PCM 编码的 WAV 文件，因为它是一种标准且简单的音频格式。

WAV 是一种容器格式，可以包含不同类型的音频编码。除了常见的 PCM 编码外，WAV 文件还可以包含其他编码格式的音频数据。

（1）浮点编码：一些 WAV 文件使用 32 位或 64 位浮点数来表示样本值。这在专业音频处理中较为常见，提供更高的动态范围。

（2）压缩格式：虽然不太常见，但 WAV 文件也可以包含压缩过的音频数据，如通过 MP3 或其他压缩算法编码的音频。

（3）非标准格式：一些特殊的 WAV 文件可能使用非标准的编码格式，这些格式可能不被所有的音频处理库支持。

由于其直接性、无损质量和简化处理的特点，PCM 编码的 WAV 文件是音频分析和处理的首选格式。其他格式的 WAV 文件可能需要额外的处理步骤，例如解码或格式转换，才能进行分析。在音频处理领域，确保使用的文件格式与所用工具和库兼容非常重要。

6.4.2 转换为 PCM 格式的 WAV 文件

如果需要进行音频分析的 WAV 文件不是 PCM 格式，而分析音频的软件或库又必须使用 PCM 格式的 WAV 文件，那么就只能进行格式转换了。

我们可以直接使用 FFmpeg 将音频文件转换为 PCM 编码的 WAV 格式，下面是具体的转换命令格式：

```
ffmpeg -i input_file -acodec pcm_s16le -ar sample_rate -ac channels output_file.wav
```

其中：

（1）-i input_file：指定输入文件。

（2）-acodec pcm_s16le：指定音频编码器为 PCM 16 位小端编码。

（3）-ar sample_rate：指定采样率，例如 44100 Hz。

（4）-ac channels：指定音频通道数，例如 2 代表立体声。

（5）output_file.wav：指定输出文件的名称。

在 Java 中，可以使用 Runtime.getRuntime().exec() 方法执行 FFmpeg 命令，具体的实现代码如下。

代码位置：src/main/java/audio/ConvertToPCM.java

```
package audio;
import java.io.IOException;
public class ConvertToPCM {
    public static void main(String[] args) {
        String sourcePath = "audio.wav";                    // 其他格式的 WAV 文件名
        String targetPath = "audio_pcm.wav";                // PCM 格式的 WAV 文件名
        try {
            convertToPCM(sourcePath, targetPath);
            System.out.println("Conversion completed successfully.");
        } catch (IOException | InterruptedException e) {
            e.printStackTrace();
        }
    }
    // 将普通格式的 WAV 文件转换为 PCM 格式的 WAV 文件
    private static void convertToPCM(String sourcePath, String targetPath)
throws IOException, InterruptedException {
        // 构造 FFmpeg 命令
        String command = "ffmpeg -i \"" + sourcePath + "\" -acodec pcm_s16le
-ar 44100 -ac 2 \"" + targetPath + "\"";

        // 执行命令
        Process process = Runtime.getRuntime().exec(command);
        // 等待转换过程完成
        process.waitFor();
    }
}
```

在运行程序之前，要确保当前目录存在 audio.wav 文件。运行程序后，会在当前目录生成一个 audio_pcm.wav 文件。

6.4.3　音高检测

音高（Pitch）是声音的一种基本属性，它是我们感知声音高低的方式。在音乐和声学领域，音高通常与声音的频率密切相关：频率较高的声音我们感知为“高音”，频率较低的声音我们感知为“低音”。

音高的科学基础如下。

（1）频率：音高通常与声波的频率相关联。频率是单位时间内振动的次数，以赫兹（Hz）为单位。人耳能感知的频率范围为 20 ～ 20 000 Hz。

（2）声音波形：声音是通过空气（或其他介质）传播的振动波形。当这些波形以一定的、稳定的频率振动时，我们听到的是一个特定的音高。

音高检测在许多领域都非常重要，尤其是在音乐、语音处理和声学研究中。

（1）音乐制作和分析：在音乐制作中，音高检测可用于音符识别、旋律提取、和声分析等。它对于音乐软件、音乐教育工具以及自动音乐转录非常重要。

（2）语音识别和处理：在语音识别系统中，音高分析可以帮助识别说话者的语调和情感，有助于改善语音理解的准确性。

（3）声学研究：在声学领域，音高分析对于研究声音如何在不同环境中传播，以及人类如何感知声音都非常关键。

（4）音乐教育：音高检测技术可以用于开发音乐教育工具，如电子调音器和音乐练习软件，帮助学习者提高演奏或唱歌的技巧。

在 Java 中，可以使用 TarsosDSP 库。这个库用于音频信号处理。它的目的是提供一个简单易用的接口，调用实际应用中常见的音频处理算法，这些算法都是纯 Java 实现的，没有任何其他外部依赖。TarsosDSP 库试图在功能强大但体积小巧的平衡点上达到最佳效果。

TarsosDSP 库有以下一些主要功能。

（1）声音起始检测：检测声音的第一下发音瞬间，可以用于乐器调音器等场景。

（2）音高检测算法：给一个音频样本，检测该样本的声音频率和音高，可以用于声音识别等领域。TarsosDSP 中提供了多种这类算法，如 YIN、Mcleod Pitch method 等。

（3）动态音高检测算法：动态的频率和音高检测算法，分析动态输入的音频流中的音高频率。

（4）解码算法：实现了 Goertzel DTMF 解码算法，用于将 MP3 等压缩格式的音频解码成 PCM 或 WAV 格式的音频。

（5）时间拉伸算法：WSOLA 算法，拉伸或变速不变调音频流中的声音。

（6）重采样算法：改变或恢复音频流中的采样率、声道数、采样大小等属性。

（7）过滤器：过滤杂音、噪声等不需要的声波。

（8）简单合成器：合成多个声波到同一个声道。

（9）音效器：为声波添加诸如混响、重低音、环绕、均衡器等效果。

（10）音高转换算法：变声器算法，将声波变为不同类型或风格。

如果读者使用 maven 管理依赖，可以在 pom.xml 文件中添加如下依赖来安装 TarsosDSP 库。

```
<dependency>
    <groupId>com.github.axet</groupId>
    <artifactId>TarsosDSP</artifactId>
    <version>2.4</version>
</dependency>
```

使用 TarsosDSP 库进行实时音高检测的步骤如下：

1. 创建 AudioDispatcher

```
AudioDispatcherFactory.fromFile(new File(audioFilePath), 1024, 0)
```

其中：

（1）fromFile 方法：用于从指定的音频文件创建一个 AudioDispatcher。

（2）new File(audioFilePath)：指定要处理的音频文件。

（3）1024：音频缓冲区的大小，单位是样本。这个大小影响音频处理的精度和性能。较小的缓冲区可以提供更高的时间分辨率，但可能增加计算负担。

（4）0：表示重叠大小。在这里，我们没有设置重叠。

2. 设置音高检测处理器

```
new PitchProcessor(PitchEstimationAlgorithm.YIN, 44100, 1024, pitchDetectionHandler)
```

其中：

（1）PitchProcessor：用于音高检测。

（2）PitchEstimationAlgorithm.YIN：音高估计算法。YIN 算法适用于音乐声音的音高检测。

（3）44100：音频的采样率。这个值应与音频文件的实际采样率相匹配。

（4）1024：音频缓冲区的大小，与 AudioDispatcher 创建时使用的大小相同。

（5）pitchDetectionHandler：处理检测到的音高的回调接口。

3. 音高检测回调处理

在 handlePitch 方法中，每当检测到音高时，这个方法就会被调用：

pitchDetectionResult.getPitch()：获取检测到的音高值（以赫兹为单位）。

完整的实现代码如下。

代码位置： src/main/java/audio/PitchDetectionExample.java

```
package audio;
import be.tarsos.dsp.AudioDispatcher;
import be.tarsos.dsp.AudioEvent;
import be.tarsos.dsp.AudioProcessor;
import be.tarsos.dsp.io.jvm.AudioDispatcherFactory;
import be.tarsos.dsp.pitch.PitchDetectionHandler;
import be.tarsos.dsp.pitch.PitchDetectionResult;
import be.tarsos.dsp.pitch.PitchProcessor;
import be.tarsos.dsp.pitch.PitchProcessor.PitchEstimationAlgorithm;
import javax.sound.sampled.UnsupportedAudioFileException;
import java.io.File;
import java.io.IOException;
public class PitchDetectionExample {
    public static void main(String[] args) throws UnsupportedAudioFileException,
IOException {
        // 指定音频文件的路径
        String audioFilePath = "audio_pcm.wav";
        AudioDispatcher dispatcher = AudioDispatcherFactory.fromFile(new
File(audioFilePath), 1024, 0);
```

```
        PitchDetectionHandler pitchDetectionHandler = new PitchDetectionHandler() {
            @Override
            public void handlePitch(PitchDetectionResult pitchDetectionResult,
AudioEvent audioEvent) {
                final float pitchInHz = pitchDetectionResult.getPitch();
                if(pitchInHz > 0){
                    System.out.printf("Pitch detected: %.2f Hz\n", pitchInHz);
                }
            }
        };
        AudioProcessor pitchProcessor = new PitchProcessor
(PitchEstimationAlgorithm.YIN, 44100, 1024, pitchDetectionHandler);
        dispatcher.addAudioProcessor(pitchProcessor);
        // 启动音频处理
        new Thread(dispatcher).start();
    }
}
```

要注意，在运行程序之前，要确保当前目录存在 audio_pcm.wav 文件。而且必须使用 PCM 格式的 WAV 文件，否则会抛出异常。运行程序，会输出类似下面的内容：

```
Pitch detected: 551.13 Hz
Pitch detected: 568.41 Hz
Pitch detected: 230.26 Hz
Pitch detected: 435.85 Hz
Pitch detected: 1245.28 Hz
Pitch detected: 616.40 Hz
Pitch detected: 655.39 Hz
Pitch detected: 704.45 Hz
Pitch detected: 725.52 Hz
...
```

在 PitchDetectionExample 程序中，每当处理一个缓冲区时，都会进行一次音高检测，并产生一个音高值。缓冲区的大小是 1024 个样本，这意味着整个缓冲区作为一个整体被用来估计一个音高值。

以采样率为 44100 Hz 和缓冲区大小为 1024 个样本为例，整个缓冲区代表的时间长度是 1024 / 44100 秒，约等于 0.023 秒。这意味着大约每 23 毫秒，程序处理一次缓冲区并输出一个音高值。每次处理都是基于这 1024 个样本的数据来估计音高。

因此，如果看到如下输出：

```
Pitch detected: 551.13 Hz
Pitch detected: 568.41 Hz
```

这表示第 1 个缓冲区的音高被估计为 551.13 Hz，第 2 个缓冲区的音高被估计为 568.41 Hz，以此类推。每个估计的音高代表了一个缓冲区时间内的音频信号的平均音高。

6.4.4 麦克风音频中的音高

这一节会直接获取从麦克风采集的音频的音高。实现原理与6.4.3节的程序类似，同样是使用TarsosDSP库，只不过需要使用AudioDispatcherFactory.fromDefaultMicrophone()方法从麦克风采集音频，并创建AudioDispatcher对象。该方法的用法如下：

```
AudioDispatcher dispatcher = AudioDispatcherFactory.fromDefaultMicrophone
(22050, 1024, 0);
```

其中：

（1）22050：采样率，表示每秒处理的样本数。22050Hz是较为常见的采样率，足以捕捉大部分听觉范围内的音频。

（2）1024：音频缓冲区的大小，单位为样本。这个值决定了处理音频时的时间分辨率。

（3）0：重叠大小，表示连续两个缓冲区之间共享的样本数。这里设置为0，意味着没有重叠。

完整的实现代码如下。

代码位置：src/main/java/audio/PitchDetectorFromMicrophone.java

```
package audio;
// 导入需要的类
import be.tarsos.dsp.AudioDispatcher;
import be.tarsos.dsp.AudioEvent;
import be.tarsos.dsp.AudioProcessor;
import be.tarsos.dsp.io.jvm.AudioDispatcherFactory;
import be.tarsos.dsp.pitch.PitchDetectionHandler;
import be.tarsos.dsp.pitch.PitchDetectionResult;
import be.tarsos.dsp.pitch.PitchProcessor;
// 定义一个类，用于实现音高检测的功能
public class PitchDetectorFromMicrophone {
    // 定义一个主方法，用于测试
    public static void main(String[] args) throws  Exception {
        // 创建一个AudioDispatcher对象，用于从麦克风获取音频数据
        AudioDispatcher dispatcher = AudioDispatcherFactory.
fromDefaultMicrophone(22050,1024,0);
        // 创建一个PitchDetectionHandler对象，用于处理音高检测的结果
        PitchDetectionHandler handler = new PitchDetectionHandler() {
            @Override
            public void handlePitch(PitchDetectionResult pitchDetectionResult,
AudioEvent audioEvent) {
                // 获取音高的频率，单位为赫兹
                double pitch = pitchDetectionResult.getPitch();
                // 打印音高的频率
                System.out.println("Pitch: " + pitch + " Hz");
            }
        };
```

```
        // 创建一个 AudioProcessor 对象，用于执行音高检测的算法
        AudioProcessor processor = new PitchProcessor(PitchProcessor.
PitchEstimationAlgorithm.YIN, 22050, 1024, handler);
        // 将 AudioProcessor 对象添加到 AudioDispatcher 对象中
        dispatcher.addAudioProcessor(processor);
        // 启动 AudioDispatcher 对象，开始音频处理
        dispatcher.run();
    }
}
```

在运行程序之前，要确保当前电脑有音频输入，并且连接了麦克风。运行程序后，对着麦克风发出声音，就会在终端输出类似下面的内容。在未发出声音时，音高是 -1.0Hz，表示没有检测到明确的音高（通常为静音）。当发出声音时，就是实际的音高。

```
Pitch: -1.0 Hz
Pitch: -1.0 Hz
Pitch: -1.0 Hz
Pitch: -1.0 Hz
Pitch: 118.38450622558594 Hz
Pitch: 120.49220275878906 Hz
Pitch: 123.86878967285156 Hz
Pitch: 128.86106872558594 Hz
Pitch: 136.3795166015625 Hz
Pitch: -1.0 Hz
Pitch: 143.2723846435547 Hz
Pitch: -1.0 Hz
Pitch: -1.0 Hz
```

6.4.5 添加噪声

本节的例子会为一个 WAV 格式的音频文件添加噪声，然后生成一个带噪声的 WAV 文件。这个例子主要涉及音频处理的基本概念，并使用 Java 中的 javax.sound.sampled 包来实现音频的读取、处理和写入。

使用的主要技术和库如下。

（1）javax.sound.sampled 包：这是 Java 标准库的一部分，专门用于处理音频数据。它提供了读取、写入音频数据以及处理音频格式的功能。

（2）音频数据处理：程序中对音频数据进行了基本的字节级操作，包括生成噪声数据和将这些数据与原始音频数据混合。

（3）文件操作：程序使用标准的 Java 文件 I/O 操作来读取和写入 WAV 文件。

下面是对技术细节和实现步骤的详细描述。

1. 读取原始音频

首先，使用 AudioSystem.getAudioInputStream 方法读取原始的 WAV 文件。这个方法

返回一个 AudioInputStream 对象，它允许我们以字节流的形式访问音频数据。同时，我们从这个流中获取 AudioFormat 对象，它包含了音频文件的格式信息，比如采样率、位深度等。

```
File originalFile = new File("path/to/your/original.wav");
AudioInputStream originalStream = AudioSystem.getAudioInputStream(originalFile);
AudioFormat format = originalStream.getFormat();
```

2. 生成噪声数据

接下来，调用 generateNoise 方法生成噪声数据。这个方法创建一个字节数组，数组的长度与原始音频流的可用字节数相同。通过遍历数组的每个元素，并为其赋值一个随机生成的字节值，这样就生成了噪声。噪声的强度可以通过调整随机值的范围来控制。

```
private static byte[] generateNoise(AudioInputStream stream) {
    byte[] noise = new byte[stream.available()];
    for (int i = 0; i < noise.length; i++) {
        noise[i] = (byte) (Math.random() * Byte.MAX_VALUE / 4);
        // 调整这里以控制噪声强度
    }
    return noise;
}
```

3. 将噪声添加到原始音频数据中

这一步是通过 mixAudio 方法实现的。首先，从原始音频流中读取所有字节到一个数组中。然后，对于音频数据的每个字节，将其与对应的噪声字节按照一定比例混合。在这个例子中，我们使用加权平均的方式来混合原始音频和噪声。原始音频数据占主导（例如 90%），而噪声数据的比重较小（例如 10%）。

```
private static byte[] mixAudio(AudioInputStream originalStream, byte[] noise)
throws IOException {
    byte[] audioBytes = originalStream.readAllBytes();
    for (int i = 0; i < audioBytes.length; i++) {
        int mixedValue = (int) (audioBytes[i] * 0.9 + noise[i] * 0.1);
        // 调整这里以控制噪声的混合比例
        audioBytes[i] = (byte) mixedValue;
    }
    return audioBytes;
}
```

4. 将混合后的音频数据写入新的 WAV 文件

最后一步是将混合后的音频数据写入一个新的 WAV 文件。这是通过创建一个新的 ByteArrayInputStream（包含混合后的音频数据）和一个新的 AudioInputStream（包含原始音频格式信息）来完成的。然后，使用 AudioSystem.write 方法将这个音频流写入一个新的文件。

```
File newFile = new File("path/to/your/newfile.wav");
ByteArrayInputStream mixedInput = new ByteArrayInputStream(mixed);
AudioInputStream mixedStream = new AudioInputStream(mixedInput, format, mixed.
length / format.getFrameSize());
AudioSystem.write(mixedStream, AudioFileFormat.Type.WAVE, newFile);
```

下面代码是添加噪声的完整实现。

代码位置：src/main/java/audio/AddNoiseToWav.java

```
package audio;
import javax.sound.sampled.*;
import java.io.*;
public class AddNoiseToWav {
    public static void main(String[] args) throws Exception {
        // 加载原始音频文件
        File originalFile = new File("audio_pcm.wav");
        AudioInputStream originalStream = AudioSystem.getAudioInputStream
(originalFile);
        AudioFormat format = originalStream.getFormat();
        // 生成噪声数据
        byte[] noise = generateNoise(originalStream);
        // 将噪声添加到原始音频数据中
        byte[] mixed = mixAudio(originalStream, noise);
        // 将混合后的音频数据写入新的 WAV 文件
        File newFile = new File("audio_pcm_noise.wav");
        ByteArrayInputStream mixedInput = new ByteArrayInputStream(mixed);
        AudioInputStream mixedStream = new AudioInputStream(mixedInput, format,
mixed.length / format.getFrameSize());
        AudioSystem.write(mixedStream, AudioFileFormat.Type.WAVE, newFile);
        originalStream.close();
        mixedStream.close();
    }
    private static byte[] generateNoise(AudioInputStream stream) throws
Exception{
        // 这里只是一个简单的噪声生成示例，读者可以根据需要实现更复杂的噪声生成算法
        byte[] noise = new byte[stream.available()];
        for (int i = 0; i < noise.length; i++) {
            noise[i] = (byte) (Math.random() * Byte.MAX_VALUE);
        }
        return noise;
    }
    private static byte[] mixAudio(AudioInputStream originalStream, byte[]
noise) throws IOException {
        byte[] audioBytes = originalStream.readAllBytes();
        for (int i = 0; i < audioBytes.length; i++) {
            // 将噪声数据混合到原始音频数据中
            // 注意：这里的混合方式可能需要根据具体需求来调整
            int mixedValue = (int) (audioBytes[i] * 0.99 + noise[i] * 0.01);
            audioBytes[i] = (byte) mixedValue;
        }
```

```
            return audioBytes;
        }
    }
```

在运行程序之前，要确保当期目录存在 audio_pcm.wav 文件，运行程序后，会发现在当前目录生成了一个名为 audio_pcm_noise.wav 的文件，该文件在保持 audio_pcm.wav 的音频效果的同时，添加了噪声效果。为了让噪声效果不至于影响原音频，本例将噪声占比调整到了 1%。

6.4.6　降低噪声

在这一节会利用 TarsosDSP 库中相关 API 降低 6.4.5 节为 WAV 音频文件添加的噪声，并生成降低噪声后的音频文件。

降低噪声是音频处理中的一个复杂任务，通常需要根据音频内容和噪声类型采取不同的策略。在这个例子中，我们使用了低通滤波器作为降低噪声的手段，这是一种简单但在某些情况下有效的方法。

低通滤波器允许频率低于特定截止点的信号通过，同时减弱或阻止更高频率的信号。在音频处理中，这可以有效减少高频噪声（例如嘶嘶声或电子干扰），同时保留低频率的音频成分，如人声或乐器声。

具体的实现步骤如下。

1. 读取音频文件

```
AudioDispatcher dispatcher = AudioDispatcherFactory.fromFile
(new File(inputFilename), 1024, 0);
```

其中：

（1）AudioDispatcher 是 TarsosDSP 库中的一个类，用于管理音频数据的读取和处理流程。它从指定的音频文件中读取数据，并将这些数据分块传递给一系列的音频处理器（AudioProcessors）。

（2）使用 AudioDispatcherFactory.fromFile 创建一个 AudioDispatcher 实例。

（3）inputFilename 指定了带噪声的音频文件路径。

（4）1024 是处理块的大小，表示每次处理 1024 个样本。这个值影响处理的精度和效率。

2. 应用低通滤波器

```
dispatcher.addAudioProcessor(new LowPassFS(1000, 44100));
```

其中：

（1）使用 LowPassFS 类创建一个低通滤波器实例，并添加到 dispatcher 的处理链中。

（2）1000 是截止频率（以赫兹为单位），只有低于此频率的信号才能通过。

（3）44100 是音频的采样率，表示每秒的样本数。

3. 写入处理后的音频

```
dispatcher.addAudioProcessor(new WaveformWriter(dispatcher.getFormat(),
outputFilename));
```

其中：

（1）使用 WaveformWriter 将处理后的音频数据写入新的文件中。

（2）outputFilename 是处理后音频的保存路径。

4. 开始处理音频

```
dispatcher.run();
```

调用 dispatcher.run() 开始音频处理流程。这个方法会按顺序执行所有添加的音频处理器。

下面的代码完整地实现了降低噪声的功能。

代码位置：src/main/java/audio/NoiseReductionExample.java

```
package audio;
import be.tarsos.dsp.AudioDispatcher;
import be.tarsos.dsp.filters.LowPassFS;
import be.tarsos.dsp.io.jvm.AudioDispatcherFactory;
import be.tarsos.dsp.io.jvm.WaveformWriter;
import javax.sound.sampled.UnsupportedAudioFileException;
import java.io.File;
import java.io.IOException;
public class NoiseReductionExample {
    public static void main(String[] args) throws UnsupportedAudioFileException,
IOException {
        String inputFilename = "audio_pcm_noise.wav"; // 输入带噪声的音频文件路径
        String outputFilename = "audio_pcm_noise_denoised.wav";
        // 输出去噪后的音频文件路径
        // 创建 AudioDispatcher 来读取音频文件
        AudioDispatcher dispatcher = AudioDispatcherFactory.fromFile
(new File(inputFilename), 1024, 0);
        // 添加低通滤波器进行噪声减少。可以调整滤波器的截止频率（这里设置为 1000Hz）
        dispatcher.addAudioProcessor(new LowPassFS(1000, 44100));
        // 将处理后的音频写入输出文件
        dispatcher.addAudioProcessor(new WaveformWriter(dispatcher.getFormat(),
outputFilename));
        // 开始处理音频
        dispatcher.run();
    }
}
```

运行程序之前，要确保当前目录存在 audio_pcm_noise.wav 文件，运行程序后，会在当前目录生成一个降低噪声后的 audio_pcm_noise_denoised.wav 文件。

6.5　音频编辑

各种音频编辑软件一直是音乐爱好者硬盘中的常客。不过这些软件往往都是别人做的。其实，读者完全可以借助 Java 强大的第三方库以及 FFmpeg 等工具，实现音频编辑自由，而且可以完全自动化,不像很多音频编辑软件,尽管功能非常强大,但不能编程,功能再强大,也需要我们手工去操作，很是麻烦。通过对本节的学习，读者完全可以用 Java 做一个相当专业的音频编辑软件。

6.5.1　音频裁剪

在这一节我们会使用 FFmpeg 工具实现音频裁剪功能，通过 Java 调用 FFmpeg 命令，可以很容易将 FFmpeg 的一切功能集成在 Java 程序中。下面是用 FFmpeg 实现音频裁剪的命令行：

```
ffmpeg -i input.mp3 -ss 00:00:05 -t 00:00:10 -acodec copy output.mp3
```

其中：

（1）FFmpeg：这是命令行调用 FFmpeg 工具的方式，FFmpeg 是一个可执行文件。

（2）-i input.mp3：-i 参数指定输入文件，这里是 input.mp3。这是要被裁剪的原始音频文件。

（3）-ss 00:00:05：-ss 参数指定了开始裁剪的起始时间。在这个例子中，它被设置为从音频的第 5 秒开始。这个参数确保 FFmpeg 从指定的时间点开始读取输入文件。

（4）-t 00:00:10：-t 参数指定了从起始点开始需要裁剪的持续时间。在这里，它设置为 10 秒。这意味着从开始时间算起，后续的 10 秒音频将被提取出来。

（5）-acodec copy：这个参数告诉 FFmpeg 在处理音频时保留原始的音频编码（即不对音频进行重新编码）。这样做通常会更快，而且可以保持原始音频质量。

（6）output.mp3：这是输出文件的名称，即裁剪后的音频将被保存为 output.mp3。

下面是用 Java 调用 FFmpeg 命令实现音频裁剪的完整代码。在这段代码中，截取了 audio.mp3 文件中 2 ～ 5 秒的音频内容，并将截取后的音频保存为 cut_audio.mp3 文件。

代码位置：src/main/java/audio/AudioCutter.java

```
package audio;
import java.io.IOException;
public class AudioCutter {
    public static void main(String[] args) {
        // 音频文件路径
        String sourceAudio = "audio.mp3";
```

```
        // 裁剪后的输出文件路径
        String outputAudio = "cut_audio.mp3";
        // 裁剪的开始时间（格式为 HH:MM:SS）
        String startTime = "00:00:02";
        // 裁剪的持续时间（格式为 HH:MM:SS）
        String duration = "00:00:05";
        try {
            // 构建 FFmpeg 命令
            String command = String.format("ffmpeg -i %s -ss %s -t %s -acodec 
copy %s",
                    sourceAudio, startTime, duration, outputAudio);
            // 执行 FFmpeg 命令
            Process process = Runtime.getRuntime().exec(command);
            // 等待 FFmpeg 命令执行完成
            process.waitFor();
            System.out.println(" 音频裁剪完成 ");
        } catch (IOException | InterruptedException e) {
            e.printStackTrace();
        }
    }
}
```

6.5.2 音频合并

在这一节我们会使用 FFmpeg 工具实现合并音频的功能，通过 Java 调用 FFmpeg 命令，可以很容易地将 FFmpeg 的一切功能集成在 Java 程序中。下面是用 FFmpeg 实现音频合并的命令行：

```
ffmpeg -f concat -safe 0 -i filelist.txt -c copy output.mp3
```

其中：

（1）-f concat：这个参数指定 FFmpeg 使用 concat（合并）格式。这告诉 FFmpeg 我们想要合并多个文件。

（2）-safe 0：这个参数用于禁用安全模式检查，允许 FFmpeg 读取任意文件路径。这通常在处理文件列表时需要，因为 FFmpeg 默认情况下不允许读取非相对路径的文件。

（3）-i filelist.txt：-i 参数后面跟着的是输入文件，这里是一个包含所有要合并音频文件的列表的文本文件 filelist.txt。

（4）-c copy：这个参数告诉 FFmpeg 在处理音频时保留原始的音频编码（即不对音频进行重新编码）。这样做可以更快地完成合并操作，同时保持原始音频的质量。

（5）output.mp3：这是输出文件的名称，即所有输入音频文件合并后的结果文件。

下面的例子使用 FFmpeg 命令合并了 filelist.txt 文件中指定的音频文件（audio.mp3 和 audio1.mp3）。

代码位置：src/main/java/audio/AudioMerger.java

```
package audio;
import java.io.BufferedWriter;
import java.io.FileWriter;
import java.io.IOException;
public class AudioMerger {
    public static void main(String[] args) {
        // 定义要合并的音频文件列表
        String[] audioFiles = {"audio.mp3", "audio1.mp3"};
        // 定义输出文件名
        String outputFileName = "merged_audio.mp3";
        try {
            // 创建一个包含所有音频文件路径的文件
            createFileList(audioFiles, "filelist.txt");

            // 构建 FFmpeg 命令来合并音频
            String command = "ffmpeg -f concat -safe 0 -i filelist.txt -c copy "
+ outputFileName;
            // 执行 FFmpeg 命令
            Process process = Runtime.getRuntime().exec(command);
            // 等待 FFmpeg 命令执行完成
            process.waitFor();
            System.out.println("成功合并音频");
        } catch (IOException | InterruptedException e) {
            e.printStackTrace();
        }
    }
    // 创建一个包含所有音频文件路径的文件
    private static void createFileList(String[] audioFiles, String fileList)
throws IOException {
        try (BufferedWriter out = new BufferedWriter(new FileWriter(fileList))) {
            for (String file : audioFiles) {
                out.write("file '" + file + "'");
                out.newLine();
            }
        }
    }
}
```

运行程序，会在当前目录生成一个名为 merged_audio.mp3 的合并后的音频文件。

6.5.3 音频混合

本节的例子会使用 FFmpeg 混合 music.mp3 文件和 music.wav 文件，并生成一个 mixed_music.mp3 文件。混合音频非常有用，在很多视频制作软件中，通常用音频混合添加背景音乐，或做一些有趣的特效。

使用 FFmpeg 混合音频的命令行如下：

```
ffmpeg -i music.mp3 -i audio.wav -filter_complex '[0:a][1:a]amix=inputs=2:
duration=longest:weights=1 2' -c:a libmp3lame output.mp3
```

其中：

（1）-i music.mp3 和 -i audio.wav：这些是输入文件选项，用于指定两个输入音频文件。

（2）-filter_complex：指定了一个复杂的过滤器图（filtergraph）。

（3）[0:a][1:a]：分别代表第 1 个 (music.mp3) 和第 2 个 (audio.wav) 输入文件的音频流。

（4）amix=inputs=2：指明要混合的音频流数量为 2。

（5）duration=longest：设置混合后的音频长度与输入音频中最长的一个相同。

（6）weights=1 2：为每个音频流设定混合权重，这里第 1 个流（music.mp3）的权重为 1，第 2 个流（audio.wav）的权重为 2。如果 audio.wav 是背景音乐，想让背景的声音不要喧宾夺主，可以为 music.mp3 设置一个比较大的权重，最好远比 audio.wav 的权重大，如 weights=5 1。

（7）-c:a libmp3lame：指定音频的编码器为 libmp3lame，它是用于 MP3 编码的。

（8）output.mp3：输出文件的名称，即混合后的音频文件。

下面的例子完整地实现了混合音频的功能。

代码位置：src/main/java/audio/AudioMixer.java

```java
package audio;
import java.io.IOException;
public class AudioMixer {
    public static void main(String[] args) {
        try {
            // 定义要混合的音频文件
            String audioFile1 = "music.mp3";
            String audioFile2 = "audio.wav";
            // 定义输出文件名
            String outputFileName = "mixed_music.mp3";
            // 构建 FFmpeg 命令作为字符串数组
            String[] command = {
                    "ffmpeg",
                    "-i", audioFile1,
                    "-i", audioFile2,
                    "-filter_complex", "[0:a][1:a]amix=inputs=2:duration=
longest:weights=1 2",
                    "-c:a", "libmp3lame",
                    outputFileName
            };
            // 执行 FFmpeg 命令
            Process process = new ProcessBuilder(command).start();
            // 等待 FFmpeg 命令执行完成
            process.waitFor();
```

```
            System.out.println("成功混合音频");
        } catch (IOException | InterruptedException e) {
            e.printStackTrace();
        }
    }
}
```

运行程序，会在当前目录生成一个 mixed_music.mp3 文件。播放该文件，两个音频文件的声音同时存在。

6.6　小结

本节介绍了一些与音频相关的 Java 第三方库，以及如何让 Java 与 FFmpeg 结合，实现更强大的音频应用。当然，这些库在音频领域中仍然是沧海一粟。就算是本章介绍的 TarsosDSP、JavaFX 等库，它们的功能远非如此。尤其是 TarsosDSP 库，可以实现非常强大的音频分析应用。不过，要想全面支持音频的各种处理任务，还是推荐直接使用 FFmpeg，当然，也可以将 FFmpeg 与其他库结合使用。总之，如果读者想实现功能强大的音频处理应用，Java 可能会成为最好的选择。

第 7 章

图像与视频

音频和视频向来不分家，在第 6 章我们介绍了 Java 与音频的相关操作，我们已经体验到了 Java 的强大，不过，对于处理视频来说，Java 也不在话下。在这一章，就让我们来体验一下如何利用 Java 与其强大的第三方模块来操控视频，这些操作主要包括获取视频信息、播放视频、截屏、拍照、录制视频、格式转换和视频编辑，利用这些功能，足可以实现一款非常强大的视频软件。

7.1 OpenCV 基础

由于本章及后面的章节会涉及 OpenCV 以及相关技术，所以本节会介绍关于 OpenCV 的一些基础知识。

OpenCV 官方只提供了 Windows、Android 和 iOS 平台的编译版本，如果要在其他平台中使用 OpenCV，就需要下载 OpenCV 源代码，然后进行编译。对于 Python 语言，可以使用 pip install opencv-Python 命令安装 cv2 模块，然后通过 cv2 模块中的 API 使用 OpenCV，非常简单。但 Java 却没有这么简单。

如果读者使用的是 Windows、Android 和 iOS 平台，那么非常幸运，因为这些平台已经将 OpenCV 编译好了，而且是静态编译，发布时只需要复制一些 jar 文件和对应的 dll 或 .so 文件即可，但其他平台（如 macOS、Linux 等），就需要自己编译了，而且需要设置一些编译选项，才能生成相应的 jar 文件。所以本节会详细介绍如何在 macOS、Linux 等平台手动编译 OpenCV 源代码，以及发布 OpenCV 应用。

7.1.1 编译 OpenCV 源代码

目前 OpenCV 只提供了 Windows、Android 和 iOS 的编译版本，所以如果读者使用的是 Windows 平台，那么可以忽略本节的内容，因为 Windows 版本的 OpenCV 不需要编译，可以直接使用。

如果读者使用的是 macOS 或 Linux，那么可以通过 https://opencv.org/releases 进入下载页面，如图 7-1 所示。

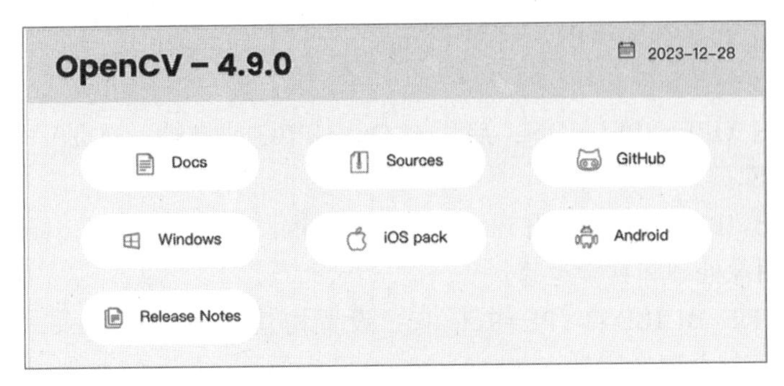

图 7-1　OpenCV 下载页面

目前 OpenCV 的最新版本是 4.9.0，可能读者在阅读本书时有更新的 OpenCV 版本，读者也可以使用它们，一般并不影响运行本书提供的案例。但建议读者使用 OpenCV 4.8.0 或以上的版本，因为这些版本提供了更多的新特性，也能尽可能保证本书提供的源代码能顺利运行。

单击 Sources 下载 OpenCV 的源代码。OpenCV 源代码是一个 ZIP 压缩文件，将其解压即可。OpenCV 不支持直接在 OpenCV 根目录编译，所以需要建立一个新目录，然后编译。读者可以建立一个与 OpenCV 根目录平级的目录，由于本节使用的是静态编译，所以可以建立一个 build-4.9.0-static 目录，然后从终端进入该目录。

接下来在终端输入如下的命令（仅适用于 macOS 和 Linux）进行配置检查，并生成一些编译必需的文件。

```
cmake \
-DBUILD_SHARED_LIBS=OFF \
-DWITH_FFMPEG=ON \
-DWITH_OPENJPEG=ON \
-DWITH_IPP=OFF \
-DCMAKE_BUILD_TYPE=Release \
-DBUILD_EXAMPLES=ON \
-DBUILD_opencv_Python2=OFF \
-DBUILD_opencv_Python3=ON \
-DBUILD_JAVA=ON \
-DJAVA_INCLUDE_PATH=$JAVA_HOME/include \
-DJAVA_INCLUDE_PATH2=$JAVA_HOME/include/darwin \
-DJAVA_AWT_INCLUDE_PATH=$JAVA_HOME/include \
-DBUILD_JS=ON \
-DCMAKE_INSTALL_PREFIX=/usr/local \
-DBUILD_OPENEXR=ON \
../opencv-4.9.0
```

命令的详细解释如下。

（1）-DBUILD_SHARED_LIBS=OFF：禁用共享库（动态链接库）的构建，意味着 OpenCV 将以静态库的形式编译。

（2）-DWITH_FFMPEG=ON：启用 FFmpeg 支持。FFmpeg 是一个用于处理视频和音频数据的库。

（3）-DWITH_OPENJPEG=ON：启用对 OpenJPEG 的支持，这是一个用于处理 JPEG 2000 格式的开源库。

（4）-DWITH_IPP=OFF：禁用 Intel Integrated Performance Primitives（IPP）的支持。IPP 用于高性能图像处理和数学运算。

（5）-DCMAKE_BUILD_TYPE=Release：设置构建类型为 Release，这意味着生成的库会被优化并且不包含调试信息。

（6）-DBUILD_EXAMPLES=ON：构建 OpenCV 示例。这将编译 OpenCV 提供的示例程序。

（7）-DBUILD_opencv_java2=OFF 和 -DBUILD_opencv_java3=ON：分别禁用 Python 2 的支持并启用 Python 3 的支持。这将构建 OpenCV 的 Python 3 绑定。

（8）-DBUILD_JAVA=ON：启用 Java 支持。这将构建 OpenCV 的 Java 绑定。

（9）-DBUILD_JS=ON：启用 JavaScript 支持。这可能涉及 OpenCV.js 的构建。

（10）-DCMAKE_INSTALL_PREFIX=/usr/local：指定安装目录。编译完成后，make install 将安装 OpenCV 到这个目录。

（11）-DBUILD_OPENEXR=ON：启用对 OpenEXR 的支持，这是一种高动态范围（HDR）图像文件格式。

（12）../opencv-4.9.0：指定 OpenCV 源代码的路径。

下面 3 个选项用于设置 Java 的相关目录，如果是 macOS 平台，应该指定 DJAVA_INCLUDE_PATH2。

```
-DJAVA_INCLUDE_PATH=$JAVA_HOME/include
-DJAVA_INCLUDE_PATH2=$JAVA_HOME/include/darwin
-DJAVA_AWT_INCLUDE_PATH=$JAVA_HOME/include
```

cmake 默认是动态编译，会生成一大堆库文件（macOS 是 .dylib 文件，Linux 是 .so 文件），所以建议静态编译 OpenCV。

执行上面的命令很快就会结束，会输出一大堆信息，由于本书只关注 Java，所以如果发现类似下面的输出信息，说明已经成功配置了 Java。否则编译 OpenCV 时不会生成 Java 相关的文件。

```
--   Java:
--     ant:                              /System/Volumes/Data/sdk/apache/apache
-ant-1.9.16/bin/ant (ver 1.9.16)
--     Java:                              YES
--     JNI:                              /Users/lining/Library/Java/JavaVirtualMachines/
openjdk-21.0.1/Contents/Home/include /Users/lining/Library/Java/JavaVirtualMachines/
openjdk-21.0.1/Contents/Home/include/darwin /Users/lining/Library/Java/
JavaVirtualMachines/openjdk-21.0.1/Contents/Home/include
```

```
--      Java wrappers:                YES (ANT)
--      Java tests:                   YES
```

如果这些信息都是 NO，那么需要进行配置。主要是执行 Apache Ant 和 Java 的相关路径。需要下载 Apache Ant，然后将 Apache Ant 的 bin 目录添加到 PATH 环境变量中。并设置 JAVA_HOME 环境变量。如果 Java wrappers 和 Java tests 都是 YES，那么在编译目录中会生成一个 java_test 子目录，在 java_test/bin 目录中会生成一个 opencv-490.jar 文件。这是 Java 使用 OpenCV 时需要引用的 jar 文件。

使用 cmake 生成完必要的文件后，可以执行 make 命令编译 OpenCV，成功编译 OpenCV 的效果如图 7-2 所示。

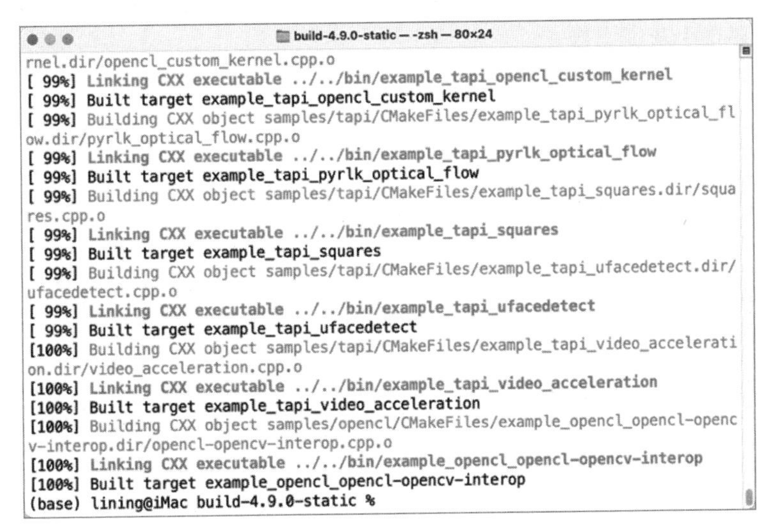

图 7-2　成功编译 OpenCV

我们要注意，这个过程比较漫长，大概需要 1 个小时。编译完后，可以直接使用编译好的库文件，也可以执行 sudo make install 命令，将编译好的 OpenCV 安装到 -DCMAKE_INSTALL_PREFIX 指定的安装目录中（本例是 /usr/local）。

如果读者希望提高 OpenCV 的编译速度，可以使用 **make -j4** 命令，这是因为 -j4 参数让 make 使用 4 个并行作业来进行编译（也可以使用更多的并行作业，如 make -j8）。这样做可以显著减少编译所需的时间，特别是在多核心的处理器上。

如果读者使用的是 macOS，那么编译后，会生成如下两个主要文件：

- build-4.9.0-static/java_test/bin/opencv-490.jar
- build-4.9.0-static/lib/libopencv_java490.dylib

如果是 Linux，那么 .dylib 文件会变成 .so 文件。

Java 使用 OpenCV 时只需要引用 opencv-490.jar 文件，并指定 libopencv_java490.dylib 文件的路径即可。

7.1.2 在 Java 中使用 OpenCV

在 Java 中使用 OpenCV，需要引用 opencv-490.jar 文件。如果读者使用的是 IntelliJ IDEA，可以将该文件复制到 IntelliJ IDEA 工程中，然后引用该文件。最后在运行当前工程的配置中添加如下的 VM options，这是因为 opencv-490.jar 只是对 OpenCV 的封装，本质上，opencv-490.jar 仍然需要调用 libopencv_java490.dylib 文件中的相关 API，所以需要使用下面的 VM options 指定 libopencv_java490.dylib 文件所在的路径。

```
-Djava.library.path=/System/Volumes/Data/sdk/opencv/build-4.9.0-static/lib
```

下面的例子演示了在 Java 中使用 OpenCV 的基本步骤。这个例子将彩色图像转换为灰度图像，并保存这个灰度图像。

代码位置：src/main/java/image/video/OpenCVExample.java

```java
package image.video;
import org.opencv.core.Core;
import org.opencv.core.Mat;
import org.opencv.imgcodecs.Imgcodecs;
import org.opencv.imgproc.Imgproc;
public class OpenCVExample {
    static {
        // 装载 libopencv_java490.dylib 文件
        System.loadLibrary("opencv_java490");
    }
    public static void main(String[] args) {
        // 读取图片文件
        Mat image = Imgcodecs.imread("screenshot.png");
        if (image.empty()) {
            System.out.println("图片读取失败！");
            return;
        }
        // 将图片转换为灰度图像
        Mat grayImage = new Mat();
        Imgproc.cvtColor(image, grayImage, Imgproc.COLOR_BGR2GRAY);
        // 保存结果
        Imgcodecs.imwrite("gray_image.jpg", grayImage);
        System.out.println("图像处理完成！");
    }
}
```

运行程序之前，要确保当前目录存在 screenshot.png 文件。程序运行后，会在当前目录生成一个 gray_image.jpg 文件。

7.1.3 发布 OpenCV 应用

如果 OpenCV 是静态编译的，那么就容易得多。本节会详细解释如何发布基于静态编译的 OpenCV 应用。

对于 Windows 来说，OpenCV 官方提供了编译好的版本。假设读者使用的是 64 位的 Windows，那么需要如下几个文件：

- <OpenCV 根目录 >\build\java\opencv-490.jar
- <OpenCV 根目录 >\build\java\x64\opencv_java490.dll
- <OpenCV 根目录 >/build/bin/opencv_videoio_ffmpeg490_64.dll

如果使用的是 32 位 Windows 系统，需要使用下面的文件：

```
<OpenCV 根目录 >\build\java\x86\opencv_java490.dll
```

发布时只需要这 3 个文件，其中 opencv-490.jar 是 Java 对 OpenCV 的封装，在 Java 中直接引用这个 jar 文件即可，然后将 opencv_java490.dll 和 opencv_videoio_FFmpeg490_64.dll 复制到系统目录或其他目录，并使用 -Djava.library.path 指定该目录即可。

对于 macOS 和 Linux 系统，如果是静态编译，也非常简单。编译完成，会在 build-4.9.0-static/lib 目录生成一个 libopencv_java490.dylib（macOS）或 libopencv_java490.so（Linux）文件。build-4.9.0-static/java_test/bin 目录会生成一个 opencv-490.jar 文件。只需要发布这两个文件即可。OpenCV 内部需要使用 FFmpeg，所以在发布时，仍然需要同时发布 FFmpeg 的库。读者可以下载 FFmpeg 源代码，直接静态编译。或者使用相关的命令下载 FFmpeg，如 macOS 可以使用 sudo brew install FFmpeg 进行安装。

7.2　获取视频信息

本节会使用 JavaCV 的相关 API 获取 MP4 视频文件的信息。JavaCV 是一个开源的 Java 接口库，它提供了对常用的计算机视觉和音视频处理库的封装，如 OpenCV、FFmpeg、libdc1394、PGR FlyCapture、OpenKinect、videoInput 等。JavaCV 使得 Java 开发者能够在不直接使用复杂本地接口的情况下，访问这些库的功能。这对于需要进行图像处理、视频处理或计算机视觉任务的 Java 项目特别有用。

JavaCV 的主要功能如下。

（1）图像处理和计算机视觉（OpenCV）：JavaCV 通过封装 OpenCV 提供了强大的图像处理和计算机视觉功能。功能包括但不限于图像的读取、写入、转换、特征检测、目标跟踪、图像分割、3D 重建、机器学习工具等。

（2）视频和音频处理（FFmpeg）：JavaCV 通过集成 FFmpeg 库提供了全面的音视频处理能力。支持多种音视频格式的读写，音视频流的编码和解码、转码、视频抓取等。

（3）摄像头和视频输入（libdc1394, PGR FlyCapture, OpenKinect, videoInput）：JavaCV 可以与多种视频采集设备接口，如 USB 摄像头、FireWire 摄像头等。支持从这些设备捕获实时视频流。

（4）实时视频处理和分析：JavaCV 可以实时处理和分析视频流，这对于视频监控、实时目标跟踪等应用场景非常有用。

（5）机器学习和人工智能：JavaCV 整合了 OpenCV 的机器学习模块，支持各种算法，如 SVM、决策树、神经网络等。

注意，由于 JavaCV 只是对本地库的封装，所以在使用 JavaCV 之前，要先安装本地库，如本节的例子，需要安装 FFmpeg。

JavaCV 在 GitHub 上的主页如下：

```
https://github.com/bytedeco/javacv
```

本节的例子会使用 FFmpegFrameGrabber 类的相关 API 获取视频文件的信息。FFmpegFrameGrabber 类是 JavaCV 库的一部分，它提供了一个接口，用于在 Java 程序中使用 FFmpeg 功能。FFmpeg 是一个非常强大的开源多媒体处理工具，支持大多数音频和视频格式。FFmpegFrameGrabber 类的作用是从视频文件、网络流或摄像头中抓取（grab）帧，并获取相关的信息。

下面是对 FFmpegFrameGrabber 类主要功能的详细解释。

（1）实例化：使用视频文件路径、网络流地址或摄像头索引来创建 FFmpegFrameGrabber 的实例。

（2）开始和停止：start() 方法在开始抓取帧之前调用，用于初始化抓取器。stop() 方法在完成抓取后调用，用于释放资源。

（3）获取视频属性：getImageWidth() 方法和 getImageHeight() 方法获取视频的宽度和高度。getLengthInTime() 方法获取视频的总长度（通常以微秒为单位）。getFrameRate() 方法获取视频的帧率。getAudioChannels() 方法用于获取媒体文件中音频流的通道数。getSampleRate() 方法用于获取音频流的采样率。

（4）帧抓取：grab() 方法抓取下一帧。返回的 Frame 对象可以是视频帧、音频样本或其他类型的帧。

（5）处理异常：FFmpegFrameGrabber 可能会抛出异常，需要进行相应的异常处理。

FFmpegFrameGrabber 类依赖 FFmpeg 工具，所以在使用 FFmpegFrameGrabber 类时，通常安装 FFmpeg 工具。

在使用 JavaCV 之前，要安装 JavaCV。如果读者使用 maven 管理依赖，需要在 pom.xml 中添加如下的依赖。

```
< dependencies>
    <dependency>
        <groupId>org.bytedeco</groupId>
        <artifactId>javacv-platform</artifactId>
        <version>1.5.9</version>
    </dependency>
</dependencies>
```

如果觉得下载依赖库比较慢，可以改成阿里或其他国内的下载源。在 pom.xml 文件中添加如下下载源即可：

```
<repositories>
    <repository>
        <id>alimaven</id>
        <name>Aliyun Central</name>
        <url>https://maven.aliyun.com/repository/central</url>
        <releases>
            <enabled>true</enabled>
        </releases>
        <snapshots>
            <enabled>true</enabled>
        </snapshots>
    </repository>
</repositories>
```

下面的代码使用 FFmpegFrameGrabber 类的相关 API 获取了 video.mp4 文件的分辨率、持续时间和帧率，并将这些信息输出到终端。

代码位置： src/main/java/image/video/VideoInfo.java

```
package image.video;
import org.bytedeco.javacv.FFmpegFrameGrabber;
import org.bytedeco.javacv.FrameGrabber.Exception;
public class VideoInfo {
    public static void main(String[] args) {
        String videoFilePath = "video.mp4";              // 确保这是视频文件的正确路径
        FFmpegFrameGrabber grabber = new FFmpegFrameGrabber(videoFilePath);
        try {
            grabber.start();
            // 获取视频分辨率
            int width = grabber.getImageWidth();
            int height = grabber.getImageHeight();
            // 获取视频持续时间（单位：微秒）
            long duration = grabber.getLengthInTime();
            // 获取视频帧率
            double frameRate = grabber.getFrameRate();
            // 打印信息
            System.out.println("分辨率：" + width + "x" + height);
            System.out.println("持续时间：" + (duration / 1_000_000) + "秒");
            System.out.println("帧率：" + frameRate);
            grabber.stop();
        } catch (Exception e) {
            e.printStackTrace();
        }
    }
}
```

运行程序，会在终端输出如下信息：

```
分辨率： 400x400
持续时间： 5 秒
帧率： 30.0
```

7.3 播放视频

本节会使用 JavaFX 的相关 API 播放 MP4 视频文件。在这个例子中会使用 JavaFX 框架中的 MediaPlayer 类和 MediaView 类，这两个类的主要功能如下。

（1）MediaPlayer 类：负责加载和控制媒体文件的播放。

（2）MediaView 类：用于在界面上显示 MediaPlayer 的视频输出。

加载、播放、暂停和停止视频的方式如下。

（1）打开视频文件：创建 Media 对象，并将视频文件名传入 Media 类的构造方法。然后创建 MediaPlayer 对象，将 Media 对象传入 MediaPlayer 类的构造方法，最后，将 MediaPlayer 与 MediaView 绑定。

（2）播放视频：调用 MediaPlayer. play() 方法开始播放视频。

（3）暂停视频：调用 MediaPlayer.pause() 方法暂停视频播放。

（4）停止视频：调用 MediaPlayer.stop() 方法停止视频播放。

下面的代码使用 JavaFX 创建了一个窗口，并在窗口的下方放置 4 个按钮：打开、播放、暂停和停止。分别用来打开视频文件、播放视频、暂停视频和停止视频。单击“打开”按钮，会弹出一个打开文件对话框，选择视频文件后，就会打开选中的视频文件，然后在按钮上方会显示视频的第一帧画面，单击“播放”按钮，开始播放视频。

代码位置：src/main/java/image/video/VideoPlayer.java

```java
package image.video;

import javafx.application.Application;
import javafx.stage.Stage;
import javafx.stage.FileChooser;
import javafx.scene.Scene;
import javafx.scene.layout.BorderPane;
import javafx.scene.layout.HBox;
import javafx.geometry.Pos;
import javafx.geometry.Insets;
import javafx.scene.control.Button;
import javafx.scene.media.Media;
import javafx.scene.media.MediaPlayer;
import javafx.scene.media.MediaView;
import java.io.File;
```

```
public class VideoPlayer extends Application {
    private MediaPlayer mediaPlayer;
    private MediaView mediaView;
    @Override
    public void start(Stage primaryStage) {
        // 创建 BorderPane 布局
        BorderPane root = new BorderPane();
        // 创建 MediaView 组件
        mediaView = new MediaView();
        mediaView.setPreserveRatio(true);
        root.setCenter(mediaView);
        // 创建按钮并设置事件处理
        Button openButton = new Button("打开");
        Button playButton = new Button("播放");
        Button pauseButton = new Button("暂停");
        Button stopButton = new Button("停止");
        // 水平布局用于放置按钮
        final HBox controlBox = new HBox(10, openButton, playButton, pauseButton,
stopButton);
        controlBox.setAlignment(Pos.CENTER);                       // 设置按钮水平居中
        controlBox.setPadding(new Insets(10, 10, 10, 10));    // 设置边距
        root.setBottom(controlBox);
        // 打开文件事件处理
        openButton.setOnAction(e -> {
            FileChooser fileChooser = new FileChooser();
            FileChooser.ExtensionFilter filter = new FileChooser.ExtensionFilter
("选择一个视频文件 (*.mp4)", "*.mp4");
            fileChooser.getExtensionFilters().add(filter);
            File file = fileChooser.showOpenDialog(primaryStage);
            if (file != null) {
                Media media = new Media(file.toURI().toString());
                if (mediaPlayer != null) {
                    mediaPlayer.stop();
                }
                mediaPlayer = new MediaPlayer(media);
                mediaView.setMediaPlayer(mediaPlayer);
                // 更新 MediaView 的尺寸绑定
                updateMediaViewSizeBindings(primaryStage, controlBox);
            }
        });
        // 播放视频事件处理
        playButton.setOnAction(e -> {
            if (mediaPlayer != null) {
                mediaPlayer.play();
            }
        });
        // 暂停视频事件处理
        pauseButton.setOnAction(e -> {
            if (mediaPlayer != null) {
                mediaPlayer.pause();
            }
```

```
        });
        // 停止视频事件处理
        stopButton.setOnAction(e -> {
            if (mediaPlayer != null) {
                mediaPlayer.stop();
            }
        });
        // 创建场景并设置舞台
        Scene scene = new Scene(root, 800, 600);
        primaryStage.setTitle(" 视频播放器 ");
        primaryStage.setScene(scene);
        primaryStage.show();
    }
    // 更新 MediaView 的尺寸绑定
    private void updateMediaViewSizeBindings(Stage stage, HBox controlBox) {
        mediaView.fitWidthProperty().bind(stage.widthProperty());
mediaView.fitHeightProperty().bind(stage.heightProperty().subtract(controlBox.
heightProperty().add(10)));
    }
    public static void main(String[] args) {
        launch(args);
    }
}
```

运行程序，打开一个视频，会看到如图 7-3 所示的效果。

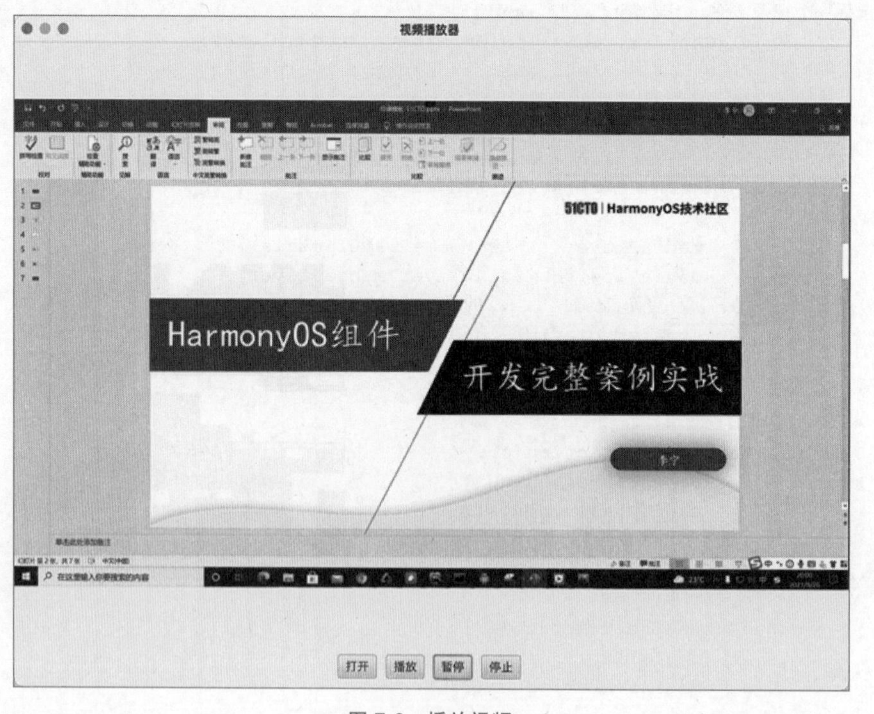

图 7-3　播放视频

注意，由于本例使用了 JavaFX 的多媒体部分，所以需要添加如下的 VM 选项：

```
--add-modules javafx.controls,javafx.media
```

7.4 截屏

本节详细介绍了如何使用 Java 截取屏幕以及 Web 页面。

7.4.1 截取屏幕

本节会使用 JavaFX 和 java.awt.Robot 实现一个截屏应用。JavaFX 的主要作用是创建截屏应用的界面，并使用下面的代码在截屏前和截屏后最小化窗口和恢复窗口。

```
// 最小化窗口
primaryStage.setIconified(true);
// 恢复窗口
primaryStage.setIconified(false);
```

java.awt.Robot 是 Java 编程语言中的一个类，它是 Java AWT（抽象窗口工具包）的一部分。Robot 类主要用于自动控制鼠标指针和键盘输入，以及捕捉屏幕上的图像。这个类提供了与图形用户界面交互的低级功能，使开发者能够编写代码来模拟用户输入和捕获屏幕内容。

Robot 类的主要功能如下。

（1）屏幕捕获：Robot 类可以对屏幕的某个区域进行截图，这是通过 createScreenCapture 方法实现的。此方法可以捕获屏幕上的任意矩形区域，并将其作为 BufferedImage 对象返回。这个功能在屏幕截图工具或者需要分析屏幕内容的软件中特别有用。

（2）模拟鼠标和键盘输入：Robot 类能够模拟鼠标移动、单击和键盘敲击等操作。通过 mouseMove、mousePress、mouseRelease、keyPress 和 keyRelease 方法，可以编写代码来自动执行这些动作。这对于自动化测试、游戏玩家的宏命令编写等场景非常实用。

（3）延迟设置：Robot 类还提供了 delay 方法，它可以在模拟输入操作之间设置延迟，以确保操作之间有足够的时间间隔。

使用 Robot 类时应注意如下几点。

（1）Robot 类操作的是整个系统的用户界面，而不仅仅是 Java 应用程序的窗口。因此，使用时需要注意对系统资源的影响。

（2）使用 Robot 类进行自动化测试时，最好确保测试环境是稳定且可预测的，因为屏幕上的任何变化都可能影响到测试结果。

（3）Robot 类在安全性方面也需要特别注意。因为它可以模拟真实用户的输入，所以在使用时应确保不会对用户的计算机安全或隐私造成威胁。

下面的例子使用 JavaFX 和 java.awt.Robot 类实现了一个用于截屏的应用，当单击“截屏”按钮后，窗口会最小化，然后截屏，将截屏图像保存为当前目录的 screenshot.png 文件，最后恢复窗口。

代码位置：src/main/java/image/video/Screenshot.java

```
package image.video;
import javafx.application.Application;
import javafx.scene.Scene;
import javafx.scene.control.Button;
import javafx.scene.layout.StackPane;
import javafx.stage.Stage;
import javax.imageio.ImageIO;
import java.awt.Rectangle;
import java.awt.Robot;
import java.awt.Toolkit;
import java.awt.image.BufferedImage;
import java.io.File;
public class Screenshot extends Application {
    private Stage primaryStage;
    @Override
    public void start(Stage primaryStage) {
        this.primaryStage = primaryStage;
        // 创建一个按钮，文本为 " 截屏 "，并充满整个窗口
        Button btnScreenshot = new Button(" 截屏 ");
        btnScreenshot.setMaxSize(Double.MAX_VALUE, Double.MAX_VALUE);
        // 设置按钮的单击事件
        btnScreenshot.setOnAction(event -> takeScreenshot());
        // 创建布局并添加按钮
        StackPane root = new StackPane();
        root.getChildren().add(btnScreenshot);
        // 设置场景和舞台
        Scene scene = new Scene(root, 200, 100);
        primaryStage.setTitle(" 截屏演示 ");
        primaryStage.setScene(scene);
        primaryStage.show();
    }
    // 定义一个截屏方法
    private void takeScreenshot() {
        try {
            // 最小化窗口
            primaryStage.setIconified(true);
            // 等待一小段时间以确保窗口已最小化
            Thread.sleep(500);
            // 使用 Robot 类进行截屏
            Robot robot = new Robot();
            Rectangle screenRect = new Rectangle(Toolkit.getDefaultToolkit().
getScreenSize());
```

```
            BufferedImage screenFullImage = robot.createScreenCapture(screenRect);
            // 保存截图到文件
            ImageIO.write(screenFullImage, "png", new File("screenshot.png"));
            // 恢复窗口
            primaryStage.setIconified(false);
        } catch (Exception ex) {
            ex.printStackTrace();
        }
    }
    public static void main(String[] args) {
        launch(args);
    }
}
```

运行程序，会显示如图 7-4 所示的窗口，单击“截屏”按钮，会最小化窗口，然后将截屏的图像保存为 screenshot.png，最后会恢复窗口。

图 7-4　截屏

7.4.2　截取 Web 页面

使用 Selenium WebDriver for Java 可以只截取打开的 Web 页面。Selenium 是一个广泛使用的开源软件，主要用于自动化网页浏览器的操作。它是自动化测试领域的一个重要工具，尤其是在 Web 应用的测试中。Selenium 提供了一套工具和库，允许开发者编写代码来自动控制浏览器，模拟用户在浏览器中的行为。

Selenium 的主要组件如下。

（1）Selenium WebDriver：WebDriver 是 Selenium 项目的核心部分，提供了一套编程接口（API），允许使用多种编程语言（如 Java、C#、Python、Ruby）来创建和执行自动化测试脚本。它直接与浏览器交互，支持多种浏览器，包括 Chrome、Firefox、Safari、Edge 等。

（2）Selenium IDE：一个浏览器扩展，可以录制、编辑和回放用户在浏览器中的操作。这对于快速创建测试脚本非常有用，尤其是对于不熟悉编程的用户。

（3）Selenium Grid：用于同时在多个环境（不同的浏览器和不同的操作系统）上运行测试脚本，可以大大提高测试的效率和覆盖率。它支持分布式测试执行，能够将测试任务分发到多个物理或虚拟机上。

Selenium 的主要特点如下。

（1）跨浏览器兼容性：支持目前所有主流的 Web 浏览器。

（2）多语言支持：提供了多种编程语言的接口，包括 Java、Python、C#、Ruby、JavaScript。

（3）广泛的社区支持：拥有一个活跃的社区和大量的文档、教程、指南。

（4）灵活性和可扩展性：可以很容易地集成到现有的测试框架中，如 JUnit、TestNG、PyTest 等。

（5）支持复杂的 Web 应用测试：可以处理弹窗、对话框、Ajax 请求、页面跳转等复杂场景。

Selenium 的使用场景如下。

（1）自动化测试：开发和测试团队使用 Selenium 来编写和执行针对 Web 应用的自动化测试脚本。

（2）爬虫和数据抓取：可以用于自动化获取网页内容，特别是在涉及 JavaScript 交互的动态网页中。

（3）自动化任务：自动执行一些重复的 Web 浏览任务，如自动化表单填写、网站监控等。

总的来说，Selenium 作为一个功能强大的 Web 浏览器自动化工具，其在测试自动化、质量保证和 Web 数据抓取方面有着广泛的应用。

本节的例子会使用 Selenium 打开 Chrome 浏览器，并在 Chrome 浏览器中打开京东商城首页，然后截取打开的页面，并保存为 jd.png，最后关闭 Chrome 浏览器。完成这些工作的主要步骤如下。

（1）设置 WebDriver 路径：使用 System.setProperty("webdriver.chrome.driver", "path/to/chromedriver"); 设置 WebDriver 驱动的路径，WebDriver 驱动是一个可执行程序。这行代码设置了系统属性 webdriver.chrome.driver，告诉 Selenium WebDriver ChromeDriver 的位置。这是必要的步骤，因为 WebDriver 需要知道如何启动和控制 Chrome 浏览器。

（2）初始化 WebDriver：使用 WebDriver driver = new ChromeDriver(); 创建了一个新的 ChromeDriver 实例，从而启动了一个新的 Chrome 浏览器窗口。ChromeDriver 是 WebDriver 接口的实现，用于控制 Chrome 浏览器。

（3）打开网页：使用 driver.get("https://www.jd.com/"); 打开指定的 URL（在本例中是京东的首页）。这个方法会导航到提供的网址，并等待页面加载完成。

（4）截取屏幕截图：使用 File screenshot = ((TakesScreenshot)driver).getScreenshotAs(OutputType.FILE); 捕获屏幕截图。getScreenshotAs 方法以文件形式获取当前页面的屏幕截图。

（5）保存截图到文件：使用 Apache Commons IO 库的 FileUtils.copyFile() 方法将截图文件保存到指定的位置（在本例中是当前目录下的 jd.png 文件）。

（6）资源清理：在 finally 块中，确保无论如何都会执行 quit() 方法来关闭浏览器并释放资源。这是自动化脚本中的重要步骤，确保不会留下未关闭的浏览器窗口。

本例完整的实现代码如下。

代码位置：src/main/java/image/video/CaptureWeb.java

```
package image.video;
import org.openqa.selenium.WebDriver;
import org.openqa.selenium.chrome.ChromeDriver;
import org.openqa.selenium.TakesScreenshot;
import org.openqa.selenium.OutputType;
import org.apache.commons.io.FileUtils;
import java.io.File;
import java.io.IOException;
public class CaptureWeb {
    public static void main(String[] args) {
        System.setProperty("webdriver.chrome.driver", "webdrivers/chromedriver");
        WebDriver driver = new ChromeDriver();
        try {
            driver.get("https://www.jd.com/");
            // 强转 WebDriver 为 TakesScreenshot
            File screenshot = ((TakesScreenshot)driver).getScreenshotAs
(OutputType.FILE);
            FileUtils.copyFile(screenshot, new File("jd.png"));
        } catch (IOException e) {
            e.printStackTrace();
        } finally {
            driver.quit();
        }
    }
}
```

在运行程序之前，需要完成下面两个工作。

（1）安装依赖。

由于本例使用了 Selenium WebDriver for Java 和 Apache Commons IO，所以需要在 pom.xml 文件中按如下形式添加相关库的依赖。

```
<dependency>
    <groupId>org.seleniumhq.selenium</groupId>
    <artifactId>selenium-java</artifactId>
    <version>4.16.1</version>
</dependency>
<dependency>
    <groupId>commons-io</groupId>
    <artifactId>commons-io</artifactId>
    <version>2.11.0</version>
</dependency>
```

（2）下载 Google Chrome 的驱动程序。

本例通过 Selenium 调用了 Google Chome 浏览器，所以需要下载 Selenium WebDriver。读者可以从下面的页面下载 Google Chrome 的驱动程序。

```
https://chromedriver.chromium.org/downloads
```

如果下载最新的 Google Chrome 驱动程序，可以从下面的页面下载。

```
https://googlechromelabs.github.io/chrome-for-testing/
```

下载完 Chrome 驱动程序，需要将 Chrome 驱动程序（一个可执行文件，读者可以根据自己使用的操作系统下载不同版本的 Chrome 驱动程序）复制到当前目录的 webdrivers 子目录中。

注意： Chrome 驱动程序版本要与 Chrome 浏览器的版本相同，否则无法启动 Chrome 浏览器。这个版本通常主要是大版本相同即可，例如，Chrome 浏览器版本是 120.0.6099.129，Chrome 驱动程序使用 120.0.6099.109 也可以访问 Chrome 浏览器。

运行程序，Chrome 浏览器会自动打开，然后自动填入网址并显示页面，接下来会截取网页，最后会关闭浏览器。在当前目录会看到生成了一个 jd.png 文件，效果如图 7-5 所示。

图 7-5 截取网页

7.5　拍照

使用 OpenCV 的 VideoCapture 类，可以从摄像头中采集图像，并使用其他 API 保存采集的图像，因此，使用 VideoCapture 类可以实现一个拍照程序。

下面是使用 VideoCapture 类实现拍照程序的主要实现原理和步骤。

（1）加载 OpenCV 库：

```
static {
   System.loadLibrary("opencv_java490");
}
```

这段静态代码块在类加载时执行，用于加载 OpenCV 的 Java 库。这里加载的是 opencv_java490，这通常对应于 OpenCV 4.9.0 版本的 Java 绑定。

（2）创建视频捕获对象：

```
VideoCapture cap = new VideoCapture(0);
```

这行代码创建了一个 VideoCapture 对象，用于从摄像头捕获视频。传入的参数 0 表示使用计算机的默认摄像头，如果计算机有多个摄像头，可以分别设置为 0、1、2、…、*n* 来切换不同的摄像头。

（3）检查摄像头状态：

```
if (!cap.isOpened()) {
   System.out.println("Error: Camera not accessible.");
   return;
}
```

这里检查摄像头是否成功打开。如果无法访问摄像头，程序会打印错误信息并终止。

（4）图像帧的处理：

```
Mat frame = new Mat();
```

创建一个 Mat 对象用来存储从摄像头读取的每一帧图像。

（5）显示视频帧：

```
org.opencv.highgui.HighGui.imshow(windowName, frame);
```

这行代码在名为 windowName 的窗口中显示捕获的视频帧。

（6）键盘输入处理：

```
int keyCode = org.opencv.highgui.HighGui.waitKey(100) & 0xff;
```

waitKey(100) 函数等待 100 毫秒的键盘输入，返回按键的 ASCII 码。& 0xff 确保结果是一个有效的字符。

（7）保存图片和退出循环：

```
if (keyCode == 83 || keyCode == 115) {
    Imgcodecs.imwrite("photo.jpg", frame);
    System.out.println("Photo saved.");
} else if(keyCode == 81 || keyCode == 113){
    break;
}
```

如果按下 's'（小写或大写），则保存当前帧为图片。如果按下 'q'（小写或大写），则退出循环。

（8）释放资源：

```
cap.release();
org.opencv.highgui.HighGui.destroyAllWindows();
```

最后，释放摄像头资源并关闭所有 OpenCV 创建的窗口。

整个程序的流程是不断从摄像头读取图像帧，将它们显示在窗口中，并检测用户的键盘输入。根据输入执行相应的操作，如保存图片或退出程序。

下面的代码是本例的完整实现。

代码位置：src/main/java/image/video/CameraCapture.java

```
package image.video;

import org.opencv.core.Core;
import org.opencv.core.Mat;
import org.opencv.videoio.VideoCapture;
import org.opencv.imgcodecs.Imgcodecs;

public class CameraCapture {
    static {
        System.loadLibrary("opencv_java490");
    }
    public static void main(String[] args) {
        // 创建 VideoCapture 对象来捕获视频，参数 0 表示使用默认摄像头
        VideoCapture cap = new VideoCapture(0);
        // 检查摄像头是否打开成功
        if (!cap.isOpened()) {
            System.out.println("Error: Camera not accessible.");
            return;
        }
        // 创建 Mat 对象来存储每一帧图像
        Mat frame = new Mat();
        // 设置窗口
```

```
        String windowName = "Camera";
        org.opencv.highgui.HighGui.namedWindow(windowName);
        while (true) {
            // 从摄像头读取一帧
            if (cap.read(frame)) {
                // 显示图像
                org.opencv.highgui.HighGui.imshow(windowName, frame);
                // 等待键盘输入
                int keyCode = org.opencv.highgui.HighGui.waitKey(100) & 0xff;
                // 获取键盘输入的 ASCII 码
                // 如果按下 's'，保存图片
                if (keyCode == 83 || keyCode == 115) {
                    Imgcodecs.imwrite("photo.jpg", frame);
                    System.out.println("Photo saved.");
                }
                // 如果按下 'q'，退出循环
                else if(keyCode == 81 || keyCode == 113){
                    break;
                }
            } else {
                System.out.println("Failed to read frame.");
            }
        }
        // 释放资源
        cap.release();
        org.opencv.highgui.HighGui.destroyAllWindows();
    }
}
```

运行程序，会显示如图 7-6 所示的图像采集窗口，按 s 键，会将当前采集的图像保存为 photo.jpg 文件。

图 7-6　图像采集窗口

7.6 制作录屏视频

要想同时录制屏幕和声音，最简单的方式是使用 FFmpeg 工具。FFmpeg 主要用于完成对音频和视频的各种处理。

如果直接用 FFmpeg 命令行工具录制带声音的视频，可以使用下面的命令。

```
ffmpeg -f avfoundation -framerate 30 -i "3:1" -vf scale=iw/2:ih/2 out.avi
```

其中 avfoundation 是 macOS 下的设备，用来捕获集成的 iSight 摄像头以及通过 USB 或 FireWire 连接的摄像头。

在 Windows 系统中，可以使用 dshow（DirectShow）输入设备，或者使用内置的 GDI 屏幕捕获器（gdigrab）。在 Linux 系统中，可以使用 video4linux2（或简称为 v4l2）输入设备来捕获实时输入，例如来自网络摄像头的输入。

上面命令行中的 30 表示帧率是 30 帧。"3:1" 表示视频设备和音频设备的索引。3 表示视频设备索引，1 表示音频设备索引。如果机器上有多个视频设备和多个音频设备，那么就需要使用下面的命令查询每个设备的索引。iw/2:ih/2 表示将屏幕尺寸缩小到原来的 50%，这一点非常重要，如果屏幕分辨率过大，FFmpeg 可能会不支持这么大分辨率的视频，所以需要等比例缩放屏幕尺寸。

- macOS

使用下面的命令查询 macOS 下的视频和音频设备：

```
ffmpeg -f avfoundation -list_devices true -i ""
```

如果执行这行命令，可能会输出如下内容：

```
[AVFoundation indev @ 0x7fefdce21e00] AVFoundation video devices:
[AVFoundation indev @ 0x7fefdce21e00] [0] USB 2.0 Camera
[AVFoundation indev @ 0x7fefdce21e00] [1] FaceTime 高清摄像头（内建）
[AVFoundation indev @ 0x7fefdce21e00] [2] EpocCam
[AVFoundation indev @ 0x7fefdce21e00] [3] Capture screen 0
[AVFoundation indev @ 0x7fefdce21e00] AVFoundation audio devices:
[AVFoundation indev @ 0x7fefdce21e00] [0] WeMeet Audio Device
[AVFoundation indev @ 0x7fefdce21e00] [1] Built-in Microphone
[AVFoundation indev @ 0x7fefdce21e00] [2] USB Digital Audio
```

方括号中的数字就是索引。本例使用的 "3:1"，表示视频使用了 Capture screen 0，音频使用了 Built-in Microphone（内建麦克风）。

- Windows

使用下面的命令查询 Windows 下的视频和音频设备：

```
ffmpeg -list_devices true -f dshow -i dummy
```

❑ Linux

使用下面的命令查询 Linux 下的视频和音频设备：

```
v4l2-ctl --list-devices
```

如果未安装 v4l2-ctl 命令，可以使用下面的命令安装：

```
sudo apt install v4l-utils
```

下面的例子使用 Java 生成并执行用于录制屏幕视频的 FFmpeg 命令行，按 Enter 键停止录制，并将视频文件保存为 output_video.avi 文件。

代码位置：src/main/java/image/video/ScreenRecorder.java

```
package image.video;
import java.io.IOException;
import java.util.Scanner;
public class ScreenRecorder {
    public static void main(String[] args) throws IOException, InterruptedException {
        String outputVideoName = "output_video.avi";
        String screenRecordCommand = "ffmpeg";
        // 使用 ProcessBuilder 并且正确分隔命令的每个部分
        ProcessBuilder processBuilder = new ProcessBuilder(
                screenRecordCommand, "-f", "avfoundation", "-framerate", "30",
"-i", "3:1",
                "-vf", "scale=iw/2:ih/2", outputVideoName
        );
        Process screenRecordProcess = processBuilder.start();
        System.out.println("按 Enter 键停止录制...");
        Scanner scanner = new Scanner(System.in);
        scanner.nextLine();
        screenRecordProcess.destroy();
        System.out.println("录制完成，视频保存为 " + outputVideoName);
    }
}
```

执行上面的程序，会立刻开始录制，按 Enter 键停止录制，output_video.avi 就是刚才录制的视频文件。

7.7　格式转换

本节会介绍如何使用 BufferedImage 类实现图像格式转换，以及使用 FFmpeg 转换视频格式。

7.7.1 图像格式转换

使用 BufferedImage 类可以实现常用图像格式的转换。BufferedImage 是一个包含图像数据的类。它可以加载图片的所有像素数据到内存中，允许程序读取、处理和写入图像数据。

图像转换的核心是 ImageIO.write 方法，该方法将 BufferedImage 对象的数据写入文件。原型如下。

```
public static boolean write(RenderedImage im, String formatName, File output)
throws IOException
```

参数含义如下。

（1）im：要写入的图像，通常是一个 BufferedImage 对象。

（2）formatName：图像的格式，如 "JPEG" "PNG" "GIF" 等。

（3）output：输出文件对象。

这里要着重介绍一下第（2）个参数 formatName，该参数指定图像的输出格式（字符串形式），这个字符串应与目标图像格式匹配。常见的格式如下。

（1）"JPEG" 或 "JPG"。

（2）"PNG"。

（3）"GIF"。

（4）"BMP"。

（5）"WBMP"。

formatName 参数值对大小写并不敏感，如 JPG、jpg 效果是一样的。这个参数决定了图像数据如何被编码和保存。例如，使用 "PNG" 会保存为 PNG 格式，而使用 "JPEG" 会保存为 JPEG 格式。

下面的例子通过命令行传入两个参数，分别表示源图像文件和目标图像文件，本例假设目标图像文件的格式是 JPG。运行程序，会将源图像文件转换为 JPG 格式的目标图像文件，如果图像文件存在，会要求确定是否覆盖目标文件，如果输入 Y，并按 Enter 键，则覆盖目标图像文件。

代码位置：src/main/java/image/video/ImageConverter.java

```
package image.video;
import javax.imageio.ImageIO;
import java.awt.image.BufferedImage;
import java.io.File;
import java.io.IOException;
import java.util.Scanner;
public class ImageConverter {
    public static void main(String[] args) {
        Scanner scanner = new Scanner(System.in);
        if (args.length < 2) {
```

```
                System.out.println("请提供源文件名和目标文件名。");
                return;
            }
            String sourceFile = args[0];
            String targetFile = args[1];
            convertImage(sourceFile, targetFile, scanner);
        }
        private static void convertImage(String sourceFile, String targetFile,
    Scanner scanner) {
            File target = new File(targetFile);
            // 检查目标文件是否存在
            if (target.exists()) {
                System.out.println("目标文件已存在，是否覆盖？(Y/N)");
                String overwrite = scanner.nextLine();
                if (!overwrite.equalsIgnoreCase("y")) {
                    return;
                }
            }
            // 打开源文件并转换格式
            try {
                BufferedImage image = ImageIO.read(new File(sourceFile));
                ImageIO.write(image, "jpg", target);
                // 这里假设目标格式为 jpg，根据需要可以修改
                System.out.println("转换成功！");
            } catch (IOException e) {
                System.out.println("无法转换文件。");
                e.printStackTrace();
            }
        }
    }
```

7.7.2　使用 FFmpeg 转换视频格式

使用 FFmpeg 可以转换常用的视频格式。FFmpeg 在转换视频文件格式时通常是通过输出文件的扩展名来确定视频的格式。FFmpeg 是一个非常强大的多媒体处理工具，它支持多种音频和视频格式。当使用 FFmpeg 命令行工具进行格式转换时，FFmpeg 会根据指定的输出文件的扩展名自动选择相应的编码器。

例如想将一个 AVI 文件转换为 MP4 文件，可能会使用以下命令：

```
ffmpeg -i video.avi -y video.mp4
```

其中，-i 指定输入文件，video.avi 表示输入文件（待转换文件），-y 选项表示覆盖输出文件，video.mp4 表示输出文件，如果文件存在，并指定了 -y，那么会覆盖已经存在的文件。

FFmpeg 能智能地处理这些扩展名和格式的映射，但用户也需要确保选择了正确的扩

展名以匹配他们期望的输出格式。在某些情况下，特别是在涉及不常见或专业格式时，可能需要显式指定所需的编码器和容器格式。

下面的例子将当前目录的 video.avi 文件转换为 video.mp4 文件。

代码位置：src/main/java/image/video/VideoConverter.java

```
package image.video;
import java.io.IOException;
public class VideoConverter {
    public static void main(String[] args) {
        String inputFileName = "video.avi";
        String outputFileName = "video.mp4";
        // 构造 FFmpeg 命令
        String ffmpegCommand = "ffmpeg -i " + inputFileName + " -y " +
outputFileName;
        // 使用 ProcessBuilder 运行 FFmpeg 命令
        ProcessBuilder processBuilder = new ProcessBuilder(ffmpegCommand.
split(" "));
        try {
            Process process = processBuilder.start();
            // 等待 FFmpeg 进程执行完毕
            int exitCode = process.waitFor();
            if (exitCode == 0) {
                System.out.println("转换完毕");
            } else {
                System.out.println("转换失败");
            }
        } catch (IOException | InterruptedException e) {
            e.printStackTrace();
        }
    }
}
```

在运行程序之前，确保当前目录存在 video.avi 文件，运行程序后，会在当前目录生成一个 video.mp4 文件。

7.8 视频编辑

本节会介绍如何对视频进行各种编辑操作，这些操作包括视频裁剪、视频合并、混合音频和视频、提取视频中的音频以及制作画中画视频。

7.8.1 视频裁剪

使用 FFmpeg 可以进行视频裁剪，可以按时间点进行裁剪，也可以按持续时间进行裁剪，下面是这两种裁剪方式的基本命令格式。

❑ 按时间点裁剪

使用“-ss”来指定裁剪的开始时间，使用“-to”来指定结束时间，命令行如下：

```
ffmpeg -i input.mp4 -ss 00:00:10 -to 00:00:20 -c:v copy -c:a copy output.mp4
```

这行命令将从 input.mp4 的第 10 秒开始裁剪，到第 20 秒结束。

❑ 按持续时间裁剪

使用“-ss”来指定开始时间，使用“-t”来指定从开始时间起的持续时间，命令行如下：

```
ffmpeg -i input.mp4 -ss 00:00:10 -t 10 -c:v copy -c:a copy output.mp4
```

这行命令将从 input.mp4 的第 10 秒开始，裁剪持续 10 秒的内容。

FFmpeg 在截取视频时，可以精确到更小的时间刻度。FFmpeg 允许指定精确到毫秒的时间点，这意味着可以精确到 1/1000 秒。

在 -ss（开始时间）和 -t（持续时间）或 -to（结束时间）参数中，可以使用小时 : 分钟 : 秒 . 毫秒的格式（例如 00:00:10.500 表示 10 秒零 500 毫秒）来指定这些时间点。

```
ffmpeg -i input.mp4 -ss 00:00:10.500 -t 2.250 -c:v copy -c:a copy output.mp4
```

这行命令将从 input.mp4 的第 10.500 秒（10 秒 500 毫秒）开始，持续 2.250 秒（2 秒 250 毫秒）。

```
ffmpeg -i input.mp4 -to 00:00:12.750 -c:v copy -c:a copy output.mp4
```

这行命令将从视频开始裁剪，一直到 12.750 秒（12 秒 750 毫秒）结束。

相关参数的详细解释如下。

（1）-i input.mp4：指定输入文件。

（2）-ss：指定从视频中的哪个时间点开始裁剪。时间格式可以是 hh:mm:ss[.xxx]，其中 hh 是小时，mm 是分钟，ss 是秒，.xxx 是小数秒（毫秒）。例如，-ss 00:00:10.500 表示从视频的 10 秒 500 毫秒处开始。

（3）-t：指定从 -ss 定义的开始时间点裁剪多长时间。时间格式同上，可以包括小数秒（毫秒）。例如，-t 2.250 表示从开始点裁剪 2 秒 250 毫秒的视频。

（4）-to：指定裁剪操作在视频的哪个时间点结束。时间格式同上，包括小时、分钟、秒和毫秒。例如，-to 00:00:15.750 表示在视频的 15 秒 750 毫秒处结束裁剪。

（5）-c:v copy 和 -c:a copy：指示 FFmpeg 使用与原视频相同的视频（v）和音频（a）编码器进行复制操作。这可以避免重新编码过程，节省时间并保持原始质量。

FFmpeg 命令中的 -ss（开始时间）和 -t（持续时间）或 -to（结束时间）参数的时间单位可以是小时、分钟、秒，以及小数形式的秒来表示毫秒。以下是这些参数的详细说明。

使用 FFmpeg 裁剪视频需要注意如下几点。

（1）确保使用的时间点在视频的总时长范围内。

（2）精确度可能受视频本身的帧率限制。例如，如果视频的帧率是每秒 30 帧，每一帧的时间间隔大约是 33.33 毫秒。这意味着裁剪时可能无法完全精确到单个毫秒。

（3）如果需要对视频进行精确的编辑，可能还需要考虑关键帧的位置。在非关键帧处开始裁剪可能导致视频开头有一小段空白或者跳跃。这可以通过重新编码（去掉 -c:v copy -c:a copy）来解决，但会增加处理时间并可能影响视频质量。

下面的例子使用 Java 执行 FFmpeg 命令裁剪 copilot.mp4 文件（从第 1 秒开始，裁剪 10 秒），将裁剪的结果保存为 clip_video.mp4 文件。

代码位置：src/main/java/image/video/VideoCutter.java

```
package image.video;
import java.io.BufferedReader;
import java.io.IOException;
import java.io.InputStreamReader;
public class VideoCutter {
    public static void main(String[] args) {
        // FFmpeg 命令，确保已经安装了 FFmpeg 并且添加到了系统路径
        String ffmpegCommand = "ffmpeg";
        // 输入视频文件路径
        String inputVideo = "copilot.mp4";
        // 输出视频文件路径
        String outputVideo = "clip_video.mp4";
        // 设置裁剪的开始时间和持续时间
        String startTime = "00:00:01";           // 开始时间
        String duration = "10";                  // 持续时间（例如，裁剪 10 秒）
        // 构建 FFmpeg 命令
        String[] command = {
                ffmpegCommand,
                "-i", inputVideo,
                "-ss", startTime,
                "-t", duration,
                "-c", "copy",
                outputVideo
        };
        // 执行命令
        ProcessBuilder processBuilder = new ProcessBuilder(command);
        try {
            Process process = processBuilder.start();
            // 读取命令的输出和错误流
            BufferedReader stdInput = new BufferedReader(new InputStreamReader
(process.getInputStream()));
            BufferedReader stdError = new BufferedReader(new InputStreamReader
(process.getErrorStream()));
            // 输出命令执行结果
            String s;
            while ((s = stdInput.readLine()) != null) {
                System.out.println(s);
```

```
            }

            // 输出错误信息
            while ((s = stdError.readLine()) != null) {
                System.err.println(s);
            }
            // 等待进程结束
            int exitVal = process.waitFor();
            if (exitVal == 0) {
                System.out.println("Success!");
                System.out.println("Output video at: " + outputVideo);
            } else {
                // 处理错误
                System.err.println("FFmpeg command failed.");
            }
        } catch (IOException | InterruptedException e) {
            e.printStackTrace();
        }
    }
}
```

7.8.2 视频合并

FFmpeg 同样也支持合并视频。不过合并视频要比视频裁剪复杂一些。首先，要检测所有待合并视频的尺寸，FFmpeg 只支持合并尺寸相同的视频，如果视频尺寸不同，那么就需要使用 FFmpeg 缩放相关的视频文件。通常以一个视频的尺寸为标杆，将其他视频的尺寸都缩放到与这个视频的尺寸完全相同。最后就是使用 FFmpeg 合并视频。下面是对这些步骤的详细描述。

1. 获取视频的尺寸

可以使用下面的命令获取视频文件的分辨率（宽度和高度）：

```
ffprobe -v error -select_streams v:0 -show_entries stream=width,height -of
csv=s=x:p=0 videoFile
```

参数含义如下。

（1）-v error：设置日志级别为 error，只显示错误信息。

（2）-select_streams v:0：选择第 1 个视频流。

（3）-show_entries stream=width,height：显示视频流的宽度和高度。

（4）-of csv=s=x:p=0：输出格式为 CSV，使用 x 作为宽度和高度的分隔符。

（5）videoFile：视频文件名。

FFprobe 和 FFmpeg 都是 FFmpeg 项目的一部分，它们在处理视频和音频文件方面有不同的用途和特点。FFprobe 是一个用于分析多媒体流的工具。它的主要功能是收集和显示有关视频、音频和其他多媒体流的信息，如格式、编解码器、持续时间、比特率、流大小、帧率、像素格式等。而 FFmpeg 是一个多功能的多媒体处理工具，主要用于视频和音频的录制、转换和流化。它可以用来编解码、转换格式、调整视频和音频的质量、合并或切割媒体流等。

2. 调整视频尺寸

可以使用下面的命令调整视频的尺寸：

```
ffmpeg -i inputFileName-vf scale=size -c:v libx264 -c:a aac newFileName
```

参数含义如下。

（1）-i inputFileName：输入视频文件名。

（2）-vf scale=size：视频过滤器，调整视频到指定的 size（尺寸）。

（3）-c:v libx264：设置视频编码器为 libx264（H.264 编码）。

（4）-c:a aac：设置音频编码器为 AAC。

（5）newFileName：输出的新视频文件名。

3. 合并视频

使用下面的命令可以合并视频文件：

```
ffmpeg -f concat -safe 0 -i tempFileList -c copy output.mp4
```

参数含义如下。

（1）-f concat：指定使用 concat（连接）过滤器。

（2）-safe 0：允许读取不安全的文件路径。

（3）-i tempFileList：输入文件列表，这里的 tempFileList 是一个文本文件，其中包含了需要合并的视频文件名，每个文件名前面要加 file 前缀，并且文件名需要用 '' 引起来，如 file 'video_path'。

（4）-c copy：复制音视频数据而不重新编码，这保证了音频和视频流在合并过程中保持原样。

在使用 FFmpeg 进行视频合并时，有两种主要方法来指定要合并的视频文件：一种是使用文本文件（如 tempFileList），另一种是直接在 FFmpeg 命令中指定要合并的视频文件。下面详细说明这两种方法。

- 使用文本文件指定视频文件（如前面提到的 tempFileList）

方法：创建一个文本文件，其中包含所有要合并的视频文件的列表，每个视频文件的路径写在一行，并以 file 开头，例如：

```
file 'video1.mp4'
file 'video2.mp4'
file 'video3.mp4'
```

这种方法尤其适用于视频文件数量较多或文件名较长的情况，因为它使得命令行操作更加简洁。

可以使用 -f concat 和 -i 参数引用这个文本文件，例如：

```
ffmpeg -f concat -safe 0 -i tempFileList.txt -c copy output.mp4
```

其中 tempFileList.txt 就是包含视频文件列表的文本文件。

❑ 直接在命令中指定要合并的视频文件

在 FFmpeg 命令中直接使用多个 -i 参数指定要合并的视频文件。这种方法在文件数量较少时比较方便。

```
ffmpeg -i video1.mp4 -i video2.mp4 -i video3.mp4 -filter_complex "[0:v][0:a][1:v]
[1:a][2:v][2:a]concat=n=3:v=1:a=1[v][a]" -map "[v]" -map "[a]" output.mp4
```

这里使用 -filter_complex 来处理多个视频流，concat 过滤器用于合并视频，n=3:v=1:a=1 表明有三个视频流和音频流需要合并。

下面的例子使用 Java 调用 FFmpeg 命令合并 videolist.txt 文件中的视频文件，为了方便，videolist.txt 文件中的视频文件并没有使用 file 前缀，而是直接指定了视频文件，如下所示：

```
copilot.mp4
video.mp4
out.mp4
```

程序会自动生成一个临时的文本文件用于指定 FFmpeg 合并文件时需要的格式（带 file 前缀的视频文件名），在使用完临时视频文件后，会删除这个临时文件。在需要修改视频文件尺寸时，也会将该视频文件复制一份，并修改副本的尺寸，合并时，会合并这个副本，最后会删除视频副本。

代码位置：src/main/java/image/video/VideoMerger.java

```java
package image.video;
import java.io.BufferedReader;
import java.io.FileReader;
import java.io.IOException;
import java.io.InputStreamReader;
import java.nio.file.Files;
import java.nio.file.Paths;
import java.util.ArrayList;
import java.util.List;
import java.util.UUID;
public class VideoMerger {
    public static void main(String[] args) {
        try {
            // 读取视频文件列表
```

```
            List<String> videoFiles = readVideoVideoList("videolist.txt");
            if (videoFiles.isEmpty()) {
                System.out.println(" 没有视频文件可以合并 ");
                return;
            }

            // 获取第一个视频的尺寸
            String size = getVideoSize(videoFiles.get(0));
            // 调整视频尺寸并获取新的文件名列表
            List<String> adjustedVideoFiles = adjustVideoSizes(videoFiles, size);
            // 合并视频
            mergeVideos(adjustedVideoFiles);
            // 删除临时视频文件
            deleteTemporaryFiles(adjustedVideoFiles, videoFiles);
        } catch (Exception e) {
            e.printStackTrace();
        }
    }
    private static List<String> readVideoVideoList(String filePath) throws IOException {
        List<String> videoFiles = new ArrayList<>();
        try (BufferedReader br = new BufferedReader(new FileReader(filePath))) {
            String line;
            while ((line = br.readLine()) != null) {
                videoFiles.add(line);
            }
        }
        return videoFiles;
    }
    private static String getVideoSize(String videoFile) throws IOException,
InterruptedException {
        String command = "ffprobe -v error -select_streams v:0 -show_entries
stream=width,height -of csv=s=x:p=0 " + videoFile;
        Process process = Runtime.getRuntime().exec(command);
        BufferedReader reader = new BufferedReader(new InputStreamReader
(process.getInputStream()));
        String size = reader.readLine();
        // 检查命令行是否执行成功
        int exitCode = process.waitFor();
        if (exitCode != 0) {
            // 如果命令执行失败，则返回 null
            return null;
        }
        return size;
    }
    private static List<String> adjustVideoSizes(List<String> videoFiles,
String size) throws IOException, InterruptedException {
        List<String> adjustedFiles = new ArrayList<>();
        for (String video : videoFiles) {
            String currentSize = getVideoSize(video);
            if (currentSize != null && !currentSize.equals(size)) {
                // 如果 currentSize 不为 null 且与目标尺寸不同，则调整视频尺寸
```

```
                String newFileName = "temp_" + UUID.randomUUID().toString() +
".mp4";
                String scaleCommand = "ffmpeg -i " + video + " -vf scale=" +
size + " -c:v libx264 -c:a aac " + newFileName;
                Runtime.getRuntime().exec(scaleCommand).waitFor();
                adjustedFiles.add(newFileName);
            } else {
                adjustedFiles.add(video);
            }
        }
        return adjustedFiles;
    }
    private static void mergeVideos(List<String> videoFiles) throws IOException,
InterruptedException {
        // 创建一个临时文件来存储用于合并的视频文件列表
        String tempFileList = "temp_videolist.txt";
        StringBuilder fileListBuilder = new StringBuilder();
        for (String video : videoFiles) {
            fileListBuilder.append("file '").append(video).append("'\n");
        }
        Files.write(Paths.get(tempFileList), fileListBuilder.toString().
getBytes());

        // 使用临时文件列表进行合并
        String mergeCommand = "ffmpeg -f concat -safe 0 -i " + tempFileList +
" -c copy output.mp4";
        Runtime.getRuntime().exec(mergeCommand).waitFor();

        // 合并完成后，删除临时文件列表
        Files.deleteIfExists(Paths.get(tempFileList));
    }
    private static void deleteTemporaryFiles(List<String> adjustedFiles,
List<String> originalFiles) throws IOException {
        for (String file : adjustedFiles) {
            if (!originalFiles.contains(file)) {
                Files.deleteIfExists(Paths.get(file));
            }
        }
    }
}
```

如果当前目录存在 videolist.txt 文件，并且该文件中指定的视频文件都存在，那么运行程序，会在当前目录生成一个合并后的 output.mp4 文件。

7.8.3　混合音频和视频

FFmpeg 支持混合音频和视频。本节的例子会使用 Java 调用 FFmpeg 命令行工具实现音频和视频混合的功能，在混合的过程中，还会输出混合的进度百分比。为了完善这个功能，首先需要获取待混合视频的总帧数，然后用当前已经处理的帧数与这个总帧数相除，就得到了当前处理的进度百分比。下面是实现这个例子的原理。

1. 获取视频总帧数

通过 getVideoTotalFrames 方法可以获取视频总帧数。该方法运行一个 FFmpeg 命令，该命令以 -i 参数打开视频文件，但不进行实际的处理。FFmpeg 会将视频的详细信息输出到标准错误流。然后该方法会从 FFmpeg 的输出中解析视频的总持续时间（时分秒）和帧率（fps）。

最后，将总持续时间（转换为秒）与帧率相乘，得到视频的总帧数，再将该值四舍五入到最接近的整数。

```
private static int getVideoTotalFrames(String videoFilePath) throws IOException {
    // ...
    while ((s = stdError.readLine()) != null) {
        if (s.contains("Duration")) {
            // 解析出视频总时长
        } else if (s.contains("fps") || s.contains("tbr")) {
            // 解析出视频帧率
        }
    }
    return (int) Math.round(durationInSeconds * frameRate);
}
```

在 getVideoTotalFrames() 方法中之所以使用标准错误流 (process.getErrorStream()) 来读取 FFmpeg 输出信息，而不是标准输入流（process.getInputStream()），是因为 FFmpeg 默认将其大部分输出，包括错误消息和信息性消息（如文件详细信息），都发送到标准错误流。

getVideoTotalFrames() 方法会通过执行下面的命令获取视频的相关信息，包括编解码信息、分辨率、帧率和时长等，然后通过时长和帧率计算出终帧数。

```
ffmpeg -i videoFilePath -hide_banner
```

命令行参数含义如下。

（1）-i videoFilePath：-i 参数后面跟着的是输入文件的路径。在这里，videoFilePath 应该替换为想要分析的视频文件的路径。

（2）-hide_banner：这个参数用于隐藏 FFmpeg 输出中的版权和版本信息的横幅（banner），使输出更清晰，只关注关键信息。

2. 混合音频和视频

本例最关键的一步就是使用 FFmpeg 命令混合音频和视频，可以使用下面的命令行完成混合工作。

```
ffmpeg -i videoFilePath -i audioFilePath -filter_complex "[1:a]volume=0.2[a1];
[0:a][a1]amerge=inputs=2[aout]" -map "0:v" -map "[aout]" -c:v copy -c:a aac
-strict experimental -progress progressFilePath outputFilePath
```

参数解释如下。

（1）-i videoFilePath：指定视频文件的路径。

（2）-i audioFilePath：指定音频文件的路径。

（3）-filter_complex "[1:a]volume=0.2[a1];[0:a][a1]amerge=inputs=2[aout]"：使用复杂的过滤器图，先将音频文件的音量调整为原来的 20%，然后将这个调整后的音频流与视频的原音频流合并。

（4）-map "0:v"：从第 1 个输入（视频）中映射视频流。

（5）-map "[aout]"：映射经过过滤器处理后的音频流。

（6）-c:v copy：在输出时直接复制视频流，不对其进行编码处理。

（7）-c:a aac：使用 AAC 编码器对音频进行编码。

（8）-strict experimental：允许使用实验性的功能，对于某些编码器可能是必需的。

（9）-progress progressFilePath：将进度信息输出到指定的文件路径。

（10）outputFilePath：指定输出文件的路径。

请确保替换 videoFilePath、audioFilePath、progressFilePath 和 outputFilePath 为实际的文件路径。这个命令将混合视频和调整音量的音频，并将结果保存到指定的输出文件中。同时，它会将进度信息输出到指定的文件，以便于跟踪处理进度。

3. 监听混合进度

本例采用异步的方式监听进度文件（progressFilePath），也就是启动 FFmpeg 进程并同时在另一个线程中监控进度文件。

在监控线程中，程序定期读取 FFmpeg 输出到 progressFilePath 的进度信息。每当读取到包含 frame= 的行时，程序解析出当前处理的帧数，并计算出处理进度的百分比。例如，下面的行表示当前正在处理第 1424 帧①。

```
frame=1423
```

下面的例子完整地演示了如何用 Java 调用 FFmpeg 命令混合音频和视频，并输出混合进度。

代码位置：src/main/java/image/video/AudioVideoMixer.java

```
package image.video;
import java.io.BufferedReader;
import java.io.File;
import java.io.FileReader;
import java.io.IOException;
import java.io.InputStreamReader;
public class AudioVideoMixer {
```

① 进度文件的当前处理帧数是从 0 开始的，所以 frame=1 其实表示正在处理第 2 帧，所以 frame=1423 表示正在处理第 1424 帧。

```
    private static int frames = 0;
    public static void main(String[] args) {
        String videoFilePath = "video.mp4";
        String audioFilePath = "audio.mp3";
        String outputFilePath = "new_video.mp4";                    // 输出视频文件
        String progressFilePath = "ffmpeg_progress.txt";            // 进度文件
        try {
            File outputFile = new File(outputFilePath);
            if (outputFile.exists()) {
                outputFile.delete();
            }
            File progressFile = new File(progressFilePath);
            if (progressFile.exists()) {
                progressFile.delete();
            }
            frames = getVideoTotalFrames(videoFilePath);
            String[] command = {
                    "ffmpeg",
                    "-i", videoFilePath,
                    "-i", audioFilePath,
                    "-filter_complex", "[1:a]volume=0.2[a1];[0:a][a1]amerge=
inputs=2[aout]",
                    "-map", "0:v",
                    "-map", "[aout]",
                    "-c:v", "copy",
                    "-c:a", "aac",
                    "-strict", "experimental",
                    "-progress", progressFilePath,
                    outputFilePath
            };
            // 执行 FFmpeg 命令混合音频和视频
            Process process = Runtime.getRuntime().exec(command);
            // 创建一个新线程来监控进度文件
            Thread thread = new Thread(() -> {
                try {
                    monitorProgress(progressFilePath);
                } catch (IOException e) {
                    e.printStackTrace();
                }
            });
            thread.start();
            int exitCode = process.waitFor();
            if (exitCode == 0) {
                thread.join();
                System.out.println("混合完成，输出文件：" + outputFilePath);
            } else {
                System.out.println("混合过程中发生错误");
            }
        } catch (IOException | InterruptedException e) {
            e.printStackTrace();
        }
```

```
    }
    // 监测进度文件
    private static void monitorProgress(String progressFilePath) throws
IOException {
        File progressFile = new File(progressFilePath);
        while (!progressFile.exists()) {
            try {
                Thread.sleep(200);
            } catch (InterruptedException e) {
                e.printStackTrace();
            }
        }
        BufferedReader reader = new BufferedReader(new FileReader(progressFile));
        String line;
        int progressFrames = 0;
        // 当前处理帧数小于总帧数减 1 时，继续监听进度文件
        while (progressFrames < frames - 1) {
            if ((line = reader.readLine()) != null) {
                if (line.startsWith("frame=")) {
                    progressFrames =  Integer.parseInt(line.substring
("frame=".length()));
                    double percentage = ((double) progressFrames / frames) * 100;
                    String formattedPercentage = String.format("%.0f%%",
percentage);
                    // 输出进度百分比
                    System.out.println("正在处理： " + formattedPercentage);
                }
            } else {
                try {
                    Thread.sleep(200);
                } catch (InterruptedException e) {
                    e.printStackTrace();
                }
            }
        }
    }
    // 获取视频总帧数
    private static int getVideoTotalFrames(String videoFilePath) throws
IOException {
        String[] command = {
                "ffmpeg",
                "-i", videoFilePath,
                "-hide_banner"
        };
        Process process = Runtime.getRuntime().exec(command);
        BufferedReader stdError = new BufferedReader(new InputStreamReader
(process.getErrorStream()));
        String s;
        double durationInSeconds = 0;
        double frameRate = 0;
        try {
```

```
            while ((s = stdError.readLine()) != null) {
                if (s.contains("Duration")) {
                    String[] parts = s.split(",")[0].split("Duration: ")
[1].trim().split(":");
                    double hours = Double.parseDouble(parts[0]);
                    double minutes = Double.parseDouble(parts[1]);
                    double seconds = Double.parseDouble(parts[2]);
                    durationInSeconds = (hours * 3600) + (minutes * 60) +
seconds;
                } else if (s.contains("fps") || s.contains("tbr")) {
                    String[] parts = s.split(",");
                    for (String part : parts) {
                        if (part.contains("fps") || part.contains("tbr")) {
                            String frameRateStr = part.trim().split(" ")[0];
                            frameRate = Double.parseDouble(frameRateStr);
                            break;
                        }
                    }
                }
            }
        } finally {
            stdError.close();
        }
        // 根据视频时长（秒）和帧率计算总帧数
        return (int) Math.round(durationInSeconds * frameRate);
    }
}
```

运行程序，如果混合成功，会输出类似下面的信息：

```
正在处理： 0%
正在处理： 9%
正在处理： 20%
正在处理： 30%
正在处理： 41%
正在处理： 52%
正在处理： 63%
正在处理： 74%
正在处理： 85%
正在处理： 100%
混合完成，输出文件：new_video.mp4
```

7.8.4 提取视频中的音频

FFmpeg 支持从视频中提取音频。下面是从视频中提取音频，并将音频保存为 MP3 文件的命令行：

```
ffmpeg -i input_video.mp4 -vn -ar 44100 -ac 2 -ab 192k -f mp3 output_audio.mp3
```

命令行参数含义如下。

（1）-i input_video.mp4：指定输入视频文件名为 input_video.mp4。可以将这个视频文件名替换成任何想使用的视频文件名。

（2）-vn：这个参数告诉 FFmpeg 在处理时忽略视频流。这意味着只处理音频内容。

（3）-ar 44100：表示音频采样率是 44100（以赫兹为单位）。44100Hz 是 CD 质量的标准采样率。

（4）-ac 2：设置音频通道数，2 表示输出的音频将是立体声（双通道）。

（5）-ab 192k：设置音频比特率。192k 指定了音频的比特率为 192kb/s，这是一个较好的音质和文件大小之间的平衡点。

（6）-f mp3：指定输出格式为 MP3。

（7）output_audio.mp3：输出的音频文件名。可以根据需要更改此文件名。

下面的例子演示如何使用 Java 调用 FFmpeg 命令提取 video.mp4 中的音频，并将音频保存为 extract_audio.mp3 文件。

代码位置：src/main/java/image/video/ExtractAudio.java

```
package image.video;
import java.io.IOException;
public class ExtractAudio {
    public static void main(String[] args) {
        // FFmpeg 的路径（根据实际路径进行修改）
        String ffmpegPath = "ffmpeg";
        // 输入视频文件的路径
        String inputVideo = "video.mp4";
        // 输出音频文件的路径
        String outputAudio = "extract_audio.mp3";
        // 构建 FFmpeg 命令
        ProcessBuilder processBuilder = new ProcessBuilder(
                ffmpegPath,
                "-i", inputVideo,                    // 指定输入文件
                "-vn",                               // 禁用视频记录
                "-acodec", "libmp3lame",             // 设置音频编码器为 libmp3lame
                "-q:a", "2",                         // 指定音频质量，值越小质量越好
                outputAudio);                        // 指定输出文件
        try {
            // 执行命令
            Process process = processBuilder.start();
            // 等待进程结束
            int exitCode = process.waitFor();
            if (exitCode == 0) {
                System.out.println("音频提取成功，保存在：" + outputAudio);
            } else {
                System.out.println("音频提取失败，退出代码：" + exitCode);
            }
        } catch (IOException | InterruptedException e) {
            e.printStackTrace();
```

```
            }
        }
    }
```

运行程序，在当前目录会生成一个 extract_audio.mp3 文件。

7.8.5 制作画中画视频

FFmpeg 可以制作画中画视频，也就是将两个或多个视频混合在一起，通常是主视频中嵌入一个或多个小视频，并且同时播放，形成一个画中画的效果。这在教学中被广泛使用，例如，用录制屏幕的方式制作教学视频，然后将讲师的讲课视频嵌入主视频的右下角。

下面的命令行将 video.mp4 作为小视频嵌入自身（video.mp4），并生成画中画视频（pip_video.mp4）：

```
ffmpeg -i video.mp4 -i video.mp4 -filter_complex "[1:v]scale=320:240,setpts=PTS-STARTPTS[pip];[0:v][pip] overlay=960:540:enable='gte(t,0)'[v]" -map "[v]" -map 0:a -c:v libx264 -c:a copy pip_video.mp4
```

命令行参数含义如下。

（1）第 1 个“-i video.mp4”：指定第 1 个输入的视频文件。

（2）第 2 个“-i video.mp4”：指定第 2 个输入的视频文件。

（3）-filter_complex"[1:v]scale=320:240,setpts=PTS-STARTPTS[pip];[0:v][pip] overlay=960:540:enable='gte(t,0)'[v]"：这是一个复杂的过滤器表达式，包含多个部分。[1:v]scale=320:240 表示将第 2 个输入视频的大小缩放到 320×240 像素。setpts=PTS-STARTPTS[pip] 表示重置小视频的时间戳，确保其从 0 开始，这在同步时很重要，这里将处理后的流标记为 pip①。[0:v][pip] overlay=960:540:enable='gte(t,0)'[v] 将 pip（小视频）覆盖到主视频流 [0:v] 上，位置在 (960,540)，即视频的右下角。enable='gte(t,0)' 表示这个覆盖从视频开始就一直有效。最后，处理后的视频流被标记为 [v]。

（4）-map "[v]"：这个选项将 filter_complex 生成的视频流 [v] 映射到输出文件。

（5）-map 0:a：将第 1 个输入源的音频流（0:a）映射到输出文件。这意味着使用原视频的音频。

（6）-c:v libx264：指定视频编码器为 libx264，这是一种常用的高效视频编码器。

（7）-c:a copy：音频编码选项，copy 表示直接复制原始音频流，而不进行转码。

（8）pip_video.mp4：指定输出文件的名称（画中画视频）。

下面的例子演示如何使用 Java 调用 FFmpeg 命令将 video.mp4 作为小视频嵌入自身，形成画中画效果，并输出为 pip_video.mp4。

① pip 是一个自定义的标签，可以根据需要给它命名。在这个例子中，我们使用 pip 作为标签，这通常代表“画中画”（Picture-in-Picture）。

代码位置：src/main/java/image/video/PictureInPictureVideo.java

```
package image.video;
import java.io.IOException;
public class PictureInPictureVideo {
    public static void main(String[] args) {
        // FFmpeg 的路径（根据实际路径进行修改）
        String ffmpegPath = "ffmpeg";
        // 输入的视频文件
        String videoFile = "video.mp4";
        // 输出的画中画视频文件
        String outputVideo = "pip_video.mp4";
        // 构建 FFmpeg 命令
        ProcessBuilder processBuilder = new ProcessBuilder(
                ffmpegPath,
                "-i", videoFile,        // 指定输入视频文件
                "-i", videoFile,        // 再次指定相同的视频文件作为画中画视频源
                "-filter_complex",
            "[1:v]scale=320:240,setpts=PTS-STARTPTS[pip];" +
            // 将画中画视频缩放并重置时间戳
                 "[0:v][pip] overlay=960:540:enable='gte(t,0)'[v]",
                 // 将画中画视频覆盖到主视频上
                "-map", "[v]",          // 确保输出视频包括上述的视频处理结果
                "-map", "0:a",          // 选择主视频的音频流
                "-c:v", "libx264",      // 使用 libx264 视频编码器
                "-c:a", "copy",         // 复制音频流，不进行转换
                outputVideo);           // 指定输出文件名

        try {
            // 执行命令
            Process process = processBuilder.start();
            // 等待进程结束
            int exitCode = process.waitFor();
            if (exitCode == 0) {
                System.out.println("视频合成成功，保存在：" + outputVideo);
            } else {
                System.out.println("视频合成失败，退出代码：" + exitCode);
            }
        } catch (IOException | InterruptedException e) {
            e.printStackTrace();
        }
    }
}
```

运行程序，会在当前目录生成一个 pip_video.mp4 文件，播放效果如图 7-7 所示。在视频中心有一个小区域，是缩小版的 video.mp4，播放视频时，由于是从 0 开始插入的小视频，所以主视频和画中画视频播放是同步的。

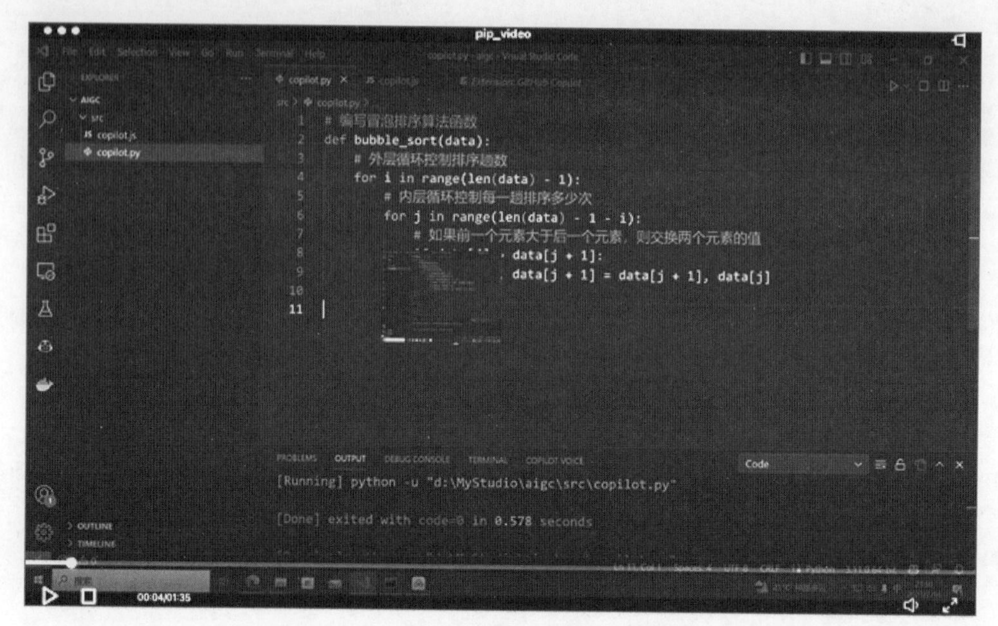

图 7-7 画中画视频

7.9 小结

本章介绍了一些与图像和视频相关的 Java 库和工具，这些库和工具主要包括 OpenCV、java.awt.Robot、BufferedImage、FFmpeg 等。这里要重点提一下 FFmpeg。FFmpeg 是一个非常强大的多媒体框架，它支持广泛的视频和音频格式，并提供了丰富的功能，如视频录制、转码、流处理等。它的这些特性使其成为开发多媒体应用程序的理想选择。因此，如果读者要开发强大的多媒体应用，那么 FFmpeg 是首选。FFmpeg 支持命令行方式、动态库方式和静态库方式调用，本章采用了最简单的命令行方式调用，其他两种方式大同小异，读者可以根据实际需要和使用场景使用不同方式与 FFmpeg 交互。

第 8 章

图像特效

Java 通过大量的库，可以实现非常酷的图像特效，效果并不亚于 Photoshop。例如，可以使用 Java 实现各种图像的滤镜、裁剪图像、翻转图像、混合图像、锐化、油画、扭曲等。本章将结合多个第三方库来详细讲解如何通过 Java 对图像完成这些复杂的处理工作。

8.1 图像滤镜

使用 OpenCV 的相关 API 可以实现各种图像滤镜，本节的例子会用 Java 编写一个带菜单的终端程序，每一个菜单项对应一个图像滤镜，用户输入菜单项对应的数字，就会将该图像滤镜应用到图像中，并保存处理后的图像。

本节的例子涉及如下几个图像滤镜。

（1）边缘增强；

（2）浮雕；

（3）查找边缘；

（4）锐化；

（5）平滑。

为了实现这几个图像特效，需要使用 4 个相关的方法：Imgproc.bilateralFilter()、Imgproc.filter2D()、Imgproc.GaussianBlur() 和 Imgproc.Canny()。这 4 个方法的详细描述如下。

❑ Imgproc.bilateralFilter 方法

bilateralFilter 方法是用于边缘保留平滑图像的非线性滤波器。它能够减少噪点而不会模糊边缘，该方法的原型如下。

```
public static void bilateralFilter(Mat src, Mat dst, int d, double sigmaColor,
double sigmaSpace)
```

参数含义如下。

（1）src：输入图像（Mat 类型）。

（2）dst：输出图像（Mat 类型），与输入图像有相同的大小和类型。

（3）d：滤波时所使用的各个像素邻域的直径。如果值是非正数，则从 sigmaSpace 计算得出。

（4）sigmaColor：颜色空间的标准方差。该参数可以视作滤波器颜色分量的空间。颜色空间越大，表示该像素邻域内有更多颜色会被混合。

（5）sigmaSpace：坐标空间中的标准方差，用于控制像素邻域的大小。空间参数越大，意味着越远的像素会相互影响，从而使更大的区域同质化。

❑ Imgproc.filter2D 方法

filter2D 方法用于将用户自定义的内核应用于图像。这个方法主要用于实现各种线性滤波器效果，例如模糊、锐化等。该方法的原型如下。

```
public static void filter2D(Mat src, Mat dst, int ddepth, Mat kernel, Point
anchor, double delta, int borderType)
```

参数含义如下。

（1）src：输入图像（Mat 类型）。

（2）dst：输出图像（Mat 类型），与输入图像有相同的大小和类型。

（3）ddepth：目标图像所需的深度。如果为 -1，则与原图像深度相同。常用值有 CvType.CV_8U, CvType.CV_16U, CvType.CV_32F 等。

（4）kernel：卷积核（或者称为滤波器），一个 Mat 类型的对象。它是一个单通道浮点矩阵，可以自定义滤波器的大小和值。

❑ Imgproc.GaussianBlur 方法

GaussianBlur 方法专用于应用高斯模糊效果。高斯模糊是一种常用的图像模糊技术，它使用正态分布（高斯函数）作为其模糊核。这种模糊技术特别有效于去除图像噪声和平滑化图像。该方法的原型如下。

```
public static void GaussianBlur(Mat src, Mat dst, Size ksize, double sigmaX)
```

参数含义如下。

（1）src：输入图像（Mat 类型）。

（2）dst：输出图像（Mat 类型），与输入图像有相同的大小和类型。

（3）ksize：高斯核的大小。核的宽度和高度可以不同，但它们都必须是正数和奇数。或者它们可以是零，然后从 sigma 中计算得出。

（4）sigmaX：X 方向上的高斯核标准差。这个参数控制核的宽度。如果 sigmaX 是 0，那么它将从 ksize.width 中计算得出。

（5）sigmaY：Y 方向上的高斯核标准差。如果这个参数是 0，那么它将从 ksize.height 中计算得出。如果 sigmaY 也是 0，它将设置成和 sigmaX 相同的值。如果两个 sigma 值都是 0，那么它们都将从 ksize 中计算得出。

❑ Imgproc. Canny 方法

Canny 方法是用于进行边缘检测的一个重要函数。它实现了 John F. Canny 在 1986 年提出的一个著名的边缘检测算法。该方法的原型如下。

```
public static void Canny(Mat image, Mat edges, double threshold1, double
threshold2, int apertureSize, boolean L2gradient)
```

参数含义如下。

（1）image：Mat 类型，表示输入图像。这应该是一个灰度图像（即单通道图像）。

（2）edges：Mat 类型，表示输出的边缘图像。它将具有相同的尺寸作为输入图像，并且是一个二值图像，边缘像素的值为 255，非边缘像素的值为 0。

（3）threshold1：双精度浮点数，表示用于边缘连接的较低阈值。

（4）threshold2：双精度浮点数，表示用于边缘检测的较高阈值。推荐的高 / 低阈值比在 2 ∶ 1 到 3 ∶ 1 之间。

（5）apertureSize：整数，表示用于计算图像梯度的 Sobel 算子的大小。默认值为 3。

（6）L2gradient：布尔值，指定梯度幅度的计算方法。如果为 True，则使用更精确的欧几里得距离计算公式，否则使用近似值。默认值为 False。

在 OpenCV 中，设置卷积核意味着创建一个 Mat 对象，并填充具有特定值的矩阵。这个矩阵定义了滤波器的特性和行为。例如，创建一个锐化的卷积核可以如下所示。

```
Mat kernel = new Mat(3, 3, CvType.CV_32F) {
    {
        put(0, 0, 0); put(0, 1, -1); put(0, 2, 0);
        put(1, 0, -1); put(1, 1, 5); put(1, 2, -1);
        put(2, 0, 0); put(2, 1, -1); put(2, 2, 0);
    }
};
```

这里，我们创建了一个 3 × 3 的卷积核，它用于锐化图像。put 方法用于在卷积核的特定位置设置值，这些值决定了如何修改图像中的每个像素。

卷积核的选择和设计对于图像处理的结果至关重要。不同的核会产生不同的效果，比如模糊、锐化、边缘检测等。下面是相应图像滤镜对应的卷积核。

浮雕效果的卷积核表示如图 8-1 所示。

此卷积核在应用时通过突出图像的高光和阴影部分产生立体感。

锐化效果的卷积核如图 8-2 所示。

$$\begin{bmatrix} -2 & -1 & 0 \\ -1 & 1 & 1 \\ 0 & 1 & 2 \end{bmatrix}$$

图 8-1　浮雕效果的卷积核

$$\begin{bmatrix} 0 & -1 & 0 \\ -1 & 5 & -1 \\ 0 & -1 & 0 \end{bmatrix}$$

图 8-2　锐化效果的卷积核

下面的例子从命令行参数读取了一个图像文件，如果没有指定命令行参数，那么使用默认的图像文件 example.png。然后输入要使用的图像滤镜序号（本例是 3），并按 Enter 键，在当前目录会生成一个 result_image.png 文件，用于保存处理结果。

代码位置：src/main/java/image/effects/ImageFilters.java

```
package image.effects;
import org.opencv.core.*;
import org.opencv.imgcodecs.Imgcodecs;
import org.opencv.imgproc.Imgproc;
import java.util.Scanner;
public class ImageFilters {
    static { System.loadLibrary("opencv_java490"); }
    public static void main(String[] args) {
        String imagePath = args.length > 0 ? args[0] : "example.png";
        // 替换为实际的图片路径
        Mat image = Imgcodecs.imread(imagePath);
        if (image.empty()) {
            System.out.println("Image not found!");
            return;
        }
        System.out.println("请选择一个滤镜：");
        System.out.println("1. 边缘增强");
        System.out.println("2. 边缘增强（更多）");
        System.out.println("3. 浮雕");
        System.out.println("4. 查找边缘");
        System.out.println("5. 锐化");
        System.out.println("6. 平滑");
        System.out.println("7. 平滑（更多）");
        Scanner scanner = new Scanner(System.in);
        int choice = scanner.nextInt();
        Mat resultImage = null;
        switch (choice) {
            case 1:
                resultImage = edgeEnhance(image);
                break;
            case 2:
                resultImage = edgeEnhanceMore(image);
                break;
            case 3:
                resultImage = emboss(image);
                break;
            case 4:
                resultImage = findEdges(image);
                break;
            case 5:
                resultImage = sharpen(image);
                break;
            case 6:
                resultImage = smooth(image);
```

```
                break;
            case 7:
                resultImage = smoothMore(image);
                break;
            default:
                System.out.println(" 无效的选择 ");
                return;
        }
        String resultPath = "result_image.png";          // 保存结果图片的路径
        Imgcodecs.imwrite(resultPath, resultImage);
        System.out.println("Image processed and saved as: " + resultPath);
    }
    private static Mat edgeEnhance(Mat image) {
        // 边缘增强
        Mat result = new Mat();
        Imgproc.bilateralFilter(image, result, -1, 15, 15);
        return result;
    }
    private static Mat edgeEnhanceMore(Mat image) {
        // 边缘增强（更多）
        Mat result = new Mat();
        Imgproc.bilateralFilter(image, result, -1, 30, 30);
        return result;
    }
    private static Mat emboss(Mat image) {
        // 浮雕效果
        Mat kernel = new Mat(3, 3, CvType.CV_32F, new Scalar(0)) {
            {
                put(0, 0, -2); put(0, 1, -1); put(0, 2, 0);
                put(1, 0, -1); put(1, 1, 1); put(1, 2, 1);
                put(2, 0, 0); put(2, 1, 1); put(2, 2, 2);
            }
        };
        Mat result = new Mat();
        Imgproc.filter2D(image, result, -1, kernel);
        return result;
    }
    private static Mat findEdges(Mat image) {
        // 查找边缘
        Mat result = new Mat();
        Imgproc.Canny(image, result, 100, 200);
        return result;
    }
    private static Mat sharpen(Mat image) {
        // 锐化
        Mat kernel = new Mat(3, 3, CvType.CV_32F, new Scalar(0)) {
            {
                put(0, 0, 0); put(0, 1, -1); put(0, 2, 0);
                put(1, 0, -1); put(1, 1, 5); put(1, 2, -1);
                put(2, 0, 0); put(2, 1, -1); put(2, 2, 0);
            }
```

```
        };
        Mat result = new Mat();
        Imgproc.filter2D(image, result, image.depth(), kernel);
        return result;
    }
    private static Mat smooth(Mat image) {
        // 平滑
        Mat result = new Mat();
        Imgproc.GaussianBlur(image, result, new Size(11, 11), 0);
        return result;
    }
    private static Mat smoothMore(Mat image) {
        // 平滑（更多）
        Mat result = new Mat();
        Imgproc.GaussianBlur(image, result, new Size(25, 25), 0);
        return result;
    }
}
```

运行程序，会看到如图 8-3 所示的图像滤镜菜单，输入 3，按 Enter 键，会在当前目录生成一个名为 result_image.png 的图像文件。

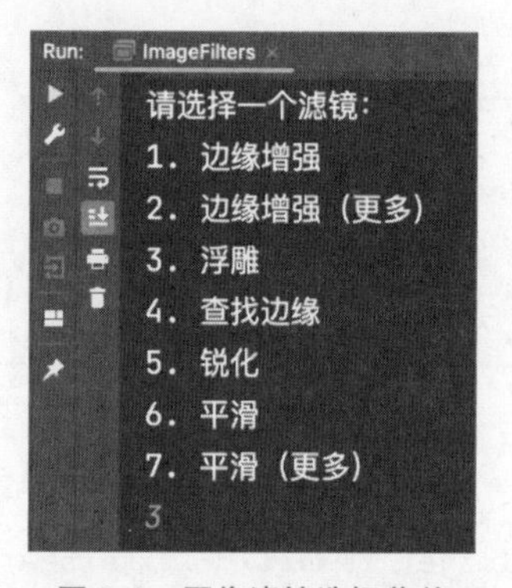

图 8-3 图像滤镜选择菜单

读者可以打开原图像和处理后的图像，对比效果如图 8-4 和图 8-5 所示。

图 8-4 原图（浮雕效果前）

图 8-5 浮雕效果的图像

8.2 缩放图像和拉伸图像

缩放图像和拉伸图像可以使用 awt 的 Graphics2D 类。Graphics2D 是 java.awt 包的一部分，它提供了一种高级接口来控制图形和文本渲染。在本节的例子中，Graphics2D 用于将一个图像从其原始尺寸调整到新的尺寸。

drawImage() 方法是 Graphics2D 类中的一个方法，用于将指定的图像绘制到 Graphics2D 的上下文中。在此程序中，drawImage 方法的重要性在于它能够对图像进行缩放和拉伸以适应新的尺寸。drawImage() 方法的原型如下。

```
public boolean drawImage(Image img, int x, int y, ImageObserver observer);
```

参数含义如下。

（1）img：要绘制的图像。

（2）x 和 y：图像在目标画布上的起始坐标。

（3）ImageObserver：通常为 null，用于接收有关图像状态的通知。

在本例中，需要使用 BufferedImage.getScaledInstance() 方法指定新图像的尺寸，然后通过 drawImage() 方法将新图像绘制到 Graphics2D 的上下文中。getScaledInstance() 方法的原型如下。

```
public Image getScaledInstance(int width, int height, int hints)
```

参数含义如下。

（1）newWidth 和 newHeight：缩放后图像的新尺寸。

（2）hints：选择标志，本例是 Image.SCALE_SMOOTH，表示使用平滑算法进行图像缩放，以优化图像质量。

缩放和拉伸的实现如下。

（1）缩放：如果新宽度和高度保持原图像的宽高比，则图像会被等比例缩放。

（2）拉伸：如果新宽度和高度改变了原图像的宽高比，则图像会被非等比例拉伸。

下面的例子将 images 目录中的 girl2.png 图像拉伸并缩放后，保存为 output.png 文件。

代码位置：src/main/java/image/effects/ImageResizerAndStretcher.java

```
package image.effects;
import javax.imageio.ImageIO;
import java.awt.image.BufferedImage;
import java.io.File;
import java.io.IOException;
import java.awt.Graphics2D;
import java.awt.Image;
public class ImageResizerAndStretcher {
    public static void main(String[] args) {
```

```
        try {
            // 加载图像
            File inputFile = new File("images/gril2.png");
            // 替换为实际的图像文件路径
            BufferedImage inputImage = ImageIO.read(inputFile);
            // 设置新的宽度和高度
            int newWidth = 1000;                    // 可以自由设置新的宽度
            int newHeight = 600;                    // 可以自由设置新的高度
            // 创建一个新的缓冲图像
            BufferedImage outputImage = new BufferedImage(newWidth, newHeight, 
inputImage.getType());
            // 使用 Graphics2D 对象来绘制调整大小后的图像
            Graphics2D g2d = outputImage.createGraphics();
            g2d.drawImage(inputImage.getScaledInstance(newWidth, newHeight, 
Image.SCALE_SMOOTH), 0, 0, null);
            g2d.dispose();
            // 保存处理后的图像
            String formatName = "png";          // 设置输出格式(如 "jpg", "png" 等)
            File outputFile = new File("images/output.png");   // 设置输出文件路径
            ImageIO.write(outputImage, formatName, outputFile);
            System.out.println("图像处理完成，保存在：" + outputFile.getAbsolutePath());
        } catch (IOException ex) {
            ex.printStackTrace();
        }
    }
}
```

运行程序，会在 images 目录生成一个 output.png 文件。对比效果如图 8-6 和图 8-7 所示。

图 8-6　原图（缩放和拉伸前）

图 8-7　缩放和拉伸后的图像

8.3　生成圆形头像

本节的例子会在图像中选择一个正方形区域，然后将这个区域制作成圆形的头像。所谓圆形头像，也就是除了正方形区域的内切圆外，其余部分都是透明的，通过 PS 裁剪，可以去掉周围的透明区域，将图像缩减为圆形区域大小。

要实现本例的功能，需要使用 Java 图形和图像处理库的相关 API，下面是对这些 API 的详细介绍。

本例的关键就是在图像中选取一个正方形区域，然后在正方形区域中按内切圆大小创建一个圆形头像，为了实现这两个功能，本例编写了 cropSquare 方法和 createCircleImage 方法。详细解释如下。

❑ cropSquare 方法

该方法的主要任务是从原始图像中裁剪出一个指定大小和位置的正方形区域。它使用 BufferedImage.getSubimage() 方法来实现。getSubimage() 方法从原始图像中提取一个子图像区域，该方法的原型如下。

```
public BufferedImage getSubimage (int x, int y, int w, int h)
```

参数含义如下。

（1）x：子区域左上角的 X 坐标。

（2）y：子区域左上角的 Y 坐标。

（3）w：子区域的宽度。

（4）h：子区域的高度。

通过指定 x、y、w 和 h，该方法可以从较大的图像中裁剪出所需的小部分。在本节的例子中，由于我们需要的是一个正方形区域，因此 w 和 h 被设定为相同的值。

❑ createCircleImage 方法

该方法的目的是将裁剪出的正方形图像转换成圆形头像。这里涉及的关键步骤包括设置抗锯齿以获得平滑的圆形边缘，绘制圆形，并通过遮罩技术将圆形区域之外的部分设为透明。在 createCircleImage 方法中主要使用了 Graphics2D 类及其方法。

Graphics2D 类是用于高级 2D 图形和文本的 Java 类。在我们的方法中，以下几个 Graphics2D 方法会被用到。

❑ setRenderingHint 方法

该方法设置渲染算法的偏好设置。这对改善图像质量非常关键，原型如下。

```
public void setRenderingHint(Key hintKey, Object hintValue)
```

参数含义如下。

（1）hintKey：指定所使用的渲染提示的键。

（2）hintValue：为该键指定的值。

在本节的例子中 hintKey 设置为 RenderingHints.KEY_ANTIALIASING，hintValue 设置为 RenderingHints.VALUE_ANTIALIAS_ON，表示开启抗锯齿，从而让圆形边缘更加平滑。

❑ fillOval 方法

该方法根据给定的坐标和尺寸绘制一个椭圆形（在我们的情况下是圆形，因为宽度和高度相等），原型如下。

```
public void fillOval(int x, int y, int width, int height)
```

参数含义如下。

（1）x 和 y：椭圆左上角的坐标。

（2）width 和 height：椭圆的宽度和高度。

在本例中绘制的是圆形，所以宽度和高度相同。

❑ setComposite 方法

该方法设置合成规则，用于定义图形间如何进行混合，原型如下。

```
public void setComposite(Composite comp)
```

其中 comp 表示合成规则。在本例中使用 AlphaComposite.SrcIn，这意味着仅在源图像（我们的圆形）与目标图像（原始正方形图像）重叠的地方绘制像素。这样，圆形之外的所有部分都将保持透明。

下面的例子从 girl2.png 中截取一个正方形，x=340，y=50，size=250，然后将正方形的内切圆制作成圆形头像，并保存结果为 circle_image.png。

代码位置：src/main/java/image/effects/CircleAvatarGenerator.java

```
package image.effects;
import javax.imageio.ImageIO;
import java.awt.*;
import java.awt.image.BufferedImage;
import java.io.File;
import java.io.IOException;
public class CircleAvatarGenerator {
    public static void main(String[] args) throws IOException {
        BufferedImage originalImage = ImageIO.read(new File("images/girl2.png"));
        // 替换为实际的图像路径
        // 指定正方形区域的起点坐标和尺寸
        int x = 340;                          // 起点 x 坐标
        int y = 50;                           // 起点 y 坐标
        int size = 250;                       // 区域的尺寸
        BufferedImage squareImage = cropSquare(originalImage, x, y, size);
        BufferedImage circleImage = createCircleImage(squareImage);
        // 保存最终的圆形图像
        ImageIO.write(circleImage, "png", new File("images/circle_image.png"));
        // 替换为想保存的路径
    }
    private static BufferedImage cropSquare(BufferedImage source, int x, int y,
int size) {
        return source.getSubimage(x, y, size, size);
    }
    private static BufferedImage createCircleImage(BufferedImage squareImage) {
```

```
        int diameter = squareImage.getWidth();
        BufferedImage circleImage = new BufferedImage(diameter, diameter,
BufferedImage.TYPE_INT_ARGB);

        Graphics2D g = circleImage.createGraphics();
        g.setRenderingHint(RenderingHints.KEY_ANTIALIASING, RenderingHints.
VALUE_ANTIALIAS_ON);
        g.fillOval(0, 0, diameter, diameter);
        g.setComposite(AlphaComposite.SrcIn);
        g.drawImage(squareImage, 0, 0, null);
        g.dispose();
        return circleImage;
    }
}
```

运行程序，会在 images 目录生成一个 circle_image.png 图像，效果如图 8-8 所示。

图 8-8　圆形头像

8.4　翻转图像

本节的例子实现了图像的翻转，包括水平翻转，垂直翻转和转置翻转（先水平翻转，再垂直翻转）。为了实现这三种翻转，本节的例子编写了三个方法，分别为 flipHorizontally 方法（实现水平翻转）、flipVertically（实现垂直翻转）方法和 transposeImage 方法（实现转置翻转），这三个方法的实现原理如下。

❑ flipHorizontally 方法（水平翻转）

水平翻转意味着将图像沿其垂直中轴线翻转。在此操作中，图像的每一行像素都会从左至右变成从右至左，即图像的左侧和右侧互换。

实现水平翻转的关键是重新定位像素点，使得每行的像素在水平方向上倒序排列。在 flipHorizontall 方法中，我们首先确定图像的宽度和高度。然后，创建一个新的 BufferedImage 对象，其尺寸与原始图像相同。接着，使用 Graphics2D.drawImage() 方法来绘制翻转后的图像。在 drawImage() 方法中，我们将源图像的右侧坐标作为目标图像的左侧坐标，源图像的左侧坐标作为目标图像的右侧坐标。这样就实现了水平翻转。

- flipVertically 方法（垂直翻转）

垂直翻转指的是将图像沿其水平中轴线翻转。在这个操作中，图像的每一列像素从上到下变成从下到上，即图像的上部和下部互换。在 flipVertically 方法中，我们同样先创建一个新的 BufferedImage 对象。使用 Graphics2D.drawImage() 方法绘制翻转后的图像时，将源图像的下边界坐标作为目标图像的上边界坐标，源图像的上边界坐标作为目标图像的下边界坐标。通过这种方式，实现了垂直翻转。

- transposeImage 方法（转置翻转）

转置翻转实际上是将图像先进行垂直翻转然后进行水平翻转，或者反过来先进行水平翻转再进行垂直翻转。在 transposeImage 方法中，我们首先对图像进行水平翻转，然后对结果进行垂直翻转。这两步操作的组合导致图像被旋转了 180°。

合并图像（mergeImages 方法）的原理涉及创建一个新的图像画布，然后将多个源图像按照指定的顺序和位置绘制到这个画布上。这个过程可以用于创建包含多个子图像的单一图像。在我们的例子中，我们合并了原图、水平翻转图、垂直翻转图和转置翻转图。

实现步骤如下。

（1）计算总宽度和最大高度：遍历所有要合并的图像。累加每个图像的宽度以计算合并后的总宽度。找出所有图像中最大的高度。

（2）创建新的 BufferedImage 对象：使用计算出的总宽度和最大高度创建一个新的 BufferedImage 实例。这个实例就是我们的画布，用于绘制所有源图像。图像类型设置为 BufferedImage.TYPE_INT_RGB，意味着使用标准的红绿蓝色彩模式，没有透明度。

（3）绘制每个源图像到画布上：创建 Graphics2D 对象用于绘制图像。遍历每个源图像，使用 drawImage() 方法将它们绘制到合并后的图像上。每个图像都绘制在当前累计宽度（currentWidth）的位置上，这保证了图像在水平方向上并排排列。每绘制完一个图像，更新 currentWidth 的值为下一个图像的起始位置。

（4）释放资源：绘制完成后，使用 g.dispose() 方法释放 Graphics2D 对象所占用的资源。

下面的例子演示了如何将 girl1.png 水平翻转，垂直翻转以及转置翻转，并将翻转后的图像合并成一张图，效果如图 8-9 所示。从左到右，分别是原图、水平翻转、垂直翻转和转置翻转。

图 8-9 图像翻转

代码位置：src/main/java/image/effects/ImageFlipper.java

```
package image.effects;
import javax.imageio.ImageIO;
import java.awt.*;
import java.awt.image.BufferedImage;
import java.io.File;
import java.io.IOException;
public class ImageFlipper {
    public static void main(String[] args) throws IOException {
        BufferedImage originalImage = ImageIO.read(new File("images/girl1.png"));
        // 替换为实际的图像路径
        BufferedImage flippedHorizontally = flipHorizontally(originalImage);
        BufferedImage flippedVertically = flipVertically(originalImage);
        BufferedImage transposeImage = transposeImage(originalImage);
        BufferedImage mergedImage = mergeImages(originalImage, flippedHorizontally,
flippedVertically, transposeImage);
        // 保存合并后的图像
        ImageIO.write(mergedImage, "png", new File("images/flipper_image.png"));
        // 替换为想保存的路径
    }
    // 水平翻转
    private static BufferedImage flipHorizontally(BufferedImage image) {
        int width = image.getWidth();
        int height = image.getHeight();
        BufferedImage flippedImage = new BufferedImage(width, height, image.getType());
        Graphics2D g = flippedImage.createGraphics();
        g.drawImage(image, 0, 0, width, height, width, 0, 0, height, null);
        g.dispose();
        return flippedImage;
    }
    // 垂直翻转
    private static BufferedImage flipVertically(BufferedImage image) {
        int width = image.getWidth();
        int height = image.getHeight();

        BufferedImage flippedImage = new BufferedImage(width, height, image.getType());
        Graphics2D g = flippedImage.createGraphics();
        g.drawImage(image, 0, 0, width, height, 0, height, width, 0, null);
        g.dispose();
        return flippedImage;
    }
    // 转置图像（180 度翻转）
    private static BufferedImage transposeImage(BufferedImage image) {
        return flipVertically(flipHorizontally(image));
    }
    // 合并 4 个图像
    private static BufferedImage mergeImages(BufferedImage... images) {
        int width = 0;
        int height = 0;
        // 计算合并后的图像的总宽度和最大高度
```

```
        for (BufferedImage image : images) {
            width += image.getWidth();
            height = Math.max(height, image.getHeight());
        }
        BufferedImage mergedImage = new BufferedImage(width, height, BufferedImage.
TYPE_INT_RGB);
        Graphics2D g = mergedImage.createGraphics();

        int currentWidth = 0;
        for (BufferedImage image : images) {
            g.drawImage(image, currentWidth, 0, null);
            currentWidth += image.getWidth();
        }
        g.dispose();

        return mergedImage;
    }
}
```

运行程序，会在 images 目录生成一个名为 flipper_image.png 文件，打开该图像，会看到如图 8-9 所示的效果。

8.5 调整图像的亮度、对比度、饱和度和锐度

本节的例子会调整图像的亮度、对比度、饱和度和锐度，下面分别给出实现这些图像特效的原理。

1. 调整亮度和对比度

调整亮度和对比度需要用到 RescaleOp 类和 RescaleOp.filter 方法，下面是对它们的详细描述。

❑ RescaleOp 类

RescaleOp 类是一个图像滤镜，它属于 java.awt.image 包。RescaleOp 类主要用于调整图像的亮度和对比度。RescaleOp 类可以通过重新缩放图像的颜色分量来改变图像的外观。这个过程通常被称为重新缩放操作。

RescaleOp 类使用一种简单的线性转换来调整图像的每个颜色分量（红色、绿色、蓝色，以及可选的 alpha 通道）。这个线性转换可以表示为：

```
output = scaleFactor * input + offset
```

其中：

（1）input：原始图像的颜色分量值。

（2）output：调整后的颜色分量值。

（3）scaleFactor：缩放因子，用于调整对比度。

（4）offset：一个偏移量，用于调整亮度。

当 scaleFactor 大于 1 时，对比度增加；小于 1 时，对比度减少。offset 会向每个颜色分量添加一个常数值，正值会使图像变亮，负值会使图像变暗。

RescaleOp 类有多个构造函数，其中最常用形式如下：

```
RescaleOp(float scaleFactor, float offset, RenderingHints hints)
```

和

```
RescaleOp(float[] scaleFactors, float[] offsets, RenderingHints hints)
```

第 1 个构造函数使用单一的缩放因子和偏移量对所有颜色分量进行相同的调整。第 2 个构造函数允许为每个颜色分量指定不同的缩放因子和偏移量。

RenderingHints 参数是可选的，它允许指定一些渲染提示，以改善图像处理的质量或性能。

❑ filter 方法

filter 方法是 RescaleOp 类的一个重要方法，用于执行实际的图像处理操作。该方法的原型如下。

```
public final BufferedImage filter (BufferedImage src, BufferedImage dst)
```

参数含义如下。

（1）src：源图像，是要调整的原始 BufferedImage。

（2）dest：目标图像，是经过滤镜处理后的图像。如果传入 null，则 filter 方法会创建一个新的 BufferedImage 作为返回值。

filter 方法会对源图像中的每个像素应用上述线性转换，然后将结果写入目标图像。如果目标图像是 null，filter 方法会创建一个具有适当类型的新 BufferedImage 对象，并将处理后的像素写入这个新图像。最后，filter 方法返回处理后的图像，可以直接显示在屏幕上，或者保存到文件中。

通过使用 RescaleOp 类和 filter 方法，可以轻松地调整图像的亮度和对比度，而无须深入了解图像处理的复杂细节。

2. 调整图像的锐度

锐度调整是通过增强图像中的边缘和细节来实现的。在数字图像处理中，锐化通常是通过应用一个卷积滤波器来完成的。卷积是一种数学运算，它通过一个卷积核（也称为滤波器或掩模）来加权平均图像的每个像素及其邻域像素。

一个常见的锐化滤波器是高通滤波器，它可以突出显示图像的高频部分，即图像中的快速变化区域，这些区域通常对应于边缘。在上面的代码示例中，我们使用了一个简单的 3×3 锐化核，如图 8-10 所示。

$$\begin{bmatrix} 0 & -1 & 0 \\ -1 & 5 & -1 \\ 0 & -1 & 0 \end{bmatrix}$$

图 8-10　3×3 锐化核

这个核的工作原理是将中心像素的权重设置得高于周围像素。中心像素的高权重（例如 5）强化了当前像素的贡献，而周围像素的负权重（例如 –1）则减少了相邻像素的影响。当这个核应用于图像的每个像素时，它会使边缘变得更加突出，从而增加了感知上的锐度。

3. 调整图像的饱和度

饱和度描述了颜色的强度或纯度。在 HSL 或 HSV 颜色空间中，饱和度是一个度量，表示颜色从灰色（0 饱和度）到全彩（100% 饱和度）的范围。在这些颜色空间中，颜色由三个组成部分定义：色相（颜色的种类）、饱和度（颜色的强度）和亮度或值（颜色的亮暗程度）。

调整饱和度通常涉及以下步骤：

（1）将 RGB 颜色空间的像素值转换为 HSL 或 HSV 颜色空间。

（2）在 HSL 或 HSV 空间中，增加或减少饱和度分量。这通常涉及乘以一个大于 1 的因子来增加饱和度，或乘以一个小于 1 的因子来减少饱和度。

（3）确保饱和度值保持在有效范围内，即 0 到 1 之间。

（4）将调整后的 HSL 或 HSV 值转换回 RGB 颜色空间，并更新图像的像素值。

通过这种方式，颜色可以变得更加鲜艳或更加柔和，而不影响图像的其他方面，如亮度或色相。这种转换对于调整图像的视觉效果非常有效，特别是在需要强调或减少颜色强度的情况下。

结合锐度和饱和度的调整，可以显著改善图像的视觉质量，使其看起来更加清晰和生动。

下面的例子演示了如何将亮度、对比度、锐度和饱和度的调整应用于一个图像，并保存处理结果的方式。

代码位置：src/main/java/image/effects/ImageAdjustment.java

```
package image.effects;
import java.awt.Color;
import java.awt.image.BufferedImage;
import java.awt.image.ConvolveOp;
import java.awt.image.Kernel;
import java.awt.image.RescaleOp;
import java.io.File;
import javax.imageio.ImageIO;
public class ImageAdjustment {
    public static void main(String[] args) throws Exception {
        // 加载图像
        File originalImageFile = new File("images/girl3.png"); // 替换为实际图像路径
        BufferedImage originalImage = ImageIO.read(originalImageFile);
        // 调整亮度和对比度
        float scaleFactor = 1.2f;                              // 对比度调整
```

```
        float offset = 20f;                                   // 亮度调整
        RescaleOp rescaleOp = new RescaleOp(scaleFactor, offset, null);
        BufferedImage adjustedImage = rescaleOp.filter(originalImage, null);
        // 调整锐度
        float[] sharpKernel = {
                0f, -1f, 0f,
                -1f, 5f, -1f,
                0f, -1f, 0f
        };
        Kernel kernel = new Kernel(3, 3, sharpKernel);
        ConvolveOp convolveOp = new ConvolveOp(kernel, ConvolveOp.EDGE_NO_OP, null);
        adjustedImage = convolveOp.filter(adjustedImage, null);
        // 调整饱和度
        adjustSaturation(adjustedImage, 1.2f);                        // 饱和度调整
        // 保存调整后的图像
        File adjustedImageFile = new File("images/adjusted_image.png");
        // 替换为实际保存路径
        ImageIO.write(adjustedImage, "png", adjustedImageFile);
    }
    // 一个用于调整图像饱和度的方法
    private static void adjustSaturation(BufferedImage image, float
saturationFactor) {
        int width = image.getWidth();
        int height = image.getHeight();
        // 遍历图像的每个像素
        for (int y = 0; y < height; y++) {
            for (int x = 0; x < width; x++) {
                int rgb = image.getRGB(x, y);
                Color color = new Color(rgb, true);
                float[] hsbVals = Color.RGBtoHSB(color.getRed(),
color.getGreen(), color.getBlue(), null);
                // 调整饱和度，确保它保持在 0 到 1 的范围内
                hsbVals[1] *= saturationFactor;
                hsbVals[1] = Math.min(1f, Math.max(0f, hsbVals[1]));
                // 将 HSB 值转换回 RGB 值
                int newRGB = Color.HSBtoRGB(hsbVals[0], hsbVals[1], hsbVals[2]);
                // 设置新的 RGB 值
                image.setRGB(x, y, newRGB);
            }
        }
    }
}
```

运行程序，会在 images 目录生成一个 adjusted_image.png 文件，对比效果如图 8-11 和图 8-12 所示。

图 8-11　原图（处理前）

图 8-12　处理后的图像

8.6　在图像上添加和旋转文字

在 Java 中，可以使用 Graphics2D.drawString() 方法将文字绘制到图像上，该方法的原型如下。

```
public void drawString(String str, int x, int y)
```

参数含义如下。

（1）str：这是要绘制的文本字符串。可以传入任何字符串，drawString() 方法会将其渲染到屏幕上。

（2）x：这是字符串的起始 x 坐标。在二维图形上，x 坐标指的是水平位置。这个坐标表示文本绘制起点的水平位置。

（3）y：这是字符串的基线 y 坐标。在二维图形上，y 坐标指的是垂直位置。重要的是，这个坐标并不是文本顶部的位置，而是文本基线的位置。在排版中，基线是大多数字符位于其上的线（例如，没有下沉部分的字母“E”会位于基线上）。

当调用 drawString() 方法时，它会在指定的 x 和 y 坐标处开始绘制文本。坐标 (x, y) 指的是文本字符串的左下角（更准确地说，是文本的基线左端点）。文本的样式（如字体、大小、颜色等）由 Graphics2D 对象的当前设置决定。

使用 drawString() 方法绘制文本的注意事项如下。

（1）在绘制文本之前，确保已经设置了所需的字体样式（如字体大小、风格等）。

（2）坐标系的原点 (0, 0) 通常位于绘图表面的左上角。

（3）文本的对齐和布局需要根据基线和字体度量进行计算，特别是当处理多行文本或需要精确布局时。

drawString() 方法只是将文本水平绘制到图像上，如果要旋转文本，需要使用二维仿射变化，也就是 AffineTransform 类和 AffineTransform.rotate 方法。

AffineTransform 类是 Java 2D API 的一部分，它提供了一种表示二维仿射变换的方式。仿射变换是一种可以在二维空间内移动、旋转、缩放和倾斜图形的数学操作。在图形处理中，AffineTransform 被广泛用于图像和文字的变换。

如果要让文字旋转，可以用 AffineTransform 创建一个新的变换，该变换定义了如何对文字进行旋转。

```
AffineTransform affineTransform = new AffineTransform();
```

这行代码创建了一个 AffineTransform 实例。初始时，它表示一个不进行任何变换的变换，即一个恒等变换。然后调用 AffineTransform.rotate 方法进行旋转变换。

```
affineTransform.rotate(Math.toRadians(45), 0, 0);                // 旋转 45 度
```

这里调用了 rotate() 方法来修改 AffineTransform 对象，使其表示一个旋转变换。这行代码意味着文字将围绕其左上角（坐标为 (0, 0)）旋转 45°。

在旋转文字时，通常需要设置字体：

```
Font rotatedFont = new Font("Serif", Font.BOLD, 60).deriveFont(affineTransform);
```

这行代码首先创建了一个新的 Font 对象，字体为 "Serif"，风格为粗体（Font.BOLD），大小为 60。然后，调用 deriveFont 方法并传入 AffineTransform 对象，以创建一个新的字体实例，该实例将包含旋转变换。这样，当使用这个字体绘制文字时，文字将被旋转。

最后，需要设置 Graphics2D 对象的当前字体为我们刚刚创建的旋转字体。Graphics2D 是一个用于在图像上绘制（如文字、形状等）的类。设置字体后，所有的绘制操作都将使用这个旋转字体进行，直到字体被更改。

```
// 设置 Graphics2D 对象的当前字体为我们刚刚创建的旋转字体
g2d.setFont(rotatedFont);
```

通过使用 AffineTransform，我们可以轻松地在 Java 中对图形和文字进行复杂的几何变换，如旋转、缩放和倾斜。在应用场景中，这意味着可以创造出视觉上吸引人的效果，比如在图片上以不同的角度绘制文字。

下面的例子会在图像的 (50,50) 位置上绘制中文文本，并以该点为圆心顺时针旋转 45°，效果如图 8-13 所示。

图 8-13　在图像上添加旋转文字

代码位置： src/main/java/image/effects/RotateTextOnImage.java

```
package image.effects;
import javax.imageio.ImageIO;
import java.awt.*;
```

```
import java.awt.geom.AffineTransform;
import java.awt.image.BufferedImage;
import java.io.File;
import java.io.IOException;
public class RotateTextOnImage {
    public static void main(String[] args) {
        try {
            // 加载图像
            BufferedImage image = ImageIO.read(new File("images/girl3.png"));
            // 创建 Graphics2D 对象，用于在图像上绘制
            Graphics2D g2d = (Graphics2D) image.getGraphics();
            // 设置字体和颜色
            g2d.setFont(new Font("Serif", Font.BOLD, 40));
            g2d.setColor(Color.RED);
            // 将文字旋转一定角度
            AffineTransform affineTransform = new AffineTransform();
            affineTransform.rotate(Math.toRadians(45), 0, 0);   // 旋转 45 度
            Font rotatedFont = new Font("Serif", Font.BOLD, 60).deriveFont
(affineTransform);
            g2d.setFont(rotatedFont);
            // 在图像上绘制文字
            g2d.drawString("旋转的文字", 50, 50);        // 在位置 (50,50) 绘制
            // 释放资源
            g2d.dispose();

            // 保存或展示图像
            ImageIO.write(image, "png", new File("images/text_image.png"));
        } catch (IOException e) {
            e.printStackTrace();
        }
    }
}
```

运行程序，会在 images 目录中生成一个 text_image.png 文件，效果如图 8-13 所示。

8.7 混合图像

在 Java 中混合两张图像通常涉及透明度的处理和像素级的图像融合。本节将详细介绍如何使用 Java 中的 BufferedImage 类和 Graphics2D 类，结合 setComposite() 方法实现图像混合。

（1）准备工作：加载图像。

首先，我们需要加载两张要混合的图像。这可以通过 ImageIO.read(File) 方法实现，它能够从文件系统中读取图像文件，并将其加载为 BufferedImage 对象。BufferedImage 是一个包含图像数据的类，提供了对图像进行操作和处理的方法。

（2）创建新的图像画布。

混合图像之前，我们需要创建一个新的 BufferedImage 实例，作为最终混合后图像的容器。这个新图像的大小通常与原始图像相同，颜色类型为 BufferedImage.TYPE_INT_ARGB，这意味着它支持 Alpha（透明度）通道。

（3）图像混合：使用 setComposite 方法。

关键步骤在于如何混合这两张图像。这是通过使用 Graphics2D 类的 setComposite 方法实现的，这个方法允许我们定义如何将图像的颜色与背景颜色融合。

（4）使用 AlphaComposite 类。

在 setComposite() 方法中，我们使用了 AlphaComposite 类。AlphaComposite 类是 Java 2D API 中用于处理 Alpha（透明度）合成规则的类。通过设置不同的 Alpha 值（0.0 ～ 1.0），我们可以控制图像的透明度级别。

例如，使用 AlphaComposite.getInstance(AlphaComposite.SRC_OVER, 0.7f) 创建一个 Alpha 合成规则，表示源图像（第一张图像）在目标图像（新创建的图像画布）上绘制时，应用 70% 的不透明度。同样，对第二张图像应用 30% 的不透明度。

这里的关键是 SRC_OVER 规则，它定义了源图像应该如何在目标图像之上绘制。简单来说，源图像的每个像素都将根据其 Alpha 值与目标图像的对应像素相结合。Alpha 值决定了源像素与目标像素的混合比例。

（5）绘制图像。

使用 Graphics2D.drawImage() 方法将每张图像绘制到新的图像画布上。因为我们已经设置了 AlphaComposite，所以每张图像都会按照之前定义的透明度规则进行绘制。

（6）保存混合后的图像。

最后一步是将混合后的图像保存到文件。这可以通过 ImageIO.write 方法实现，它允许将 BufferedImage 对象写入文件系统。

通过上述步骤，我们可以在 Java 中有效地混合两张图像。关键在于理解 Alpha 合成的概念和如何使用 Graphics2D 类的 setComposite 方法来控制图像的透明度。通过调整 Alpha 值，可以实现从完全透明到完全不透明的各种混合效果。

下面的例子完整地演示了将两个图像（girl1.png 和 girl2.png）按一定比例混合，然后生成混合后的 blendedImage.png 文件的过程，混合后的效果如图 8-14 所示。

图 8-14　混合图像

代码位置：src/main/java/image/effects/ImageBlender.java

```
package image.effects;
import javax.imageio.ImageIO;
import java.awt.*;
import java.awt.image.BufferedImage;
import java.io.File;
```

```
import java.io.IOException;
public class ImageBlender {
    public static void main(String[] args) {
        try {
            // 加载第 1 张图像
            BufferedImage image1 = ImageIO.read(new File("images/girl1.png"));

            // 加载第 2 张图像
            BufferedImage image2 = ImageIO.read(new File("images/girl2.png"));
            // 创建一个新的图像，用于存放混合后的图像
            BufferedImage blendedImage = new BufferedImage(
                    image1.getWidth(),
                    image1.getHeight(),
                    BufferedImage.TYPE_INT_ARGB);
            // 创建 Graphics2D 对象
            Graphics2D g2d = blendedImage.createGraphics();
            // 设置透明度并绘制第 1 张图像
            AlphaComposite alphaComposite = AlphaComposite.getInstance
(AlphaComposite.SRC_OVER, 0.7f);                          // 70% 的透明度
            g2d.setComposite(alphaComposite);
            g2d.drawImage(image1, 0, 0, null);
            // 设置透明度并绘制第 2 张图像
            alphaComposite = AlphaComposite.getInstance(AlphaComposite.
SRC_OVER, 0.3f);                                          // 30% 的透明度
            g2d.setComposite(alphaComposite);
            g2d.drawImage(image2, 0, 0, null);
            // 释放资源
            g2d.dispose();
            // 保存混合后的图像
            ImageIO.write(blendedImage, "png", new File("images/blendedImage.png"));
        } catch (IOException e) {
            e.printStackTrace();
        }
    }
}
```

注意：本例混合的两个图像的尺寸必须相同。在混合两个图像时，严格来说，两个图像的尺寸并不必须完全相同，但处理起来会更复杂。以下是几种情况和对应的处理方法。

（1）相同尺寸：如果两个图像的尺寸完全相同，那么混合它们将非常直接。可以简单地将一个图像的像素与另一个图像的对应像素按照指定的比例混合。

（2）不同尺寸，缩放其中一个图像：如果两个图像的尺寸不同，一种常见的方法是将其中一个图像缩放到与另一个图像相同的尺寸。这可以通过 BufferedImage.getScaledInstance 方法或通过创建一个新的 BufferedImage 并使用 Graphics2D.drawImage 方法来实现。

（3）不同尺寸，部分混合：另一种方法是只混合两个图像重叠的部分。例如，如果有一个较小的图像和一个较大的图像，可以将小图像放置在大图像的某个位置上，然后只混

合重叠的区域。

（4）扩展画布：也可以创建一个新的 BufferedImage，其尺寸足以容纳两个原始图像的尺寸（例如，取两者的较大值）。然后，可以将每个原始图像绘制到这个新画布的适当位置上，进行混合。

在选择方法时，需要考虑到最终图像的预期用途和视觉效果。例如，缩放图像可能会影响其质量，而部分混合则可能导致图像内容的丢失。因此，选择最佳方法时需要根据具体情况和需求来决定。

8.8　油画

实现图像的油画效果，要使用到 org.opencv.photo.Photo 类的 stylization 方法，下面是具体的实现步骤。

（1）灰度转换：将原始彩色图像转换为灰度图像，这通常是图像处理中的第一步，特别是在需要突出图像结构而不是颜色信息的情况下。

（2）高斯模糊：对灰度图像应用高斯模糊，用于减少图像噪声和细节，这也是准备进行边缘检测的常见预处理步骤。

（3）Canny 边缘检测：使用 Canny 算法检测图像中的边缘。这个算法能够找到图像中亮度变化明显的地方，即边缘。

（4）边缘图像三通道化：将单通道的边缘图像转换成与原始图像相同的三通道格式，以便能够与油画效果的图像合并。

（5）油画效果 (stylization)：使用 OpenCV 的 Photo.stylization 方法对原始图像应用油画效果。这个方法基于一种叫做 smoothing filter-based edge-preserving 的技术。它能够在保持边缘的同时平滑图像的颜色，模仿油画中颜色的融合和流动，但同时保持边缘和线条的清晰度。

（6）图像混合：将边缘检测后的图像与应用了油画效果的图像混合，这样可以在油画风格化的图像中添加更多的细节和对比度，使得效果更加逼真。

下面的例子演示了如何使用 OpenCV 的相关 API 实现油画效果，并将处理结果保存为 images/oilify.png 文件，效果如图 8-15 所示。

图 8-15　油画效果

代码位置：src/main/java/image/effects/OilPaintEffect.java

```
package image.effects;
import org.opencv.core.*;
import org.opencv.imgcodecs.Imgcodecs;
```

```
import org.opencv.imgproc.Imgproc;
import org.opencv.photo.Photo;
public class OilPaintEffect {
    static {
        // 加载 OpenCV 本地库，确保添加了正确的版本号
        System.loadLibrary("opencv_java490");
    }
    public static void main(String[] args) {
        // 读取图像文件
        Mat src = Imgcodecs.imread("images/girl3.png");
        Mat gray = new Mat();
        Mat blur = new Mat();
        Mat edges = new Mat();
        Mat dst = new Mat();

        // 转换图像到灰度
        Imgproc.cvtColor(src, gray, Imgproc.COLOR_BGR2GRAY);
        // 应用高斯模糊
        Imgproc.GaussianBlur(gray, blur, new Size(5, 5), 0);
        // 使用 Canny 算法检测边缘
        Imgproc.Canny(blur, edges, 50, 150);
        // 将边缘图像转换到三通道
        Imgproc.cvtColor(edges, edges, Imgproc.COLOR_GRAY2BGR);
        // 应用油画效果
        Photo.stylization(src, dst, 60, 0.45f);
        // 结合边缘图像和油画风格化图像
        Core.addWeighted(dst, 0.9, edges, 0.1, 0, dst);
        // 保存最终的结果图像
        Imgcodecs.imwrite("images/oilify.png", dst);
    }
}
```

8.9 波浪扭曲

本节的例子实现了图像的波浪扭曲效果，核心代码是自定义的 applyWaveDistortion 方法，下面是该方法的实现原理。

applyWaveDistortion 方法的核心目的是对给定的图像应用一种波浪扭曲效果。它通过遍历原始图像的每一个像素，并基于数学公式来调整这些像素的位置，从而实现波浪效果。

（1）波浪扭曲的原理：这种扭曲效果是通过在图像的每一行（或列）上应用正弦波形变化来实现的。在处理过程中，每个像素的水平位置（x 坐标）保持不变，而垂直位置（y 坐标）则根据正弦波的公式进行调整。

（2）遍历像素：方法开始时，我们遍历图像中的每个像素。对于图像中的每个像素点 (x, y)，我们将保持其 x 坐标不变，而 y 坐标则根据波浪的正弦函数进行调整。

（3）应用波浪效果：对于每个像素点，我们计算新的 y 坐标（ny），方法是在原始 y 坐

标的基础上加上一个偏移量。这个偏移量是通过将 x 坐标代入正弦函数得到的，正弦函数由两个关键参数控制：波浪频率（waveFrequency）和波浪振幅（waveAmplitude）。

- 波浪频率

定义：double waveFrequency = 0.3;

作用：这个参数定义了波浪的频率。频率决定了波浪的密集程度，即在图像上出现的波浪的数量。频率较高意味着波浪更加紧密，频率较低则波浪更加稀疏。

影响：调整这个参数会改变波浪的“波峰”和“波谷”在图像上的间距，从而影响图像扭曲的整体效果。

- 波浪振幅

定义：double waveAmplitude = 12;

作用：振幅定义了波浪的高度。振幅较大意味着波浪的峰值和谷值更加极端，从而使图像的扭曲效果更加明显。

影响：增加振幅会使图像的扭曲更加夸张，而减少振幅则会使扭曲效果更加微妙。

applyWaveDistortion 方法的执行流程是：对于原始图像中的每一个像素，我们计算出一个新的 y 坐标，然后在新图像的对应位置上放置原始像素。通过这种方式，原始图像上的直线或边缘在新图像中会呈现出波浪状的扭曲效果，从而达到了波浪效果的视觉表现。

下面的例子演示了如何实现图像的波浪扭曲效果，并将处理结果保存为 images/distorted_image.png 文件。效果如图 8-16 所示。

图 8-16　波浪扭曲

代码位置：src/main/java/image/effects/ImageWaveDistortion.java

```
package image.effects;

import java.awt.image.BufferedImage;
import java.io.File;
import javax.imageio.ImageIO;
public class ImageWaveDistortion {
    public static void main(String[] args) {
        try {
            // 读取图片文件
            BufferedImage originalImage = ImageIO.read(new File("images/girl2.
png"));
            // 创建一个新的图像用于存放扭曲效果的图像
            BufferedImage distortedImage = new BufferedImage(originalImage.
getWidth(), originalImage.getHeight(), BufferedImage.TYPE_INT_ARGB);
            // 应用波浪扭曲效果
            applyWaveDistortion(originalImage, distortedImage);
            // 保存扭曲后的图像
```

```
                File outputfile = new File("images/distorted_image.png");
                ImageIO.write(distortedImage, "png", outputfile);
                System.out.println("图像扭曲效果已保存");
            } catch (Exception e) {
                e.printStackTrace();
            }
        }
        private static void applyWaveDistortion(BufferedImage originalImage,
BufferedImage distortedImage) {
            // 获取图像宽度和高度
            int width = originalImage.getWidth();
            int height = originalImage.getHeight();
            // 波浪效果参数
            double waveFrequency = 0.3;                // 波浪的频率
            double waveAmplitude = 12;                 // 波浪的振幅
            // 对每个像素应用波浪扭曲效果
            for (int y = 0; y < height; y++) {
                for (int x = 0; x < width; x++) {
                    int nx = x;
                    int ny = (int) (y + waveAmplitude * Math.sin(x * waveFrequency));
                    // 计算扭曲后的 y 坐标
                    // 确保坐标在图像范围内
                    if (ny >= 0 && ny < height) {
                        // 获取并设置像素值
                        distortedImage.setRGB(nx, ny, originalImage.getRGB(x, y));
                    }
                }
            }
        }
    }
```

8.10 挤压扭曲

挤压扭曲的原理是通过计算挤压扭曲后的坐标映射矩阵，实现图像从左右向中间的挤压扭曲效果。所以挤压扭曲同样是坐标点的映射。

坐标点坐标是 x 和 y，表示转换后 x 序列的是 map_x，表示转换后 y 序列的是 map_y，所以有如下公式：

```
map_x[y, x] = w / 2 + (x - new_w / 2) / (1 - squeeze_factor * abs(y - h / 2) / (h / 2))
map_y[y, x] = y
```

这两个公式的推导过程如下：

首先考虑 y 坐标。由于这个效果只是在 x 方向上实现挤压，y 方向并未发生变化。所以 y 坐标的映射关系很简单，直接映射不变：map_y[y, x] = y。

然后考虑 x 坐标。我们首先需要确定几个基准点：

（1）输入图像的宽度 w 和高度 h。

（2）挤压因子 squeeze_factor，它控制挤压的强度，取值范围为 0 ～ 1。

（3）输出图像的宽度 new_w = w * (1 - squeeze_factor)。

（4）输入图像的中点坐标 (w/2, h/2)。

我们要实现的效果是图像从两边向中间挤压。所以输出图像的中点 new_w/2 需要映射到输入图像的中点 w/2。两边被挤压的部分需要映射到输入图像更宽的区域。

由于我们实现的是一个上下对称的效果，所以上下位置也需要考虑在映射关系中。上下靠近中点的部分挤压幅度更大，上下两端挤压幅度更小。

根据以上考虑，我们可以推导出 x 坐标的映射关系：

```
map_x[y, x] = w / 2 + (x - new_w / 2)
```

我们要实现的效果是：上下位置越靠近中点，挤压越强烈。这要求挤压系数与 y 的值有关，需要一个函数来表达这种关系。首先考虑使用挤压幅度 squeeze_factor 直接作为挤压系数。但是，这样无法实现我们要的和 y 相关的效果。所以，需要对 squeeze_factor 进行修正。修正的方式是将 squeeze_factor 与 y 的距离比值相乘。y 的距离使用 abs(y – h/2) 表示。为了归一化（某个变量的值范围映射到 0 ～ 1 之间），我们把它除以 h/2。所以，修正后的挤压系数为 **squeeze_factor * abs(y – h / 2) / (h / 2)**。然后从 1 减去这个值，得到最后的挤压系数表达式 **1 – squeeze_factor * abs(y – h / 2) / (h / 2)**。这个表达式的值与 y 成反比，实现了我们想要的效果：y 越靠近中点，表达式的值越小，挤压系数越大，映射程度越高。

接下来，为了得到映射关系，我们需要以这个挤压系数为分母。x 坐标与此成正比，所以选择作为分母。

最后，通过 remap 函数根据 map_x 和 map_y 实现图像的像素重映射，就可以得到挤压扭曲的效果。

下面的代码使用上述的方式挤压扭曲图像，并显示扭曲结果，以及将扭曲结果保持为 squeezed_image.png。图 8-17 是原图，图 8-18 是被挤压的效果。

图 8-17　原图（挤压前）

图 8-18　挤压扭曲后

代码位置：src/main/java/image/effects/ImageSqueezeDistortion.java

```
package image.effects;
import javax.imageio.ImageIO;
import java.awt.image.BufferedImage;
import java.io.File;
import java.io.IOException;
public class ImageSqueezeDistortion {
    public static void main(String[] args) throws IOException {
        // 读取图像
        BufferedImage image = ImageIO.read(new File("images/girl4.png"));
        // 应用挤压扭曲效果
        BufferedImage result = squeezeWarp(image, 0.3);
        // 保存挤压扭曲后的图像
        ImageIO.write(result, "png", new File("images/squeezed_image.png"));
    }
    private static BufferedImage squeezeWarp(BufferedImage image, double 
squeezeFactor) {
        int h = image.getHeight();
        int w = image.getWidth();
        int newW = (int) (w * (1 - squeezeFactor));
        // 创建一个新的图像用于存放挤压效果的图像
        BufferedImage result = new BufferedImage(newW, h, BufferedImage.TYPE_
INT_RGB);
        for (int y = 0; y < h; y++) {
            for (int x = 0; x < newW; x++) {
                // 计算挤压扭曲后的 x 坐标
                int mapX = (int) (w / 2.0 + (x - newW / 2.0) / (1.0 - 
squeezeFactor * Math.abs(y - h / 2.0) / (h / 2.0)));
                // 确保新坐标在图像范围内
                if (mapX >= 0 && mapX < w) {
                    // 获取并设置像素值
                    result.setRGB(x, y, image.getRGB(mapX, y));
                }
            }
        }
        return result;
    }
}
```

8.11 3D 浮雕效果

本节的例子会实现图像的 3D 浮雕效果。3D 浮雕效果是通过直接计算当前像素与其相邻像素（通常是右侧或下方像素）的差异来实现的，而不是使用传统的卷积核矩阵。尽管没有显式定义一个卷积核，但计算相邻像素差的操作本质上与应用一个卷积核的效果相似。本例的核心是一个自定义的 applyEnhancedEmbossEffect 方法，该方法的详细描述如下：在

applyEnhancedEmbossEffect() 方法中，实现 3D 浮雕效果的核心步骤是计算相邻像素之间的颜色差，并对结果加上一个固定的偏移量。这个过程模拟了一个光源从图像一侧照射的效果，其中像素颜色的变化给人一种凹凸的感觉。

❑ 颜色差计算

具体来说，对于图像中的每个像素，我们取其 RGB 颜色值（R, G, B），并从每个颜色分量中减去右侧相邻像素的对应颜色分量。这样，如果一个像素比它右侧的像素亮，那么在浮雕图像中，这个像素会被显示得更亮；反之，如果它比右侧的像素暗，那么它会被显示得更暗。这种对比效果模拟了光与物体表面相互作用的方式，从而产生了浮雕效果。

❑ 固定偏移量

为了保证颜色值在合理的范围内（0 ～ 255），并且在没有明显差异的平坦区域创造出一种中性的浮雕效果，我们在计算完颜色差之后加上了一个固定的偏移量 128。这个偏移量将颜色范围从可能的负值区间重新映射到可显示的颜色范围内，同时让没有变化的区域呈现为中性的灰色，从而增强了浮雕的 3D 效果。

❑ 边界处理

在图像的边界，我们没有更多的相邻像素可以用来计算差值。为了处理这种情况，在图像的最后一列和最后一行，我们将像素设置为中性灰色（RGB 值为 128,128,128），这样可以在视觉上保持浮雕效果的连续性，避免边缘产生不自然的颜色断层。

综上所述，尽管代码中没有使用传统的卷积核矩阵，但通过计算相邻像素之间的颜色差异并加上偏移量的方法，我们实现了类似卷积效果的 3D 浮雕效果。这种处理方式相对简单直接，且不需要进行复杂的卷积运算，可以快速地对图像进行浮雕效果处理。

下面的例子根据前面描述的理论对图像实现 3D 浮雕效果，并将处理后的文件保存为 embossed_image.png 文件。3D 浮雕效果如图 8-19 所示。

图 8-19　3D 浮雕效果

代码位置：src/main/java/image/effects/ImageEmbossEffect.java

```
package image.effects;
import java.awt.image.BufferedImage;
import java.io.File;
```

```
import javax.imageio.ImageIO;
public class ImageEmbossEffect {
    public static void main(String[] args) throws Exception {
        // 读取图像文件
        BufferedImage srcImage = ImageIO.read(new File("images/girl3.png"));
        // 应用浮雕效果
        BufferedImage embossImage = applyEnhancedEmbossEffect(srcImage);
        // 保存处理后的图像
        ImageIO.write(embossImage, "png", new File("images/embossed_image.png"));
    }
    private static BufferedImage applyEnhancedEmbossEffect(BufferedImage srcImage) {
        // 获取图像的宽度和高度
        int width = srcImage.getWidth();
        int height = srcImage.getHeight();
        // 创建新的图像对象，用于存放带有浮雕效果的图像
        BufferedImage embossImage = new BufferedImage(width, height,
BufferedImage.TYPE_INT_RGB);

        // 遍历原图像的每个像素除了最后一列和最后一行
        for (int i = 0; i < height - 1; i++) {
            for (int j = 0; j < width - 1; j++) {
                // 获取当前像素和其右侧及下方像素的颜色值
                int pixel = srcImage.getRGB(j, i);
                int rightPixel = srcImage.getRGB(j + 1, i);
                int bottomPixel = srcImage.getRGB(j, i + 1);
                // 计算新的 RGB 值
                int r = ((pixel >> 16) & 0xff) - ((rightPixel >> 16) & 0xff) + 128;
                int g = ((pixel >> 8) & 0xff) - ((rightPixel >> 8) & 0xff) + 128;
                int b = (pixel & 0xff) - (rightPixel & 0xff) + 128;
                // 调整 RGB 值，确保它们在 0 到 255 的范围内
                r = Math.min(Math.max(r, 0), 255);
                g = Math.min(Math.max(g, 0), 255);
                b = Math.min(Math.max(b, 0), 255);
                // 将计算后的 RGB 值组合回一个整数
                int newPixel = (r << 16) | (g << 8) | b;
                // 在新图像上设置新的像素值
                embossImage.setRGB(j, i, newPixel);
            }
        }
        // 处理图像的最后一列和最后一行，这里简单地将其设为灰色（128,128,128）
        int edgeColor = (128 << 16) | (128 << 8) | 128;
        for (int i = 0; i < width; i++) {
            embossImage.setRGB(i, height - 1, edgeColor);
        }
        for (int i = 0; i < height; i++) {
            embossImage.setRGB(width - 1, i, edgeColor);
        }
        return embossImage;
    }
}
```

8.12　小结

相信读者看完本章一定会激情澎湃，原来 Java 这么强大，不，其实大家看到的只是 Java 的冰山一角，Java 远比你想象的强大得多。尽管本章主要介绍了 Java 如何实现图像特效，但 Java 几乎可以做任何事情，甚至很多底层的操作。这得益于 Java 可以与 C++ 等语言交互，有很多第三方库底层使用的都是 C++ 语言，所以，理论上，只要 C++ 语言能做的，Java 就能做。就拿本章介绍的图像特效来说，通过 JavaFX、OpenCV、Java 图形库等第三方库，完全可以用 Java 做一个与 Photoshop 相匹敌的图像处理系统，相信广大读者已经蠢蠢欲动了，那么就让我们开始吧！

第 9 章

视 频 特 效

Java 不仅可以实现丰富的图像特效，还可以实现炫酷的视频特效。其实视频特效就是图像特效的升级版。因为视频也是由图像组成的，理论上，任何图像特效都可以用在视频特效中。不过 Java 通过 FFmpeg 等工具，让我们可以不单独操作视频中的图像，就可以实现丰富多彩的视频特效。除此之外，如果遇到非常复杂的视频特效，例如，在视频中插入动态图像，就需要 OpenCV、Java2D API、FFmpeg 等库和工具配合才能实现。本章会深入讲解如何将 Java 与多种工具和库深度结合来制作各种炫酷的视频特效。

9.1 旋转视频

使用 FFmpeg 可以实现旋转视频的功能，假设要将视频的每一帧画面顺时针旋转 90°，可以使用下面的命令。

```
ffmpeg -i input.mp4 -vf "transpose=1" output.mp4
```

参数含义如下。

（1）-i input.mp4：-i 参数用于指定输入文件。在这里，input.mp4 是要旋转的视频文件。

（2）-vf "transpose=1"：这是视频过滤器（-vf）的参数。在这里，我们使用 transpose 过滤器来旋转视频。transpose=1 指定旋转的方式。transpose 参数可以接收 0、1、2、3。它们的含义是：0 表示逆时针旋转 90° 并垂直翻转（相当于顺时针旋转 270°），1 表示顺时针旋转 90°，2 表示逆时针旋转 90°，3 表示顺时针旋转 90° 并垂直翻转（相当于逆时针旋转 270°）。

（3）output.mp4：指定了输出文件的名称。在这个例子中，处理后的视频将被保存为 output.mp4。

关于旋转的方向，FFmpeg 中的 transpose 过滤器不直接使用角度来定义旋转，而是使用上述的预设值。如果需要进行不同角度的旋转（如 180° 或任意角度），可以使用 rotate 过滤器，它允许指定旋转的弧度。例如，旋转 45°（π/4 弧度）的命令如下：

```
ffmpeg -i input.mp4 -vf "rotate=PI/4" output.mp4
```

正值表示顺时针旋转，负值表示逆时针旋转。例如，要逆时针旋转 45°，可以使用 -PI/4。但请注意，使用 rotate 过滤器可能导致输出视频的尺寸改变或出现黑边。

下面的例子用 Java 调用 FFmpeg 命令将指定视频的所有帧画面顺时针旋转 90°，并将新视频保存为 rotate_video.mp4，视频截图如图 9-1 所示。

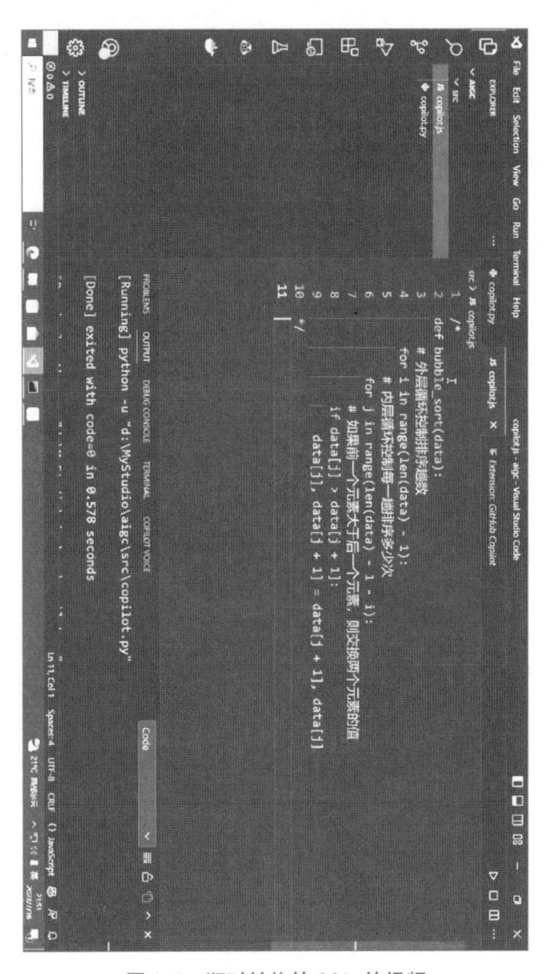

图 9-1　顺时针旋转 90° 的视频

代码位置： src/main/java/video/effects/VideoRotator.java

```
package video.effects;
import java.io.BufferedReader;
import java.io.InputStreamReader;
public class VideoRotator {
    // 使用 FFmpeg 旋转视频
    public void rotateVideo(String inputPath, String outputPath, int
rotationDegree) throws Exception {
        // 构建 FFmpeg 命令
```

```
        String cmd = buildFFmpegRotateCommand(inputPath, outputPath,
rotationDegree);
        // 执行命令
        Process process = Runtime.getRuntime().exec(cmd);
        BufferedReader reader = new BufferedReader(new InputStreamReader(process.
getInputStream()));
        String line;
        while ((line = reader.readLine()) != null) {
            System.out.println(line);              // 打印 FFmpeg 的输出，用于调试
        }
        // 等待 FFmpeg 进程结束
        process.waitFor();
    }
    // 构建用于旋转视频的 FFmpeg 命令
    private String buildFFmpegRotateCommand(String inputPath, String
outputPath, int rotationDegree) {
        String transposeValue = "";
        switch (rotationDegree) {
            case 90:
                transposeValue = "1";
                break;
            case 180:
                transposeValue = "2";
                break;
            case 270:
                transposeValue = "3";
                break;
            default:
                transposeValue = "0";
                break;
        }
        // FFmpeg 命令格式
        return "ffmpeg -i " + inputPath + " -vf transpose=" + transposeValue +
" " + outputPath;
    }
    public static void main(String[] args) {
        VideoRotator rotator = new VideoRotator();
        try {
            // 旋转视频示例：输入路径，输出路径，旋转角度
            rotator.rotateVideo("video.mp4", "rotate_video.mp4", 90);
        } catch (Exception e) {
            e.printStackTrace();
        }
    }
}
```

9.2 镜像视频

使用 FFmpeg 可以实现镜像视频的功能，假设要将视频的每一帧画面水平镜像，可以使用下面的命令。

```
ffmpeg -i input.mp4 -vf "hflip" output.mp4
```

参数含义如下。

（1）-i input.mp4：-i 参数用于指定输入文件。在这里，input.mp4 是想要镜像处理的视频文件。

（2）-vf "hflip"：这是视频过滤器（-vf）的参数。在这里，我们使用 hflip 过滤器来实现水平镜像，vflip 表示垂直镜像。hflip 会水平翻转视频中的每一帧画面，vflip 会垂直翻转视频中的每一帧画面。

（3）output.mp4：指定了输出文件的名称。在这个例子中，处理后的视频将被保存为 output.mp4。

下面的例子用 Java 调用 FFmpeg 命令将视频水平翻转，并保存为 mirrored.mp4 文件。效果如图 9-2 所示。

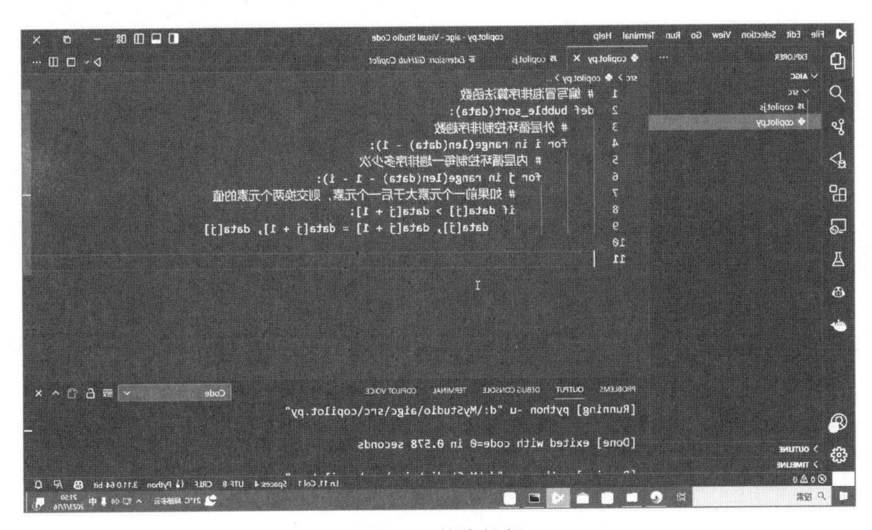

图 9-2　镜像视频

代码位置：src/main/java/video/effects/VideoMirror.java

```
package video.effects;
import java.io.BufferedReader;
import java.io.InputStreamReader;
public class VideoMirror {
    // 使用 FFmpeg 实现视频的水平镜像
    public void mirrorVideo(String inputPath, String outputPath) throws Exception {
        // 构建 FFmpeg 命令
        String cmd = buildFFmpegMirrorCommand(inputPath, outputPath);
        // 执行命令
        Process process = Runtime.getRuntime().exec(cmd);
        BufferedReader reader = new BufferedReader(new InputStreamReader
(process.getInputStream()));
```

```
            String line;
            while ((line = reader.readLine()) != null) {
                System.out.println(line);          // 打印 FFmpeg 的输出，用于调试
            }
            // 等待 FFmpeg 进程结束
            process.waitFor();
        }
        // 构建用于实现视频水平镜像的 FFmpeg 命令
        private String buildFFmpegMirrorCommand(String inputPath, String outputPath) {
            // FFmpeg 命令格式，使用 hflip 过滤器进行水平镜像
            return "ffmpeg -i " + inputPath + " -vf hflip " + outputPath;
        }
        public static void main(String[] args) {
            VideoMirror mirror = new VideoMirror();
            try {
                // 镜像视频示例：输入路径，输出路径
                mirror.mirrorVideo("video.mp4", "mirrored.mp4");
            } catch (Exception e) {
                e.printStackTrace();
            }
        }
    }
```

9.3 变速视频

使用 FFmpeg 可以变速视频，例如，下面的命令将视频的速度变为原来的 0.5 倍，并将变速结果保存为 speed_video.mp4 文件。

```
ffmpeg -i video.mp4 -filter_complex "[0:v]setpts=2.00*PTS[v];[0:a]atempo=0.50[a]"
-map "[v]" -map "[a]" speed_video.mp4
```

参数含义如下。

（1）-i video.mp4：-i 选项用于指定输入文件，这里 video.mp4 是要处理的源视频文件。

（2）-filter_complex "[0:v]setpts=2.00*PTS[v];[0:a]atempo=0.50[a]"：这个参数用于应用复杂的过滤器链。其中 [0:v] 表示输入文件中的视频流（0 表示第一个输入文件，v 表示视频流）。setpts=2.00*PTS 用于更改视频的播放时间戳（PTS），2.00*PTS 表示将视频的速度减半（使视频变慢）。PTS（Presentation Time Stamp）是决定视频帧显示时间的标记。[v] 是处理后的视频流的标签，用于后续的 -map 选项。[0:a] 表示输入文件中的音频流。atempo=0.50 用于调整音频的播放速度，0.50 表示音频速度减半（与视频同步）。[a] 是处理后的音频流的标签。

（3）-map "[v]" -map "[a]"：这些选项用于选择输出文件中应包含的流。其中 -map "[v]" 表示选择标签为 [v] 的视频流作为输出流。-map "[a]" 表示选择标签为 [a] 的音频流作

为输出流。

（4）speed_video.mp4：处理后的视频将要保存的文件名。

下面的例子使用 Java 调用 FFmpeg 命令将 video.mp4 视频的播放速度变慢为原来的 0.5 倍，并将结果保存为 speed_video.mp4 文件。

代码位置：src/main/java/video/effects/SpeedControl.java

```
package video.effects;
import java.io.BufferedReader;
import java.io.File;
import java.io.IOException;
import java.io.InputStreamReader;
public class SpeedControl {
    public static void main(String[] args) {
        String inputVideo = "video.mp4";                // 输入视频文件路径
        String outputVideo = "speed_video.mp4";         // 输出视频文件路径
        double speedFactor = 0.5;                       // 变速因子
        try {
            File file = new File(outputVideo);
            if (file.exists()) {
                file.delete();
            }
            processVideo(inputVideo, outputVideo, speedFactor);
        } catch (Exception e) {
            e.printStackTrace();
        }
    }
    private static void processVideo(String inputVideo, String outputVideo,
double speedFactor) throws IOException, InterruptedException {
        String[] command = buildFfmpegCommand(inputVideo, outputVideo,
speedFactor);
        Process process = Runtime.getRuntime().exec(command);
        // 捕获错误输出
        try (BufferedReader stdError = new BufferedReader(new InputStreamReader
(process.getErrorStream()))) {
            String line;
            while ((line = stdError.readLine()) != null) {
                System.out.println(line);
            }
        }
        int exitCode = process.waitFor();
        if (exitCode != 0) {
            throw new IOException("FFmpeg exited with error code: " + exitCode);
        } else {
            System.out.println("Video processing completed successfully.");
        }
    }
    private static String[] buildFFmpegCommand(String inputVideo, String
```

```
outputVideo, double speedFactor) {
        // 根据速度因子计算 setpts 和 atempo 的值
        String setptsValue = String.format("%.2f*PTS", 1 / speedFactor);
        String atempoValue = String.format("%.2f", speedFactor);

        return new String[]{"ffmpeg", "-i", inputVideo, "-filter_complex",
                String.format("[0:v]setpts=%s[v];[0:a]atempo=%s[a]", setptsValue,
atempoValue),
                "-map", "[v]", "-map", "[a]", outputVideo};
    }
}
```

9.4 为视频添加水印

使用 FFmpeg 可以为视频添加水印，包括文本水印和图像水印。例如，下面的命令同时为 video.mp4 在左上角添加了文本水印，在右下角添加了图像水印，并将带水印的视频保存为 watermark_video.mp4 文件。

```
ffmpeg -i video.mp4 -i images/watermark.png -filter_complex "[1][0]scale2ref=w
='iw/10':h='ow/mdar'[wm][vid];[vid][wm]overlay=W-w-10:H-h-10,drawtext=fontfile=
msyh.ttc:text=' 文本水印 ':fontcolor=black:fontsize=24:x=10:y=10" -codec:a copy
watermark_video.mp4
```

参数含义如下。

（1）-i video.mp4：指定输入视频文件。

（2）-i images/watermark.png：指定图像水印文件。

（3）-filter_complex：用于定义复杂的过滤流程。

（4）[1][0]scale2ref=w='iw/10':h='ow/mdar'[wm][vid]：这部分的命令用于调整水印图像的大小。其中 [1][0] 表示将第 2 个输入（水印图像）和第 1 个输入（视频）作为参考。scale2ref 表示调整水印图像大小为视频宽度的 1/10，同时保持宽高比。

（5）[vid][wm]overlay=W-w-10:H-h-10：表示将调整后的水印图像放在视频的右下角。W 和 H 是视频的宽度和高度，w 和 h 是水印图像的宽度和高度。

（6）drawtext=fontfile=msyh.ttc:text=' 文本水印 ':fontcolor=black:fontsize=24:x=10:y=10`：使用 drawtext 过滤器在视频左上角添加文本水印，指定字体文件、文本内容、颜色、大小和位置。

（7）-codec:a copy：复制音频流而不重新编码。

（8）watermark_video.mp4：输出文件名。

下面的例子演示了如何用 Java 调用 FFmpeg 为 video.mp4 添加文本水印（左上角）和图像水印（右下角），效果如图 9-3 所示。

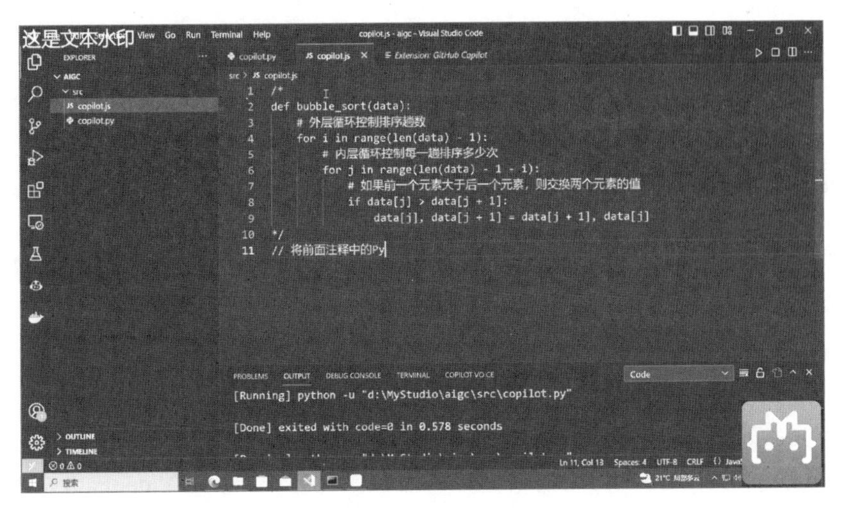

图 9-3　添加了水印的视频

代码位置：src/main/java/video/effects/VideoWatermark.java

```
package video.effects;
import java.io.IOException;
public class VideoWatermark {
    public static void main(String[] args) {
        String inputVideo = "video.mp4";
        String outputVideo = "watermark_video.mp4";
        String watermarkImage = "images/watermark.png";
        String watermarkText = " 这是文本水印 ";
        String fontFile = "msyh.ttc";                    // 替换为实际的字体文件路径
        String[] command = buildFfmpegCommand(inputVideo, outputVideo,
watermarkText, fontFile, watermarkImage);
        try {
            Process process = new ProcessBuilder(command).inheritIO().start();
            process.waitFor();
        } catch (IOException | InterruptedException e) {
            e.printStackTrace();
        }
    }
    private static String[] buildFfmpegCommand(String inputVideo, String
outputVideo, String watermarkText, String fontFile, String watermarkImage) {
        return new String[]{
                "ffmpeg", "-i", inputVideo,
                "-i", watermarkImage,
                "-filter_complex",
                String.format("[1][0]scale2ref=w='iw/10':h='ow/mdar'[wm][vid];" +
                               "[vid][wm]overlay=W-w-10:H-h-10," +
"drawtext=fontfile=%s:text='%s':fontcolor=white:fontsize=60:x=10:y=10",
                        fontFile, watermarkText),
                "-codec:a", "copy",
                outputVideo
        };
```

```
        }
    }
```

9.5 缩放和拉伸视频

使用 FFmpeg 可以缩放和拉伸视频，下面分别介绍 FFmpeg 的这两种操作的命令行格式。

❑ 缩放视频

```
ffmpeg -i inputVideoPath -vf scale=640:480 scaleOutputPath
```

参数含义如下。

（1）-i inputVideoPath：指定输入文件的路径。在这里，inputVideoPath 应被替换为实际的视频文件路径。

（2）-vf scale=640:480：使用视频过滤器（-vf）对视频进行缩放。scale=640:480 设置输出视频的分辨率为 640×480 像素。

（3）scaleOutputPath：指定输出文件的路径。在这里，scaleOutputPath 是输出视频文件的名称。

❑ 拉伸视频

```
ffmpeg -i inputVideoPath -vf scale=640:360 stretchOutputPath
```

拉伸视频和缩放视频的命令行格式完全相同，唯一不同的是拉伸比例。如果拉伸比例与原视频相同，那么就是缩放，否则就是拉伸。

下面的例子是缩放和拉伸 video.mp4，并分别生成了两个视频文件：scaled_video.mp4 和 stretched_video.mp4。

代码位置：src/main/java/video/effects/VideoTransformations.java

```
package video.effects;
import java.io.BufferedReader;
import java.io.IOException;
import java.io.InputStreamReader;
public class VideoTransformations {
    // 方法：执行 FFmpeg 命令
    public static void executeFFmpegCommand(String command) {
        Process process;
        try {
            process = Runtime.getRuntime().exec(command);
            // 读取命令的输出信息
            BufferedReader reader = new BufferedReader(new InputStreamReader
(process.getErrorStream()));
            String line;
```

```
            while ((line = reader.readLine()) != null) {
                System.out.println(line);                   // 打印 FFmpeg 命令的输出信息
            }
            // 等待 FFmpeg 命令执行完毕
            process.waitFor();
        } catch (IOException | InterruptedException e) {
            e.printStackTrace();
        }
    }
    public static void main(String[] args) {
        String inputVideoPath = "video.mp4";                // 输入视频文件的路径
        String scaleOutputPath = "scaled_video.mp4";   // 缩放后的视频文件路径
        String stretchOutputPath = "stretched_video.mp4";   // 拉伸后的视频文件路径
        // 缩放视频命令
        String scaleCommand = "ffmpeg -i " + inputVideoPath + " -vf scale=
640:480 " + scaleOutputPath;
        // 拉伸视频命令
        String stretchCommand = "ffmpeg -i " + inputVideoPath + " -vf scale=
640:360 " + stretchOutputPath;
        // 执行缩放视频命令
        executeFFmpegCommand(scaleCommand);
        // 执行拉伸视频命令
        executeFFmpegCommand(stretchCommand);
    }
}
```

9.6　视频 3D 透视变换

FFmpeg 本身并没有能力对视频做 3D 透视变换，但可以利用 OpenCV 的 Imgproc.warpPerspective 方法，对视频的每一帧做 3D 透视变换，然后生成一个新的视频文件，但用 OpenCV 生成的这个视频文件是不带音频的，所以仍然需要用 FFmpeg 将原视频的音频添加到新视频中，形成最终的带音频的 3D 透视变换视频。在《奇妙的 Python：神奇代码漫游之旅》一书中第 9.5 节的例子使用 moviepy 模块对视频进行 3D 透视变换，基本原理与本例类似，只是 Java 中没有类似 moviepy 的库，所以我们只能手动来完成这个过程。

本节案例主要分两部分实现 :3D 透视变换和音频合成。下面分别介绍这两部分的实现原理。

1. 3D 透视变换

透视变换矩阵的关键是定义透视变换的参考框架。本例使用 srcPoints（源点）和 dstPoints（目标点）定义了透视变换的参考框架。源点通常是视频帧的四个角，目标点则根据所需的 3D 透视效果进行调整。本例中，右上角和右下角的点被向上和向下移动，以创建一种深度感。

Imgproc.getPerspectiveTransform() 方法用于计算从源点到目标点的透视变换矩阵，该矩阵随后被用于每一帧的变换。

要处理视频中的每一帧，需要使用循环。在循环中，每次读取一帧，对其应用 warpPerspective() 方法进行透视变换，然后使用 VideoWriter.write() 方法，将进行了透视变换后的帧图像写入新的视频文件。当所有帧都被处理后，释放 VideoCapture（源视频）和 VideoWriter（目标视频）资源，完成视频的变换部分。

2. 音频合成

合成音频使用了 FFmpeg 命令行工具，具体的命令如下。

```
ffmpeg -i video_temp.mp4 -i video.mp4 -c:v copy -c:a aac -strict experimental
perspective_video.mp4
```

参数含义如下：

（1）-i video_temp.mp4：指定变换后无声的视频文件为输入。

（2）-i video.mp4：指定原始视频文件为第 2 个输入，用于提取音频流。

（3）-c:v copy：指示 FFmpeg 复制视频流而不重新编码。

（4）-c:a aac：指示 FFmpeg 使用 AAC 编解码器对音频进行编码。

（5）-strict experimental：某些 FFmpeg 版本要求此选项以启用某些功能，如 AAC 编码。

（6）perspective_video.mp4：指定输出文件的名称。

这行命令的作用是将 video_temp.mp4 的视频流和 video.mp4 的音频流合并到 perspective_video.mp4 输出文件中。

整个程序的工作流程是首先读取原始视频，逐帧进行 3D 透视变换，然后将这些变换后的帧写入一个新的无声视频文件。最后，使用 FFmpeg 将原始视频的音频流合并到变换后的视频中，生成最终的带音频的视频文件。这种处理方式可以利用多种图像处理工具制作出非常炫酷的视频特效。

下面的例子演示了如何使用 Java，通过 OpenCV 和 FFmpeg 完成对视频的 3D 透视变换渲染，效果如图 9-4 所示。

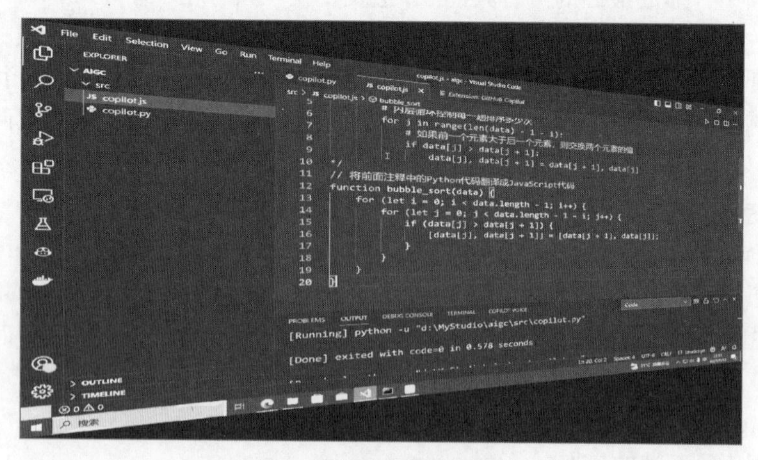

图 9-4 视频 3D 透视变换

代码位置：src/main/java/video/effects/VideoFramePerspectiveTransform.java

```
package video.effects;
import org.opencv.core.*;
import org.opencv.imgproc.Imgproc;
import org.opencv.videoio.VideoCapture;
import org.opencv.videoio.Videoio;
import org.opencv.videoio.VideoWriter;
import java.io.File;
public class VideoFramePerspectiveTransform {
    static { System.loadLibrary("opencv_java490"); }
    public static void main(String[] args) {
        String inputVideoPath = "video.mp4";            // 输入视频路径
        String outputVideoPath = "video_temp.mp4";      // 输出视频路径
        // 打开视频文件
        VideoCapture videoCapture = new VideoCapture(inputVideoPath);
        if (!videoCapture.isOpened()) {
            System.out.println("视频文件打开失败。");
            return;
        }
        // 获取视频属性
        Size size = new Size(videoCapture.get(Videoio.CAP_PROP_FRAME_WIDTH),
                videoCapture.get(Videoio.CAP_PROP_FRAME_HEIGHT));
        int fourcc = VideoWriter.fourcc('H', '2', '6', '4');
        double fps = videoCapture.get(Videoio.CAP_PROP_FPS);
        // 创建视频写入器
        VideoWriter writer = new VideoWriter(outputVideoPath, fourcc, fps,
 size, true);
        Mat frame = new Mat();
        Mat transformedFrame = new Mat(size, CvType.CV_8UC3);
        double w = size.width;
        double h = size.height;
        // 定义透视变换的源点和目标点
        MatOfPoint2f srcPoints = new MatOfPoint2f(
                new Point(0, 0),
                new Point(w, 0),
                new Point(0, h),
                new Point(w, h)
        );
        MatOfPoint2f dstPoints = new MatOfPoint2f(
                new Point(0, 0),                  // 左上角点保持不变
                new Point(w, h * 0.2),            // 右上角点向上移动
                new Point(0, h),                  // 左下角点保持不变
                new Point(w, h * 0.8)             // 右下角点向下移动
        );
        // 计算透视变换矩阵
        Mat perspectiveTransform = Imgproc.getPerspectiveTransform(srcPoints,
 dstPoints);
        while (true) {
            // 读取帧
            boolean hasFrame = videoCapture.read(frame);
```

```
                if (!hasFrame) {
                    break;
                }
                // 应用透视变换
                Imgproc.warpPerspective(frame, transformedFrame, perspectiveTransform,
size);
                // 写入新的视频文件
                writer.write(transformedFrame);
            }
            // 释放资源
            videoCapture.release();
            writer.release();
            // 添加音频
            addAudio(inputVideoPath, outputVideoPath);
        }
        // 使用 FFmpeg 命令行工具添加音频
        public static void addAudio(String inputVideoPath, String outputVideoPath) {
            try {
                String outputPath = "perspective_video.mp4"; // 最终带有音频的视频文件
                String[] command = {
                        "ffmpeg",
                        "-i", outputVideoPath,
                        "-i", inputVideoPath,
                        "-c:v", "copy",
                        "-c:a", "aac",
                        "-strict", "experimental",
                        outputPath
                };
                Process process = new ProcessBuilder(command).start();
                process.waitFor();
                // 删除无声视频文件
                new File(outputVideoPath).delete();
            } catch (Exception e) {
                e.printStackTrace();
            }
        }
    }
```

注意： 在透视变换中，点的移动取决于它们在目标图像中的位置。原始图像中的点（在 srcPoints 中定义）通常是视频帧的四个角，这些点在未经变换的状态下代表了图像的实际角落。

目标点（在 dstPoints 中定义）则确定了这些角落在变换后的图像中应该出现在哪里。当改变这些点的位置时，会在图像中创建出透视的效果。

（1）如果减小右上角点的 y 坐标（例如，使用 h * 0.2），那么这个点在图像中实际上是向上移动了，因为图像坐标系中 y 值是从上往下增加的（在 OpenCV 坐标系中，左上角是坐标原点）。

（2）如果增大右下角点的 y 坐标（例如，使用 h * 0.8），那么这个点在图像中是向下移动了。

所以，h * 0.2 意味着右上角的点被移动到了其原始高度的 20% 处，看起来像是向上移动，因为它离图像的顶部更近了。相反，h * 0.8 意味着右下角的点被移动到了其原始高度的 80% 处，看起来像是向下移动，因为它离图像的顶部更远了。

9.7 视频局部高斯模糊

对视频进行高斯模糊与 3D 透视变换类似，仍然需要 FFmpeg 与 OpenCV 配合。也就是利用 OpenCV 对视频的每一帧画面进行高斯模糊，并生成一个经过高斯模糊处理的视频文件，但这个视频文件是没有声音的，所以需要利用 FFmpeg 将原视频的音频与新视频合并，形成最终带音频的高斯模糊视频。

1. 高斯模糊

对视频中的每一帧画面进行高斯模糊，需要在循环中完成。在这个循环中，使用 capture.read(frame) 方法逐帧读取视频。对于每一帧，定义一个矩形区域 Rect roi = new Rect(0, 0, 120, 200)，这个区域指定了要进行高斯模糊的视频帧的左上角部分。使用 GaussianBlur(subMat, subMat, new Size(27, 27), 0) 对该区域应用高斯模糊。这里使用了 27 × 27 的核和 0 的标准差（让 OpenCV 自动计算标准差）。最后，使用 writer.write(frame) 将处理后的帧写入新的视频文件。

2. 音频和视频合并

这一部分需要两步完成，首先会从原视频中提取出音频，然后会将音频与高斯模糊后的视频合并。

（1）提取音频：

```
ffmpeg -i input.mp4 -vn -acodec copy audio.aac
```

① -i input.mp4：-i 指定输入文件，这里是原始视频文件 input.mp4。

② -vn：这个选项指示 FFmpeg 忽略视频流，即不处理视频，只处理音频。

③ -acodec copy：这里 -acodec 指定音频编解码器。copy 表示直接复制音频流，不进行任何转码。这意味着原始视频的音频将被保留在输出文件中，没有经过任何质量上的改变。

④ audio.aac：指定输出文件的名称和格式，这里是 audio.aac。

此命令的目的是从原视频中提取未经修改的音频流，并将其保存为 AAC 格式的文件。

（2）合并音频与视频：

```
ffmpeg -i blurred_video.mp4 -i audio.aac -c:v copy -c:a aac -strict experimental
final_output.mp4
```

① -i blurred_video.mp4：指定第 1 个输入文件，这里是经过高斯模糊处理的视频文件 blurred_video.mp4。

② -i audio.aac：指定第 2 个输入文件，这里是之前提取的音频文件 audio.aac。

③ -c:v copy：-c:v 指定视频编解码器。copy 表示直接复制视频流，不进行转码。这意味着处理后的视频流将直接被复制到最终的输出文件中。

④ -c:a aac：-c:a 指定音频编解码器，这里将音频转码为 AAC 格式。即使原音频已经是 AAC 格式，此选项也确保输出文件的音频编码的一致性。

⑤ -strict experimental：这是一个额外的参数，用于启用实验性功能，有时需要用于支持某些编解码器。

⑥ final_output.mp4：指定输出文件的名称和格式，这里是 final_output.mp4。

此命令的目的是将经过高斯模糊处理的视频和提取的原始音频合并为一个新的 MP4 文件。通过这种方式，能够得到一个既包含原始音频又包含经过视觉处理视频的输出文件。

下面的例子使用 OpenCV 和 FFmpeg，将视频左上角的区域 (0,0,120,200)进行高斯模糊，其他区域正常，效果如图 9-5 所示。

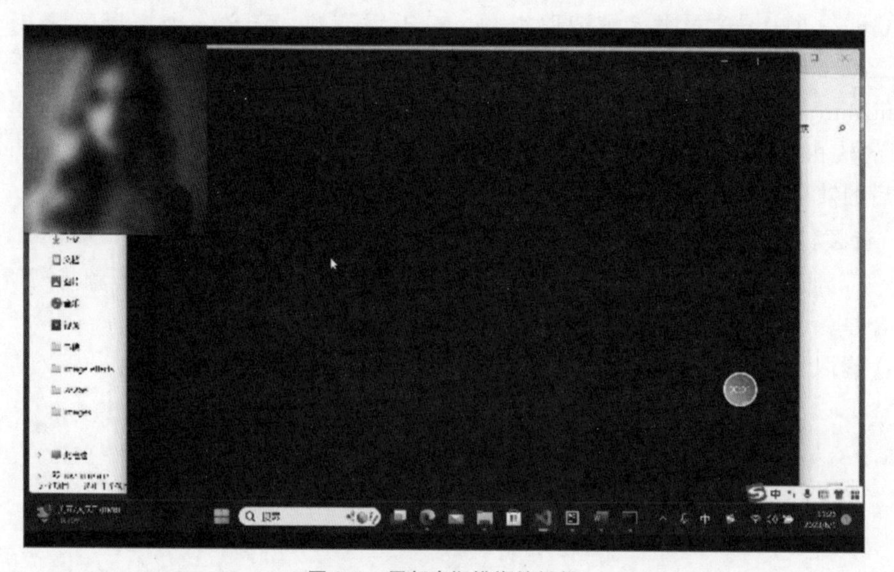

图 9-5 局部高斯模糊的视频

代码位置：src/main/java/video/effects/VideoGaussianBlur.java

```
package video.effects;
import org.opencv.core.Mat;
import org.opencv.core.Rect;
import org.opencv.core.Size;
import org.opencv.videoio.VideoCapture;
import org.opencv.videoio.VideoWriter;
import org.opencv.videoio.Videoio;
```

```java
import java.io.File;
import java.io.IOException;
import static org.opencv.imgproc.Imgproc.GaussianBlur;
public class VideoGaussianBlur {
    static {
        System.loadLibrary("opencv_java490");
    }
    public static void main(String[] args) throws Exception {
        String inputVideoPath = "input.mp4";                    // 输入视频路径
        String outputVideoPath = "blurred_video.mp4";           // 输出视频路径
        String audioPath = "audio.aac";                         // 音频路径
        File file = new File(outputVideoPath);
        if (file.exists()) {
            file.delete();
        }
        file = new File(audioPath);
        if (file.exists()) {
            file.delete();
        }
        file = new File("final_output.mp4");
        if (file.exists()) {
            file.delete();
        }
        // 提取音频
        executeCommand("ffmpeg -i " + inputVideoPath + " -vn -acodec copy " +
audioPath);
        // 应用高斯模糊
        VideoCapture capture = new VideoCapture(inputVideoPath);
        Size frameSize = new Size((int) capture.get(Videoio.CAP_PROP_FRAME_WIDTH),
                (int) capture.get(Videoio.CAP_PROP_FRAME_HEIGHT));
        int fourcc = VideoWriter.fourcc('M', 'P', '4', 'V');
        VideoWriter writer = new VideoWriter(outputVideoPath, fourcc, capture.
get(Videoio.CAP_PROP_FPS), frameSize, true);
        Mat frame = new Mat();
        while (capture.read(frame)) {
            // 对左上角区域进行高斯模糊
            Rect roi = new Rect(0, 0, 120, 200);                // 模糊区域
            Mat subMat = frame.submat(roi);
            GaussianBlur(subMat, subMat, new Size(27, 27), 0);
            writer.write(frame);
        }
        capture.release();
        writer.release();
        // 合并音频和视频
        executeCommand("ffmpeg -i " + outputVideoPath + " -i " + audioPath +
" -c:v copy -c:a aac -strict experimental final_output.mp4");
    }
    private static void executeCommand(String command) throws IOException,
InterruptedException {
        Process process = Runtime.getRuntime().exec(command);
        int exitCode = process.waitFor();
```

```
            if (exitCode != 0) {
                throw new IOException("Command execution failed: " + command);
            }
        }
    }
```

9.8 视频转场淡入淡出动画

FFmpeg 支持视频的各种转场动画，其中淡入淡出是最常见的转场动画。如果要实现两个视频之间的淡入淡出转场动画，就需要第 1 个视频在结尾的一段视频实现淡出效果，第 2 个视频开始的一段视频实现淡入的效果。命令行如下。

```
ffmpeg -i a.mp4 -i b.mp4 -filter_complex "[0:v]fade=t=out:st=4:d=2,setpts=PTS-STARTPTS[v0];[1:v]fade=t=in:st=0:d=2,setpts=PTS-STARTPTS[v1];[v0][0:a][v1][1:a]concat=n=2:v=1:a=1[v][a]" -map "[v]" -map "[a]" fade_in_out_video.mp4
```

参数含义如下。

（1）-i a.mp4：指定第 1 个输入文件，即 "a.mp4"。

（2）-i b.mp4：指定第 2 个输入文件，即 "b.mp4"。

（3）-filter_complex：开始一个复杂的过滤器图。这里用于定义多个视频流之间的操作。

（4）[0:v]fade=t=out:st=4:d=2,setpts=PTS-STARTPTS[v0];：对第 1 个视频应用淡出效果。[0:v] 表示第 1 个视频的视频流；fade=t=out 表示应用淡出效果；st=4 表示开始时间，这里是从第 1 个视频的总长度减去淡出时间（例如，如果视频时长是 6 秒，淡出时间是 2 秒，则开始时间是 4 秒）；d=2 表示淡出持续时间，这里是 2 秒；setpts=PTS-STARTPTS[v0] 表示重设视频帧的展示时间戳，确保从 0 开始，并将处理后的视频标记为 [v0]。

（5）[1:v]fade=t=in:st=0:d=2,setpts=PTS-STARTPTS[v1];：对第 2 个视频应用淡入效果。[1:v] 表示第 2 个视频的视频流；fade=t=in 表示应用淡入效果；st=0 表示淡入开始时间，从第 2 个视频的开始处；d=2 表示淡入持续时间，这里是 2 秒；setpts=PTS-STARTPTS[v1] 表示类似于前面的 setpts，为第 2 个视频重设时间戳，并标记为 [v1]。

（6）[v0][0:a][v1][1:a]concat=n=2:v=1:a=1[v][a]：将处理后的视频和音频流进行拼接。concat=n=2:v=1:a=1 表示拼接两个视频和音频流；[v][a] 表示拼接后的视频和音频流分别标记为 [v] 和 [a]。

（7）-map "[v]" -map "[a]"：选择拼接后的视频流 [v] 和音频流 [a] 作为输出。

（8）fade_in_out_video.mp4：指定输出文件名。

下面的例子使用 Java 调用 FFmpeg 将 a.mp4 和 b.mp4 合并，添加淡入淡出转场动画，并将合并后的视频保存为 fade_in_out_video.mp4。

代码位置：src/main/java/video/effects/VideoFadeInOut.java

```
package video.effects;
import java.io.BufferedReader;
import java.io.File;
import java.io.IOException;
import java.io.InputStreamReader;
import org.bytedeco.javacv.FFmpegFrameGrabber;
import org.bytedeco.javacv.FrameGrabber.Exception;
public class VideoFadeInOut {
    public static void main(String[] args) {
        // 输入视频文件路径
        String video1Path = "a.mp4";
        String video2Path = "b.mp4";
        // 输出视频文件路径
        String outputVideoPath = "fade_in_out_video.mp4";
        File file = new File(outputVideoPath);
        if (file.exists()) {
            file.delete();
        }
        // 淡入淡出的时间长度（秒）
        int fadeDuration = 2;
        // 构建 FFmpeg 命令
        String[] ffmpegCommand = buildFFmpegCommand(video1Path, video2Path,
outputVideoPath, fadeDuration);
        // 执行命令
        executeCommand(ffmpegCommand);
    }
    // 构建 FFmpeg 命令
    private static String[] buildFFmpegCommand(String video1Path, String
video2Path, String outputVideoPath, int fadeDuration) {
        long duration1 = getDuration(video1Path);
        return new String[]{
                "ffmpeg", "-i", video1Path, "-i", video2Path,
                "-filter_complex",
                "[0:v]fade=t=out:st=" + (duration1 - fadeDuration) + ":d=" +
fadeDuration + ",setpts=PTS-STARTPTS[v0];" +
                        "[1:v]fade=t=in:st=0:d=" + fadeDuration + ",setpts=PTS-
STARTPTS[v1];" +
                        "[v0][0:a][v1][1:a]concat=n=2:v=1:a=1[v][a]",
                "-map", "[v]", "-map", "[a]", outputVideoPath
        };
    }
    // 获取视频时长
    private static long getDuration(String videoPath) {
        try {
            FFmpegFrameGrabber grabber = new FFmpegFrameGrabber(videoPath);
            grabber.start();
            // 获取视频持续时间（单位：微秒）
            long duration = grabber.getLengthInTime() / 1000000L;
            grabber.stop();
```

```
            return duration;
        } catch (Exception e) {
            e.printStackTrace();
        }
        return 0;
    }
    // 执行命令行命令
    private static void executeCommand(String[] command) {
        try {
            ProcessBuilder processBuilder = new ProcessBuilder(command);
            Process process = processBuilder.start();
            BufferedReader stdInput = new BufferedReader(new InputStreamReader
(process.getInputStream()));
            BufferedReader stdError = new BufferedReader(new InputStreamReader
(process.getErrorStream()));
            // 读取命令输出
            String line;
            while ((line = stdInput.readLine()) != null) {
                System.out.println(line);
            }
            // 读取错误输出
            while ((line = stdError.readLine()) != null) {
                System.err.println(line);
            }
            process.waitFor();
        } catch (IOException e) {
            e.printStackTrace();
        } catch (InterruptedException e) {
            e.printStackTrace();
        }
    }
}
```

9.9 向视频中添加动态图像

向视频中添加动态图像，其实对于每一帧，都是静态图像，只是不同帧的图像，插入的图像有不同的状态，所以一帧一帧播放视频时，图像就会变成动态图像，就像小时候快速翻动引用不同状态的孙悟空插画书一样，这样孙悟空就会不断翻跟头了。

本节的例子会让一个 logo.png 图像从视频的左上角逐渐向视频的右下角移动，在移动的同时，会不断顺时针旋转。要完成这个例子，需要使用多种技术，如通过 AWT 中的相关 API 移动和旋转 logo.png。使用 OpenCV 从视频中读取每一帧图像，使用 FFmpeg 将没有声音的视频与原视频的音频部分合并，生成最终添加了动态图像的视频文件。下面是本例关键技术的实现原理。

1. 初始化和视频读取

使用 OpenCV 的 VideoCapture 类来打开视频文件，并从中读取帧。VideoWriter 类则用于创建和写入新的视频文件。初始化时，需要设置 logo 的开始位置在视频的左上角，且初始旋转角度为 0。

2. 缩放 logo 图像

logo 图像首先被缩放到 100 × 100 像素。这是通过创建一个新的 BufferedImage 对象并使用 Graphics2D 来将原始 logo 图像绘制到这个新的尺寸来实现的。

3. 逐帧处理

在循环中，每读取一帧视频，就对该帧进行以下操作。

（1）转换帧格式：将 OpenCV 的 Mat 对象转换为 Java AWT 的 BufferedImage 对象，以便使用 Java 2D API 进行图像处理。

（2）设置图像变换：使用 AffineTransform 对象来定义 logo 图像的旋转和平移操作。这是通过 translate() 和 rotate() 方法实现的。平移操作移动 logo 图像到当前计算出的新位置，而旋转操作则以当前的角度旋转图像。

（3）绘制变换后的图像：在设置好变换后，使用 Graphics2D 的 drawImage() 方法将 logo 绘制到当前帧上。

（4）将帧转换回 Mat 格式：完成绘制后，将修改后的 BufferedImage 对象转换回 OpenCV 的 Mat 格式，以便写入新视频文件。

4. 更新 logo 位置和角度

在每次循环迭代后，logo 的位置根据预定义的 deltaX 和 deltaY 值更新，这样 logo 就能够在视频播放时逐渐从左上角移动到右下角。同时，旋转角度增加，这使得 logo 在移动的同时绕其中心旋转。

5. FFmpeg 音视频合并

音视频合并的命令如下。

```
ffmpeg -i output_temp.mp4 -i input.mp4 -c:v copy -c:a aac -strict experimental
dynamic_image_video.mp4
```

参数含义如下。

（1）-i output_temp.mp4：指定输入文件。在这里，output_temp.mp4 是添加了 logo 的视频文件，但没有音频。

（2）-i input.mp4：再次使用 -i 指定第 2 个输入文件。在这个案例中，input.mp4 是原始视频文件，用于提供音频轨道。

（3）-c:v copy：指定视频编码器的设置。-c:v 是指视频编码器 (codec)，copy 表示直接复制视频流而不重新编码。这样做的优点是处理速度快，因为它避免了重新编码过程。

（4）-c:a aac：设置音频编码器。-c:a 是指音频编码器，aac 是音频编码格式。这里表示将音频重新编码为 AAC 格式。

（5）-strict experimental：这是一个额外的参数，用于启用实验性功能，特别是当使用较新的编码器或特殊选项时。在旧版本的 FFmpeg 中，某些编码器（如 AAC）被标记为实验性的，需要这个标志才能使用。

（6）dynamic_image_video.mp4：指定输出文件的名称。这个文件将是合并了音频和视频的最终结果。

在执行这行命令时，FFmpeg 将 output_temp.mp4（只有视频，无音频）和 input.mp4（原始视频的音频）合并成一个新文件 dynamic_image_video.mp4，这个文件既包含了视频（含 logo），也包含了音频。

下面的例子使用 Java 以及 FFmpeg、OpenCV 和 Java 2D API 在 input.mp4 中插入动态图像，该图像是一个 logo，从视频左上角开始向视频右下角匀速移动的同时不断顺时针旋转（每播放一帧，旋转一度）。效果如图 9-6 所示。

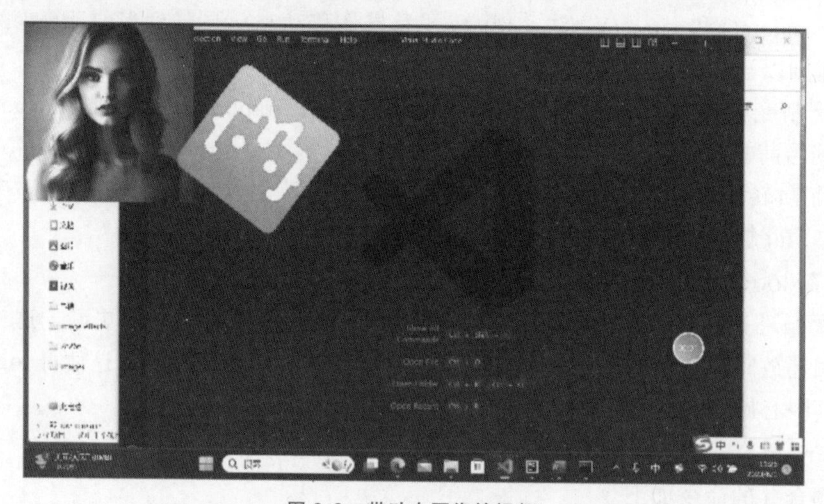

图 9-6　带动态图像的视频

代码位置：src/main/java/video/effects/DynamicImageVideo.java

```
package video.effects;
import org.opencv.core.*;
import org.opencv.videoio.VideoCapture;
import org.opencv.videoio.VideoWriter;
import org.opencv.videoio.Videoio;
import org.opencv.imgcodecs.Imgcodecs;
import javax.imageio.ImageIO;
import java.awt.*;
import java.awt.Point;
import java.awt.geom.AffineTransform;
import java.awt.image.BufferedImage;
import java.awt.image.DataBufferByte;
```

```
import java.io.ByteArrayInputStream;
import java.io.File;
import java.io.IOException;
public class DynamicImageVideo {
    static {
        System.loadLibrary("opencv_java490");
    }
    public static void main(String[] args) throws IOException {
        // 视频文件路径
        String videoFilePath = "input.mp4";
        String logoFilePath = "images/logo.png";
        String outputVideoPath = "output_temp.mp4";
        String finalVideoPath = "dynamic_image_video.mp4";
        BufferedImage originalLogo = ImageIO.read(new File(logoFilePath));
        // 缩放 logo 图像到 100×100
        BufferedImage scaledLogo = new BufferedImage(100, 100, BufferedImage.
TYPE_INT_ARGB);
        Graphics2D g2dLogo = scaledLogo.createGraphics();
        g2dLogo.setRenderingHint(RenderingHints.KEY_INTERPOLATION,
RenderingHints.VALUE_INTERPOLATION_BILINEAR);
        g2dLogo.drawImage(originalLogo, 0, 0, 100, 100, null);
        g2dLogo.dispose();
        // 打开视频文件
        VideoCapture capture = new VideoCapture(videoFilePath);
        if (!capture.isOpened()) {
            System.out.println("Error: Cannot open video file.");
            return;
        }
        // 获取视频信息
        Size videoSize = new Size(capture.get(Videoio.CAP_PROP_FRAME_WIDTH),
capture.get(Videoio.CAP_PROP_FRAME_HEIGHT));
        int fourcc = VideoWriter.fourcc('M', 'J', 'P', 'G');
        double fps = capture.get(Videoio.CAP_PROP_FPS);
        // 创建视频写入器
        VideoWriter writer = new VideoWriter(outputVideoPath, fourcc, fps,
videoSize, true);
        int frameNumber = 0;
        double angle = 0.0;                          // 初始化角度为 0
        Point position = new Point(0, 0);            // 初始化位置为左上角
        double deltaX = videoSize.width / (capture.get(Videoio.CAP_PROP_FRAME_
COUNT) - 1);                                         // 每帧在 X 方向上移动的距离
        double deltaY = videoSize.height / (capture.get(Videoio.CAP_PROP_FRAME_
COUNT) - 1);                                         // 每帧在 Y 方向上移动的距离
        Mat frame = new Mat();
        int frameCount = 0;
        while (capture.read(frame)) {
            // 每帧增加的旋转角度
            angle += 1.0;                            // 每帧旋转 1 度
            // 将 Mat 转换为 BufferedImage
```

```
                BufferedImage currentFrame = matToBufferedImage(frame);
                Graphics2D g2d = currentFrame.createGraphics();
                // 设置旋转参数
                AffineTransform transform = new AffineTransform();
                transform.translate(position.x, position.y);                // 移动图像
                transform.rotate(Math.toRadians(angle), 50, 50);            // 旋转图像
                // 执行旋转
                g2d.setTransform(transform);
                g2d.drawImage(scaledLogo, 0, 0, null);
                g2d.dispose();
                // 将 BufferedImage 转换回 Mat
                Mat logoFrame = bufferedImageToMat(currentFrame);
                // 将处理后的帧写入输出视频
                writer.write(logoFrame);
                // 更新位置
                position.x += deltaX;
                position.y += deltaY;
                // 如果 logo 到达右下角，则停止移动
                if (position.x >= videoSize.width - scaledLogo.getWidth() ||
position.y >= videoSize.height - scaledLogo.getHeight()) {
                    break;
                }
                frameNumber++;
            }
            // 释放资源
            capture.release();
            writer.release();
            try {
                // 这里添加调用 FFmpeg 合并音频的代码
                mergeAudioWithVideo(videoFilePath, outputVideoPath, finalVideoPath);
            }catch (Exception e) {
            }
        }
    private static void mergeAudioWithVideo(String originalVideoPath, String
videoNoAudioPath, String finalVideoPath) throws IOException, InterruptedException {
        // 构建 FFmpeg 命令
        String command = "ffmpeg -i " + videoNoAudioPath + " -i " +
originalVideoPath +
                " -c:v copy -c:a aac -strict experimental " + finalVideoPath;
        // 执行命令
        Process process = Runtime.getRuntime().exec(command);
        // 等待 FFmpeg 命令执行完成
        process.waitFor();
        // 获取命令执行的返回值
        int exitValue = process.exitValue();
        if (exitValue != 0) {
            throw new IOException("Error during merging audio and video with
FFmpeg. Exit code: " + exitValue);
        }
```

```
    }
private static BufferedImage matToBufferedImage(Mat mat) throws IOException{
        // 对图像进行编码
        MatOfByte matOfByte = new MatOfByte();
        Imgcodecs.imencode(".jpg", mat, matOfByte);
        // 将编码后的 Mat 存储在字节数组中
        byte[] byteArray = matOfByte.toArray();
        // 准备 BufferedImage
        BufferedImage bufImage = ImageIO.read(new ByteArrayInputStream(byteArray));
        return bufImage;
    }
    private static Mat bufferedImageToMat(BufferedImage bi) {
        Mat mat = new Mat(bi.getHeight(), bi.getWidth(), CvType.CV_8UC3);
        byte[] data = ((DataBufferByte) bi.getRaster().getDataBuffer()).getData();
        mat.put(0, 0, data);
        return mat;
    }
}
```

9.10　将视频转换为动画 GIF

用 FFmpeg 可以将视频转换为动画 GIF，命令行如下。

```
ffmpeg -i input.mp4 -vf fps=10 -c:v gif output.gif
```

参数含义如下。

（1）-i input.mp4：这部分指定输入文件。-i 是输入文件的选项，input.mp4 是要转换的视频文件的名称和路径。

（2）-vf fps=10：这部分是视频过滤器（-vf）的设置。fps=10 表示输出的 GIF 将以每秒 10 帧的速度播放。调整此值可以改变 GIF 的平滑程度和文件大小。

（3）-c:v gif：这部分指定视频编码器（-c:v）。gif 表示 FFmpeg 使用 GIF 编码器。这意味着输出文件将被编码为 GIF 格式。

（4）output.gif：输出文件的名称和路径。在这里，它被命名为 output.gif，但可以根据需要更改它。

这行命令将 input.mp4 视频文件转换为一个帧率为 10 帧 / 秒的 GIF 文件，并保存为 output.gif。注意，如果原始视频的帧率远高于 10 帧 / 秒，可能会导致动画看起来不够平滑。另外，由于没有指定尺寸调整，所以这个命令将保留视频的原始尺寸。如果视频尺寸很大，生成的 GIF 文件可能会非常大。

下面的例子用 Java 调用 FFmpeg，将 input.mp4 转换为 output.gif。图 9-7 是在 Chrome 浏览器中查看 output.gif 文件的效果。

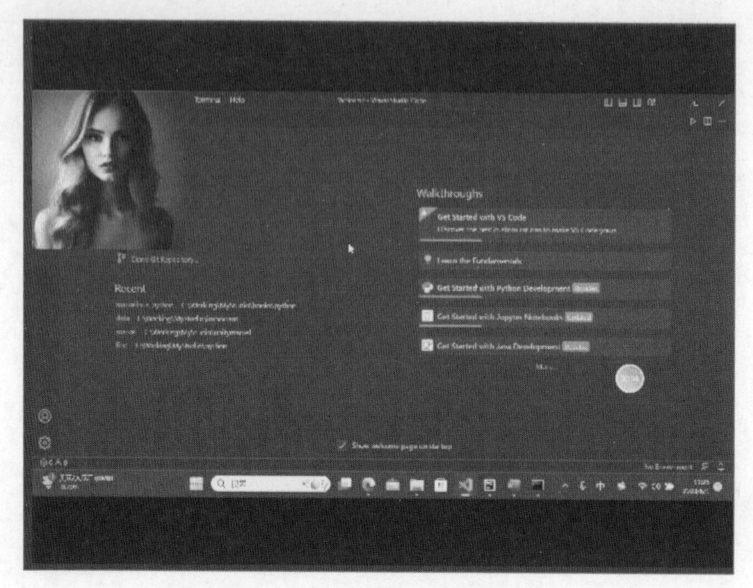

图 9-7　output.gif 的显示效果

代码位置：src/main/java/video/effects/Video2Gif.java

```
package video.effects;
import java.io.BufferedReader;
import java.io.File;
import java.io.IOException;
import java.io.InputStreamReader;
public class Video2Gif {
    // 执行 FFmpeg 命令
    public static void executeFFmpegCommand(String[] command) {
        Process process;
        try {
            process = new ProcessBuilder(command).start();
            // 读取命令的输出信息
            BufferedReader reader = new BufferedReader(new InputStreamReader
(process.getErrorStream()));
            String line;
            while ((line = reader.readLine()) != null) {
                System.out.println(line);              // 打印 FFmpeg 命令的输出信息
            }
            // 等待 FFmpeg 命令执行完毕
            process.waitFor();
        } catch (IOException | InterruptedException e) {
            e.printStackTrace();
        }
    }
    public static void main(String[] args) {
        // 输入文件路径
        String inputPath = "input.mp4";
```

```
        // 输出文件路径
        String outputPath = "output.gif";
        File file = new File(outputPath);
        if (file.exists()) {
            file.delete();
        }
        // FFmpeg 命令，直接将视频转换为 GIF，保留原始尺寸
        String[] command = {"ffmpeg", "-i", inputPath, "-vf", "fps=10", "-c:v",
"gif", outputPath};
        // 执行 FFmpeg 命令
        executeFFmpegCommand(command);
    }
}
```

9.11 为视频添加字幕

字幕通常是文本形式，并且以一段时间作为显示特定字幕的一句，例如，在第 4 秒到第 8 秒的时间间隔内，视频下方会显示一段文本，这称为字幕。

为视频添加字幕的基本方法就是视频与视频或视频与图片的混合。通常将字幕做成一组透明的图片（除了文字部分，其他部分都是透明的），这组透明的图片可以制作成字幕视频，每一个透明图片（特定字幕）会根据对应视频的时长决定展示的时间。最后将主视频与字幕视频混合，就形成了带字幕的视频。

尽管我们已经了解了为视频添加字幕的基本原理，但通常并不需要这么麻烦，使用 FFmpeg 可以直接将 srt 文件中的字幕添加到视频中，命令如下。

```
ffmpeg -i video.mp4 -vf subtitles=subtitles.srt -codec:a copy subtitle_output.mp4
```

参数含义如下。

（1）-i video.mp4：指定输入视频文件。

（2）-vf subtitles=subtitles.srt：应用视频过滤器来添加 subtitles.srt 文件中的字幕。

（3）-codec:a copy：复制原视频的音频流到输出文件，不进行转码。

（4）subtitle_output.mp4：指定输出文件名。

srt 文件的标准格式如下。

```
1
00:00:00,000 --> 00:00:02,000
<font color="red" size="20">first<font color="blue" size="120">subtitle</font></font>

2
00:00:02,000 --> 00:00:5,000
second subtitle
```

```
3
00:00:5,000 --> 00:00:7,000
<font color="green">third<font color="blue" size="120">subtitle</font><</font>
```

每条字幕分为如下 3 部分，每部分占一行：

（1）字幕序号从 1 开始。

（2）字幕对应的时间段，格式为 hh:mm:ss:xxx，表示时（hh）、分 (mm)、秒（ss）和毫秒（xxx）。

（3）字幕内容。可以是普通文本，也可以是带 <font> 标签的文本。通过 color 属性和 size 属性，可以设置字幕的颜色和尺寸。

下面的例子在 video.mp4 文件的底部添加了不同字体和颜色的字幕，并将新视频保存为 subtitle_video.mp4，效果如图 9-8 所示。

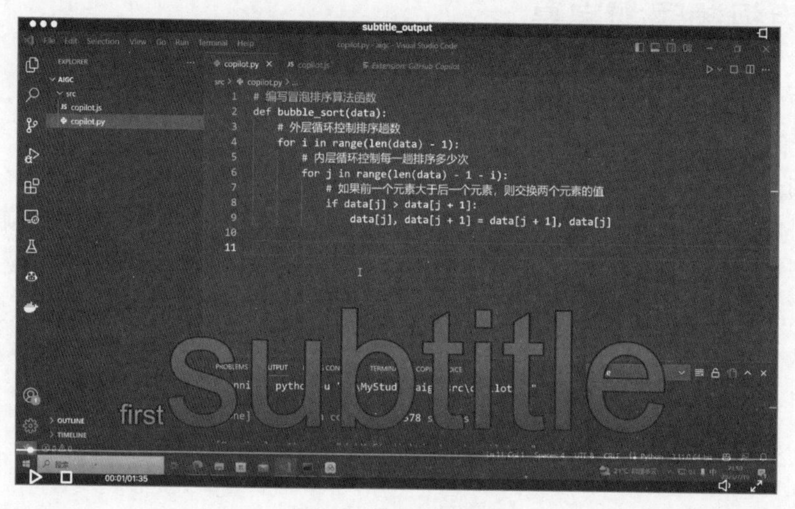

图 9-8 带字幕的视频

代码位置：src/main/java/video/effects/VideoSubtitle.java

```
package video.effects;
import java.io.IOException;
public class VideoSubtitle {
    public static void main(String[] args) {
        try {
            String videoPath = "video.mp4";                // 替换为视频文件的路径
            String subtitlePath = "subtitles.srt";         // 替换为 SRT 字幕文件的路径
            String outputPath = "subtitle_output.mp4"; // 替换为输出视频文件的路径
            addSubtitlesToVideo(videoPath, subtitlePath, outputPath);
        } catch (IOException | InterruptedException e) {
            e.printStackTrace();
        }
    }
```

```
    private static void addSubtitlesToVideo(String videoPath, String subtitlePath,
String outputPath) throws IOException, InterruptedException {
        // 构建 FFmpeg 命令
        String command = "ffmpeg -i " + videoPath + " -vf subtitles=" +
subtitlePath + " -codec:a copy " + outputPath;
        // 执行命令
        Process process = Runtime.getRuntime().exec(command);
        // 等待 FFmpeg 命令执行完成
        process.waitFor();
        // 获取命令执行的返回值
        int exitValue = process.exitValue();
        if (exitValue != 0) {
            throw new IOException("Error during adding subtitles with FFmpeg.
Exit code: " + exitValue);
        }
    }
}
```

9.12　将彩色视频变为灰度视频

使用 FFmpeg 可以将彩色视频转换为灰度视频，命令如下。

```
ffmpeg -i input.mp4 -vf format=gray -c:a copy output.mp4
```

参数含义如下。

（1）-i input.mp4：-i 代表输入文件，这里的 input.mp4 是想要转换的视频文件。需要替换 input.mp4 为实际视频文件的路径。

（2）-vf format=gray：这部分是视频过滤器（Video Filter）的命令。-vf 表示视频过滤器。format=gray 表示使用 format 过滤器将视频格式转换为灰度。gray 是指定的目标格式，它会将视频的每个帧转换为灰度图像。

（3）-c:a copy：这部分指定了对音频流的操作。

（4）-c:a：指音频编码器（codec for audio）。

（5）copy：表示直接复制音频流，不进行转码或修改。这确保了原始视频的音频部分被保留。

（6）output.mp4：这是输出文件的名称。转换后的视频将保存为这个文件。可以根据需要替换为不同的文件名或路径。

这条命令将输入的彩色视频转换为灰度视频，并保持音频流不变，输出为指定的文件。这是一种常见的操作，特别适用于需要减小视频文件大小或进行特定视觉效果处理的场景。

下面的例子将 input.mp4 文件转换为灰度视频文件，并保存为 gray_input.mp4。

代码位置：src/main/java/video/effects/ConvertToGray.java

```
package video.effects;
import java.io.IOException;
public class ConvertToGray {
    public static void main(String[] args) {
        try {
            String inputVideoPath = "input.mp4";          // 替换为输入视频文件的路径
            String outputVideoPath = "gray_output.mp4"; // 替换为输出视频文件的路径
            convertToGrayscale(inputVideoPath, outputVideoPath);
        } catch (IOException | InterruptedException e) {
            e.printStackTrace();
        }
    }
    private static void convertToGrayscale(String inputVideoPath, String
outputVideoPath) throws IOException, InterruptedException {
        // 构建 FFmpeg 命令
        String command = "ffmpeg -i " + inputVideoPath + " -vf format=gray
-c:a copy " + outputVideoPath;
        // 执行命令
        Process process = Runtime.getRuntime().exec(command);
        // 等待 FFmpeg 命令执行完成
        process.waitFor();
        // 获取命令执行的返回值
        int exitValue = process.exitValue();
        if (exitValue != 0) {
            throw new IOException("Error during converting video to grayscale
with FFmpeg. Exit code: " + exitValue);
        }
    }
}
```

9.13 小结

本节主要介绍了如何利用 FFmpeg 以及 OpenCV、Java2D API 等库实现各种视频特效。其实有了 FFmpeg，再配合各种图像处理库，如 OpenCV、OpenGL 等。几乎可以实现各种视频特效。本章只是抛砖引玉，通过将 FFmpeg、OpenCV 和 Java2D API 结合的方式，让读者了解 Java 是如何借用各种工具和库实现炫酷视频特效的。其实 FFmpeg 和 OpenCV 是跨语言的，不光 Java 可以用，Python、JavaScript 也可以用。如果读者想深入了解 OpenCV，可以看我的另外一本书《Python OpenCV 从菜鸟到高手》。

第 10 章

IntelliJ IDEA 插件

在本章中，我们将深入探讨 IntelliJ IDEA 插件开发的精髓，这是一项能够极大提升开发效率和个性化 IDE 体验的技术。通过学习本章，我们将了解到如何利用 IntelliJ IDEA 的强大扩展性，为开发环境量身定制功能。插件开发不仅能够满足个人或团队的特定需求，还能为更广泛的开发者社区贡献价值。我们将从插件的基本概念入手，逐步深入插件的类型、开发流程，以及如何将插件集成到 IntelliJ IDEA 中。通过实际案例分析，可学会如何创建菜单项、定制工具栏、上下文菜单、创建视图等多样化的插件，从而丰富我们的开发工具箱。

10.1 IntelliJ IDEA 插件简介

IntelliJ IDEA 插件是扩展 IntelliJ IDEA 集成开发环境（Integrated Development Environment，IDE）功能的组件。通过安装插件，用户可以定制和增强 IDE 的功能，以适应特定的开发需求和工作流程。插件可以添加新的语言支持、框架集成、代码分析工具、视觉主题和更多的实用功能。JetBrains 提供了一个强大的插件开发工具包（Software Development Kit SDK），使得开发者能够使用 Java 或 Kotlin 语言来创建插件。

从技术角度看，IntelliJ IDEA 主要支持如下的插件类型。

（1）添加菜单项和工具栏按钮：插件可以通过定义动作（AnAction）来添加新的菜单项或工具栏按钮。这些动作可以被绑定到特定的代码逻辑，如打开一个对话框、执行一个操作或修改编辑器内容。

（2）对话框和工具窗口：插件可以创建自定义的对话框（JDialog）和工具窗口（ToolWindow）来提供交互式的用户界面，展示信息或接收用户输入。

（3）语言支持：插件可以为新的或现有的编程语言提供额外的支持，包括语法高亮、代码补全、重构工具等。这通常涉及定义语言的语法和语义分析器。

（4）代码检查和修正：通过实现代码检查（InspectionTool）和快速修复（QuickFix）功能，插件可以帮助用户发现并修正代码中的问题。

（5）项目事件监听：插件可以监听和响应项目生命周期事件（如项目打开、关闭等）和文件系统事件（如文件添加、修改、删除等）。

（6）版本控制系统集成：插件可以扩展或添加对版本控制系统的支持，提供如分支管理、变更历史查看等功能。

（7）状态栏小部件：插件可以在状态栏添加自定义的小部件，用于显示信息或提供快速访问某些功能的入口。

（8）自定义主题和颜色方案：插件可以提供新的 IDE 主题或颜色方案，改变编辑器、工具窗口等的外观样式。

（9）构建和运行任务：插件可以定义新的运行 / 调试配置类型，允许用户配置和执行特定的构建任务或应用程序。

IntelliJ IDEA 插件开发利用了 IntelliJ Platform 的强大扩展性，几乎没有什么是不可能实现的。通过使用提供的 API 和扩展点，插件开发者可以深度集成并扩展 IDE 的核心功能，从而为用户带来更加丰富和个性化的开发体验。

JetBrains 提供了一系列基于 IntelliJ Platform 的 IDE，包括 PyCharm（针对 Python）、WebStorm（针对 JavaScript 和前端开发）、RubyMine、PhpStorm 等。许多 IntelliJ IDEA 插件可以在这些 IDE 中使用，前提是它们基于相同的平台特性和 API。这种兼容性使得开发者能够在不同的 JetBrains IDE 之间享受相似的工具和功能，尽管有时候需要插件作者确保他们的插件在不同 IDE 中都能正常工作。

然而，并非所有的 IntelliJ IDEA 插件都能在其他 JetBrains IDE 中使用。插件的兼容性取决于它所依赖的特定平台组件和 API。如果插件使用了仅在 IntelliJ IDEA 中可用的特定功能，那么这个插件就可能不兼容其他基于 IntelliJ Platform 的 IDE。因此，插件的通用性和兼容性取决于插件开发时的目标和依赖选择。

总的来说，IntelliJ IDEA 插件极大地丰富了开发者的开发体验，提供了无限的定制和扩展可能。通过 JetBrains 插件市场，开发者可以轻松地发现、安装和使用成千上万的插件，以适应他们的开发需求。

10.2 IntelliJ IDEA 插件的开发步骤

开发 IntelliJ IDEA 插件是一个创造性和技术性相结合的过程，涉及理解 IntelliJ Platform 的架构、API、扩展点以及插件开发工具包 (SDK)。以下是开发 IntelliJ IDEA 插件的一般步骤。

1. 环境准备

（1）安装 IntelliJ IDEA：确保安装了最新版本的 IntelliJ IDEA，包括社区版或旗舰版。

（2）配置 JDK：插件开发需要 JDK，确保开发环境配置了合适版本的 JDK。

2. 创建插件项目

（1）在 IntelliJ IDEA 中，单击 File → New → Project 菜单项，然后在如图 10-1 所示的 New Project 对话框中选择左侧的 IDE Plugin 项。

（2）配置项目设置，如项目名称、位置和使用的 JDK。其中 Group 将作为 Java 代码中包名的前一部分，而 Artifact 将作为包名的后一部分。对于图 10-1 所示的配置，Java 包名是 com.unitymarvel.codeline。

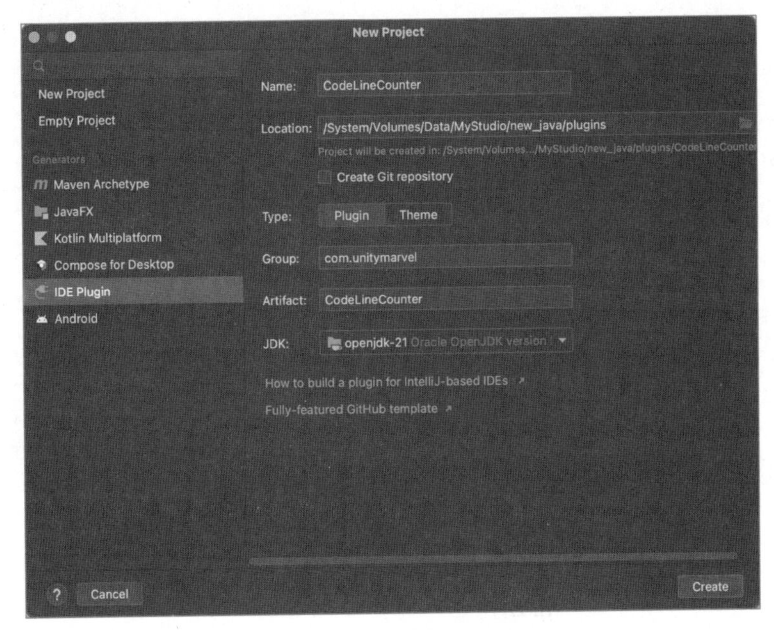

图 10-1　创建插件工程

3. 配置 plugin.xml 文件

plugin.xml 是插件的主要配置文件，位于 src/main/resources/META-INF 目录下。该文件定义了插件的基本信息、扩展点和依赖。

4. 编写插件代码

根据插件的功能需求，使用 Java 或 Kotlin 编写代码。例如，定义一个继承自 AnAction 的类来实现一个菜单项动作。

5. 运行和调试

在工具栏上的运行下拉列表中选择如图 10-2 所示的 Run Plugin 列表项，然后单击运行列表右侧的运行按钮（绿色箭头）。这时会新启动一个 IntelliJ IDEA 实例，插件会自动安装。如果要调试插件，单击运行按钮右侧的调试按钮即可。

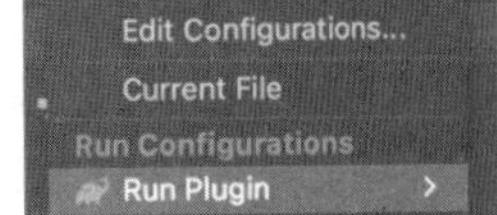

图 10-2　运行和调试插件

6. 打包和发布

运行插件后，会在 < 插件工程根目录 >/build/libs 中生成一个

JAR 文件，可以直接在 Settings 对话框中选择左侧的 Plugins，在右侧界面中单击配置菜单按钮，并单击 Install Plugin from Disk 菜单项安装 JAR 形式的插件，如图 10-3 所示。插件也可以通过 JetBrains Plugin Repository 发布，这样在 Marketplace 上就可以搜索到了。

图 10-3 从本地安装插件

10.3 plugin.xml 文件解读

plugin.xml 是 IntelliJ IDEA 插件的核心配置文件，其中主要的配置解释如下。

（1）<id>：插件的唯一标识符，通常使用反向域名格式，如 com.example.MyPlugin。

（2）<name>：插件的显示名称。

（3）<vendor>：插件的供应商信息，包括电子邮件和网址。

（4）<description>：插件的详细描述，支持基本的 HTML 标记。

（5）<version>：插件版本。

（6）<idea-version>：定义插件兼容的 IntelliJ IDEA 版本范围。

（7）<depends>：插件依赖的其他模块或插件。

（8）<extensions>：插件实现的扩展点。

（9）<actions>：插件添加的动作，可以指定添加到哪个菜单或工具栏中。

开发 IntelliJ IDEA 插件时，理解 plugin.xml 的结构和配置是至关重要的，因为它决定了插件的行为、外观和与 IDE 的集成方式。通过精心设计 plugin.xml，开发者可以确保插件能够顺利地集成到 IntelliJ IDEA 中，为用户提供有价值的功能和改进。

10.4　统计代码行数的插件

本章的例子会实现一个用于统计 Java 文件代码行数的 IntelliJ IDEA 插件。该插件会统计当前工程内所有 Java 文件去除注释和空行后的代码行数总和。开发这个插件需要考虑如下几个技术问题。

（1）在 Code 菜单中添加一个子菜单。

（2）弹出对话框。

（3）获取当前工程的目录。

（4）统计 Java 文件中的代码行数。

下面将详细介绍如何完成这些工作。

（1）在 Code 菜单中添加一个子菜单。

要在 IntelliJ IDEA 的 Code 菜单中添加一个子菜单项，首先需要在插件的 plugin.xml 配置文件中声明一个新的 action，并指定它应该被添加到 Code 菜单的某个组中。通过指定 <add-to-group> 标签的 group-id 属性为 CodeMenu，我们可以确保插件的菜单项会出现在 Code 菜单中。anchor 属性用于指定该菜单项相对于组中其他项的位置，last 表示将其放在菜单的末尾。核心代码如下：

```
<actions>
    <action id="CountCodeLinesAction" class="com.unitymarvel.codeline.
CountCodeLinesAction" text="统计代码行数" description="Counts code lines in Java
files.">
        <add-to-group group-id="CodeMenu" anchor="last"/>
    </action>
</actions>
```

在本例中，将新的菜单项添加到 Code 菜单的末尾是为了避免干扰到该菜单中已有的更常用的功能项，同时也使得这个新增的功能容易被找到。

（2）弹出对话框。

当用户单击菜单项时，我们使用 Swing 库中的 JDialog 类来弹出一个对话框。对话框的创建和显示是通过在 actionPerformed 方法中编写相应的 Swing 代码来完成的。对话框中包含一个按钮和两个标签，用于触发代码行数的统计操作和展示结果。

```
JDialog dialog = new JDialog();
dialog.setTitle("Code Lines Statistics");
dialog.setSize(300, 150);
dialog.setLayout(new BorderLayout());
// 添加按钮和标签到对话框
dialog.setLocationRelativeTo(null);
dialog.setVisible(true);
```

（3）获取当前工程的目录。

为了获取当前工程的目录，我们可以使用 ProjectUtil.guessProjectDir(project) 方法。

```
VirtualFile projectDir = ProjectUtil.guessProjectDir(project);
if (projectDir == null) return;
```

这里，projectDir 将是我们用于开始扫描项目中 Java 文件的 VirtualFile 对象。如果无法确定项目目录（即 guessProjectDir 返回 null），则操作终止。

（4）在线程中扫描所有 Java 文件并统计代码行数。

为了避免阻塞 UI 线程，需要在一个新的线程中执行文件扫描和代码行数统计的操作。使用 VirtualFile 类的递归遍历方法，我们能够访问项目中的每一个文件。对于每个文件，我们首先检查文件扩展名是否为 java，然后读取并处理文件内容以统计代码行数。

```
new Thread(() -> {
    AtomicInteger codeLines = new AtomicInteger();
    scanFiles(baseDir, codeLines);
    SwingUtilities.invokeLater(() -> {
        codeLinesLabel.setText("Pure code lines: " + codeLines.get());
    });
}).start();
```

在 scanFiles 方法中，我们读取文件内容并分隔成单独的行。通过逐行检查，我们排除了空行和注释行，使用 AtomicInteger 来累计纯代码行数。对于多行注释的处理，我们设置了一个标志变量 inBlockComment 来标志当前是否处于多行注释块中。

下面的例子完整地演示了统计 Java 代码行数插件的实现过程。该插件会在 Code 菜单项最后添加一个“统计代码行数”的菜单项，如图 10-4 所示。

单击该菜单项，会弹出一个对话框。单击 Start Counting 按钮，开始统计当前工程中所有扩展名为 java 的文件中的 Java 代码行数，最后会将结果显示在按钮的下方，如图 10-5 所示。

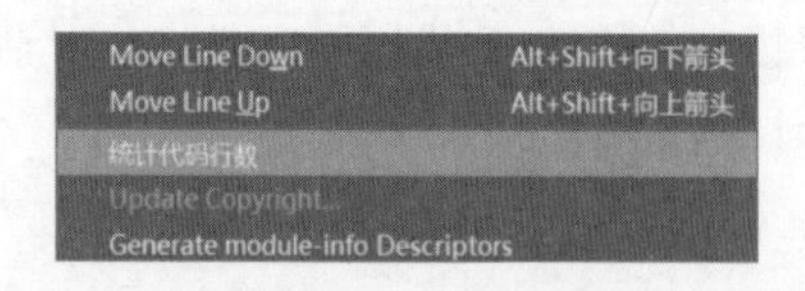

图 10-4　插件添加的菜单项

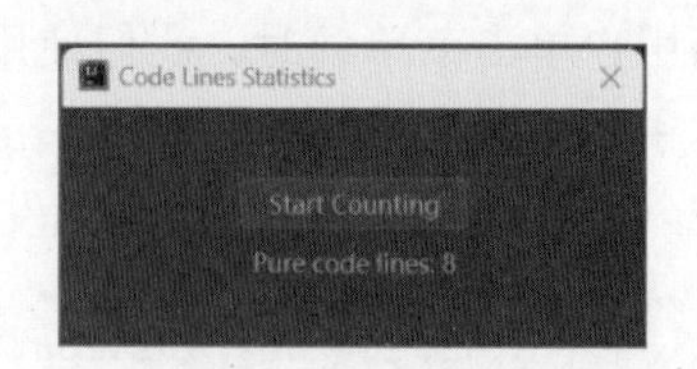

图 10-5　插件对话框

工程目录：src/plugins/CodeLineCounter

本例的 Action 类是 CountCodeLinesAction，代码如下：

```
package com.unitymarvel.codeline;
import com.intellij.openapi.actionSystem.AnAction;
```

```
import com.intellij.openapi.actionSystem.AnActionEvent;
import com.intellij.openapi.project.Project;
import com.intellij.openapi.project.ProjectUtil;
import com.intellij.openapi.vfs.VirtualFile;
import javax.swing.*;
import java.awt.*;
import java.nio.charset.StandardCharsets;
import java.util.concurrent.atomic.AtomicInteger;
public class CountCodeLinesAction extends AnAction {
    @Override
    public void actionPerformed(AnActionEvent e) {
        Project project = e.getProject();
        if (project == null) return;                    // 如果项目为空，直接返回
        JDialog dialog = new JDialog();
        dialog.setTitle("Code Lines Statistics");       // 设置对话框标题
        dialog.setSize(300, 150);                       // 设置对话框大小
        dialog.setLayout(new BorderLayout());           // 设置对话框布局
        // 主面板，使用 BoxLayout 进行垂直对齐
        JPanel mainPanel = new JPanel();
        mainPanel.setLayout(new BoxLayout(mainPanel, BoxLayout.Y_AXIS));
        // 创建按钮，用于开始计数
        JButton button = new JButton("Start Counting");
        button.setAlignmentX(Component.CENTER_ALIGNMENT);
        // 创建标签，用于显示结果
        JLabel codeLinesLabel = new JLabel("Pure code lines: 0");
        codeLinesLabel.setAlignmentX(Component.CENTER_ALIGNMENT);
        // 将组件添加到主面板，并通过 Glue 实现垂直居中
        mainPanel.add(Box.createVerticalGlue());        // 在组件前添加 Glue
        mainPanel.add(button);
        mainPanel.add(Box.createRigidArea(new Dimension(0, 5)));
        // 添加按钮和标签之间的空间
        mainPanel.add(codeLinesLabel);
        mainPanel.add(Box.createVerticalGlue());        // 在组件后添加 Glue
        // 为按钮添加事件监听器
        button.addActionListener(actionEvent -> new Thread(() -> {
            AtomicInteger codeLines = new AtomicInteger();
            scanFiles(ProjectUtil.guessProjectDir(project), codeLines);
            SwingUtilities.invokeLater(() -> {
             codeLinesLabel.setText("Pure code lines: " + codeLines.get());
             // 在标签中显示统计结果
            });
        }).start());
        // 将主面板添加到对话框中
        dialog.add(mainPanel, BorderLayout.CENTER);
        dialog.setLocationRelativeTo(null);
        dialog.setVisible(true);
    }
    private void scanFiles(VirtualFile file, AtomicInteger codeLines) {
        if (file.isDirectory()) {
            for (VirtualFile child : file.getChildren()) {
                scanFiles(child, codeLines);            // 递归扫描子文件夹
```

```
                }
            } else if ("JAVA".equalsIgnoreCase(file.getExtension())) {
                try {
                    String content = new String(file.contentsToByteArray(),
StandardCharsets.UTF_8);
                    codeLines.addAndGet(countPureCodeLines(content));
                    // 计算纯代码行数
                } catch (Exception e) {
                    e.printStackTrace();
                }
            }
        }
    private int countPureCodeLines(String content) {
        String[] lines = content.split("\n");
        int count = 0;
        boolean inBlockComment = false;
        for (String line : lines) {
            line = line.trim();
            if (line.startsWith("/*") && !line.endsWith("*/")) {
                inBlockComment = true;          // 进入块注释模式
                continue;
            }
            if (inBlockComment && line.endsWith("*/")) {
                inBlockComment = false;         // 退出块注释模式
                continue;
            }
            if (inBlockComment || line.startsWith("//") || line.isEmpty()) {
                continue;                       // 忽略块注释内、单行注释和空行
            }
            count++;                            // 累加有效代码行数
        }

        return count;
    }
}
```

然后向 plugin.xml 文件中添加如下的 action：

```
<actions>
    <group id="CodeLineCounterGroup" text="Code Line Counter" description=
"Code Line Counter Tools">
        <add-to-group group-id="CodeMenu" anchor="last"/>
        <action id="CountCodeLinesAction" class="com.unitymarvel.codeline.
CountCodeLinesAction" text="统计代码行数" description="Counts code lines in Java
files.">
        </action>
    </group>
</actions>
```

现在运行插件，在新启动的 IDE 中的 Code 菜单的最后，就会看到“统计代码行数”

菜单项，单击即可统计 Java 代码行数。

在默认情况下，会在 <插件工程根目录>/build/libs 目录生成 CodeLineCounter-1.0-SNAPSHOT.jar 文件，可以修改 build.gradle.kts 文件中的 version 属性，该属性的默认值是 1.0-SNAPSHOT，将该属性的值改成 1.0，那么再次运行插件，就会在 <插件工程根目录>/build/libs 目录重新生成一个 CodeLineCounter-1.0.jar 文件。该插件不仅可以在 Intellij IDEA 中安装，还可以在 PyCharm、WebStorm 等类似的 IDE 中安装。

如果使用的 Intellij IDEA 版本较低，那么在 build.gradle.kts 文件中 patchPluginXml 属性设置的最高兼容版本号可能比较低，如果 Intellij IDEA 推出更高的版本，将无法安装这个插件。不过不用担心，只要插件没有调用 Intellij IDEA 最新的 API 和库，那么只需要将 untilBuild.set(...) 的参数指定的版本号设置得高一点，就可以解决这个问题。

```
patchPluginXml {
    sinceBuild.set("213")
    untilBuild.set("233.*")
}
```

10.5　圆形头像插件

本节会实现一个将图像文件转换为圆形头像的 IntelliJ IDEA 插件。主要涉及在工具栏上添加图像按钮，以及获取当前选中的图像文件路径，如何将矩形图像转换为圆形图像以及刷新工程树等技术问题。下面将详细介绍这些技术的具体原理和实现。

1. 插件动作类的定义

首先，需要定义一个继承自 AnAction 的 GenerateCircleAvatarAction 类。这个类负责响应用户的操作，即单击工具栏上的按钮。

```
public class GenerateCircleAvatarAction extends AnAction {
    public GenerateCircleAvatarAction() {
        super("制作大头像", "将一个图像文件变为大头像", IconLoader.getIcon
("/images/tray.png", GenerateCircleAvatarAction.class));
    }
    ...
}
```

在构造函数中，调用了 super 方法，设置了按钮的名称、描述和图标。这里的图标路径需要是相对于资源目录的路径，确保图标文件已经放在了正确的位置。

2. 检测图像文件

本例会编写一个 isImageFile 方法，该方法用于检测当前选中的文件是否为图像文件。这个方法通过检查文件扩展名是否符合常见的图像格式来判断选中的是否为图像文件。

```
private boolean isImageFile(VirtualFile file) {
    String extension = file.getExtension();
    return "png".equalsIgnoreCase(extension) || "jpg".equalsIgnoreCase(extension) ||
                                              "jpeg".equalsIgnoreCase(extension);
}
```

3. 响应按钮单击事件

actionPerformed 方法是按钮单击事件的响应方法。在这个方法中，我们实现了从项目中选择图像文件、处理图像并保存为新文件的逻辑。

```
@Override
public void actionPerformed(AnActionEvent e) {
    Project project = e.getProject();
    ...
    if (selectedFile == null || !isImageFile(selectedFile)) {
        Messages.showErrorDialog(" 请选择一个图像文件来应用效果。", " 未选择图像 ");
        return;
    }
    ...
}
```

如果选中的是图像文件，则读取该文件，调用 createCircleAvatar 方法进行处理，并将结果保存为新文件。如果选中的不是图像文件，则弹出错误对话框。

4. 图像处理逻辑

本例会编写一个用于处理裁剪图像动作的 createCircleAvatar 方法，该方法接收原始图像，首先将其裁剪为正方形，然后创建一个内切圆的图像，外围透明。

使用 cropSquare 方法从原始图像中裁剪出中心的最大正方形区域。再通过 createCircleImage 方法使用 Graphics2D 对象在正方形图像基础上绘制一个圆形区域。

5. 刷新工程树

生成圆形头像文件后，工程树并不会立刻刷新，如果要立刻刷新工程树，并显示出新创建的图像文件，需要使用下面的代码：

```
VirtualFileManager.getInstance().asyncRefresh(null);
```

6. 配置 plugin.xml 文件

plugin.xml 文件配置了插件的元数据和扩展点。在这个文件中，我们通过 <action> 标签将 GenerateCircleAvatarAction 动作添加到工具栏上。

```
<actions>
    <action id="GenerateCircleAvatarAction" class="com.unitymarvel.image_
effects.GenerateCircleAvatarAction">
        <add-to-group group-id="ToolbarRunGroup" anchor="last"/>
    </action>
</actions>
```

<add-to-group> 标签指定了动作被添加到的工具栏组（ToolbarRunGroup），这通常是运行 / 调试相关工具的组。

下面的例子会在工具栏创建一个按钮，如图 10-6 中间的图标就是创建的按钮。单击该按钮，会将当前工程选中的图像文件转换为圆形头像图像，并生成 image_avatar.png。其中 image 是图像的原名，效果如图 10-7 所示。

图 10-6　创建的按钮

图 10-7　圆形头像

工程目录：src/plugins/ImageEffects

```
package com.unitymarvel.image_effects;
import com.intellij.openapi.actionSystem.AnAction;
import com.intellij.openapi.actionSystem.AnActionEvent;
import com.intellij.openapi.actionSystem.CommonDataKeys;
import com.intellij.openapi.fileChooser.FileChooser;
import com.intellij.openapi.fileChooser.FileChooserDescriptor;
import com.intellij.openapi.project.Project;
import com.intellij.openapi.ui.Messages;
import com.intellij.openapi.util.IconLoader;
import com.intellij.openapi.vfs.VirtualFile;
import com.intellij.openapi.vfs.VirtualFileManager;
import javax.imageio.ImageIO;
import java.awt.*;
import java.awt.image.BufferedImage;
import java.io.File;
import java.io.IOException;
    // 插件动作类，用于生成圆形头像
public class GenerateCircleAvatarAction extends AnAction {
    // 构造函数：设置动作的名称、描述和图标
    public GenerateCircleAvatarAction() {
```

```
        super(" 制作大头像 ", " 将一个图像文件变为大头像 ", IconLoader.getIcon
("/images/tray.png", GenerateCircleAvatarAction.class));
    }
    // 判断文件是否为图像文件
    private boolean isImageFile(VirtualFile file) {
        String extension = file.getExtension();
        return "png".equalsIgnoreCase(extension) || "jpg".equalsIgnoreCase
(extension) || "jpeg".equalsIgnoreCase(extension);
    }
    // 动作执行时的方法
    @Override
    public void actionPerformed(AnActionEvent e) {
        Project project = e.getProject();
        // 文件选择器描述符，设置为只能选择文件
        FileChooserDescriptor descriptor = new FileChooserDescriptor
(true, false, false, false, false, false);
        // 获取当前选中的文件
        VirtualFile selectedFile = e.getData(CommonDataKeys.VIRTUAL_FILE);

        // 如果没有选择文件或选择的不是图像文件，则弹出错误对话框
        if (selectedFile == null || !isImageFile(selectedFile)) {
            Messages.showErrorDialog(" 请选择一个图像文件来应用效果。", " 未选择图像 ");
            return;
        }
        if (project != null && selectedFile != null) {
            try {
                // 读取原始图像
                BufferedImage originalImage = ImageIO.read(new File
(selectedFile.getPath()));
                // 创建圆形头像
                BufferedImage avatarImage = createCircleAvatar(originalImage);
                // 构建新文件路径，添加 "_avatar" 后缀
                String newPath = selectedFile.getPath().replaceFirst
("[.][^.]+$", "") + "_avatar.png";
                // 保存圆形头像为新文件
                ImageIO.write(avatarImage, "PNG", new File(newPath));
                // 刷新项目视图，使新文件显示
                VirtualFileManager.getInstance().asyncRefresh(null);
            } catch (IOException ex) {
                ex.printStackTrace();
            }
        }
    }
    // 从原始图像中裁剪出正方形区域
    private static BufferedImage createCircleAvatar(BufferedImage originalImage) {
        int size = Math.min(originalImage.getWidth(), originalImage.getHeight());
        int x = (originalImage.getWidth() - size) / 2;
        int y = (originalImage.getHeight() - size) / 2;
        BufferedImage squareImage = cropSquare(originalImage, x, y, size);
        return createCircleImage(squareImage);
    }
```

```
    // 裁剪出正方形区域
    private static BufferedImage cropSquare(BufferedImage source, int x, int y,
int size) {
        return source.getSubimage(x, y, size, size);
    }
    // 将正方形图像转换为圆形图像
    private static BufferedImage createCircleImage(BufferedImage squareImage) {
        int diameter = squareImage.getWidth();             // 假设正方形图像已裁剪
        BufferedImage circleImage = new BufferedImage(diameter, diameter,
BufferedImage.TYPE_INT_ARGB);
        Graphics2D g = circleImage.createGraphics();
        // 设置抗锯齿渲染提示
        g.setRenderingHint(RenderingHints.KEY_ANTIALIASING, RenderingHints.
VALUE_ANTIALIAS_ON);
        // 绘制圆形
        g.fillOval(0, 0, diameter, diameter);
        // 设置合成规则，只在源图像和目标图像重叠的地方绘制源图像
        g.setComposite(AlphaComposite.SrcIn);
        // 绘制原始正方形图像
        g.drawImage(squareImage, 0, 0, null);
        g.dispose();                                        // 释放资源
        return circleImage;
    }
}
```

plugin.xml 的配置：

```
<actions>
    <action id="GenerateCircleAvatarAction"
            class="com.unitymarvel.image_effects.GenerateCircleAvatarAction">
        <!-- 添加到工具栏 -->
        <add-to-group group-id="ToolbarRunGroup" anchor="last"/>
    </action>
</actions>
```

由于代码中包含中文，如果在运行插件时出现乱码，可以在 <IntelliJ IDEA 安装目录 >/bin 中找到 idea64.exe.vmoptions 文件，打开该文件，在最后添加下面一行内容，然后重启 IntelliJ IDEA 即可。

```
-Dfile.encoding=UTF-8
```

10.6　视频转动画 GIF 插件

本节的例子是用于将视频（MP4）转换为动画 GIF 的插件。通过在工程树上下文菜单中添加“转换为 GIF”菜单项，然后单击菜单项，会将当前选中的 MP4 文件转换为动

画 GIF 文件。这其中涉及了如何在工程树上下文菜单中添加新的菜单项，如何只有在选中 MP4 文件时显示这个菜单项，如何获取选中的 MP4 文件，如何利用 FFmpeg 将 MP4 文件转换为动画 GIF 文件。下面会详细介绍这些技术的实现过程。

1. 在工程树上下文中添加菜单项

IntelliJ IDEA 平台提供了一套丰富的 API 来与 IDE 的各个组成部分交互，包括项目视图（工程树）的上下文菜单。要向上下文菜单添加自定义菜单项，需要创建一个继承自 AnAction 的类。这个类代表了用户可以执行的动作。

```
public class ConvertToGifAction extends AnAction {
    ...
}
```

然后需要在 plugin.xml 文件中添加相应的动作。

```
<actions>
    <!-- 定义一个新的动作 -->
    <action id="ConvertToGifAction" class="com.unitymarvel.convert_to_gif.
ConvertToGifAction"
            text="转换为GIF" description="将MP4视频转换为GIF格式">
        <!-- 添加到项目视图的上下文菜单 -->
        <add-to-group group-id="ProjectViewPopupMenu" anchor="last"/>
    </action>
</actions>
```

2. 控制菜单项的可见性

为了确保“转换为 GIF”菜单项仅在 MP4 文件上可见和可用，需要重写 AnAction 类的 update 方法，它负责更新动作的状态，包括它是否可见或可用。

```
@Override
public void update(AnActionEvent e) {
    final VirtualFile file = e.getData(CommonDataKeys.VIRTUAL_FILE);
    boolean visible = file != null && "mp4".equals(file.getExtension());
    e.getPresentation().setEnabledAndVisible(visible);
}
```

这段代码检查当前选中的文件是否为 MP4 格式。这是通过获取事件的数据上下文中的 VIRTUAL_FILE，然后检查文件扩展名是否为 MP4 来实现的。

3. 获取被选中的 MP4 文件

当用户单击菜单项时，会触发 actionPerformed 方法。在这个方法内，可以获取到被选中的文件，并执行相应的操作。

```
@Override
public void actionPerformed(AnActionEvent e) {
```

```
    final VirtualFile file = e.getData(CommonDataKeys.VIRTUAL_FILE);
    ...
}
```

这段代码利用 e.getData(CommonDataKeys.VIRTUAL_FILE) 来获取当前选中的文件。

4. 通过 FFmpeg 将 MP4 转换为动画 GIF 文件

转换过程是通过执行 FFmpeg 命令行工具来完成的。这里使用了 ProcessBuilder 来构造和执行命令。

```
private void executeFFmpegCommand(String[] command) {
    try {
        Process process = new ProcessBuilder(command).start();
        ...
    } catch (IOException | InterruptedException e) {
        e.printStackTrace();
    }
}
```

FFmpeg 命令通过指定输入文件、输出文件以及转换参数（如帧率）来配置。在 actionPerformed 方法中，构建了这个命令并传递给了 executeFFmpegCommand 方法。

```
String[] command = {"ffmpeg", "-i", inputPath, "-vf", "fps=10", "-c:v", "gif", outputPath};
executeFFmpegCommand(command);
```

完成转换后，使用 VirtualFileManager 的 asyncRefresh 方法刷新项目视图，使新生成的 GIF 文件在 IDE 中可见。

这个插件展示了如何通过 IntelliJ 平台 API 和外部工具（如 FFmpeg）扩展 IDE 的功能。通过仅几个关键步骤——定义动作、控制可见性、获取选中文件以及执行外部命令——创建强大的插件来增强开发环境。这个过程不仅限于文件转换，同样的模式可以应用于其他类型的文件操作或集成外部工具的需求。

下面的代码完整地展示了本例的实现过程。

工程目录：src/plugins/ConvertToGif

```
package com.unitymarvel.convert_to_gif;
import com.intellij.openapi.actionSystem.AnAction;
import com.intellij.openapi.actionSystem.AnActionEvent;
import com.intellij.openapi.actionSystem.CommonDataKeys;
import com.intellij.openapi.vfs.VirtualFile;
import com.intellij.openapi.project.Project;
import com.intellij.openapi.vfs.VirtualFileManager;
import javax.swing.*;
import java.io.BufferedReader;
import java.io.File;
import java.io.IOException;
```

```
import java.io.InputStreamReader;
public class ConvertToGifAction extends AnAction {
    @Override
    public void update(AnActionEvent e) {
        // 只有当选中的文件扩展名为 MP4 时，菜单项才可见和可用
        final VirtualFile file = e.getData(CommonDataKeys.VIRTUAL_FILE);
        boolean visible = file != null && "mp4".equals(file.getExtension());
        e.getPresentation().setEnabledAndVisible(visible);
    }
    @Override
    public void actionPerformed(AnActionEvent e) {
        final VirtualFile file = e.getData(CommonDataKeys.VIRTUAL_FILE);
        if (file != null) {
            Project project = e.getProject();
            String inputPath = file.getPath();
            String outputPath = inputPath.substring(0, inputPath.
lastIndexOf('.')) + ".gif";
            File outputFile = new File(outputPath);
            if (outputFile.exists()) {
                outputFile.delete();
            }
            String[] command = {"ffmpeg", "-i", inputPath, "-vf", "fps=10",
"-c:v", "gif", outputPath};
            executeFFmpegCommand(command);
            // 刷新项目视图，使新文件显示
            VirtualFileManager.getInstance().asyncRefresh(null);

            JOptionPane.showMessageDialog(null, "转换完成。GIF 已保存至：" +
outputPath);
        }
    }
    // 使用 ProcessBuilder 执行 FFmpeg 命令
    private void executeFFmpegCommand(String[] command) {
        try {
            Process process = new ProcessBuilder(command).start();
            BufferedReader reader = new BufferedReader(new InputStreamReader
(process.getErrorStream()));
            String line;
            while ((line = reader.readLine()) != null) {
                System.out.println(line);                  // 打印 FFmpeg 命令的输出信息
            }
            process.waitFor();
        } catch (IOException | InterruptedException e) {
            e.printStackTrace();
            JOptionPane.showMessageDialog(null, "转换失败：" + e.getMessage());
        }
    }
}
```

运行插件，在工程树中选中一个 MP4 文件，就会在右键菜单的最后看到“转换为 GIF”，菜单项，如图 10-8 所示。单击该菜单项，会将 MP4 文件转换为动画 GIF 文件，假设视频文件为 input.mp4，那么生成的 GIF 动画文件就是 input.gif。

图 10-8 “转换为 GIF”菜单项

10.7 绘图插件

本节的例子会实现一个可以绘图的插件。该插件会创建一个视图，并运行在视图中自由绘图，然后单击 Save 按钮，会弹出一个对话框，输入文件名后，会将绘制的图像保存在工程目录的 images 子目录中，如果文件存在，则会弹出对话框询问是否需要覆盖图像文件，如果选择“是”，那么就会覆盖图像文件。在这个例子中主要涉及如何创建一个 IntelliJ IDEA 视图，并在视图上添加画布，以及绘制图像和保存图像，下面会详细介绍这些技术的实现原理和步骤。

1. 添加视图

在插件开发中，首先需要在 plugin.xml 中配置一个新的视图（ToolWindow），这样用户就可以在 IDE 的界面中访问这个视图。这通过 <toolWindow> 标签完成，其中 id 属性定义了工具窗口的唯一标识符，同时也决定了视图的标题文本。

```
<toolWindow id="My Drawing" anchor="right" factoryClass=
"...DrawingToolWindowFactory" />
```

2. 实现 ToolWindowFactory

ToolWindowFactory 接口负责初始化工具窗口的内容。在 createToolWindowContent 方法中，我们创建插件 UI 的根面板，并将其添加到工具窗口中。这个根面板通常包含了绘图面板和其他控件（如保存按钮）。

```
public class DrawingToolWindowFactory implements ToolWindowFactory {
    @Override
    public void createToolWindowContent(@NotNull Project project, @NotNull
ToolWindow toolWindow) {
        // 创建并添加内容到 toolWindow
    }
}
```

3. 绘图面板

绘图面板是一个自定义的 JPanel，它重写了 paintComponent 方法以支持画图操作。此外，它监听鼠标事件来允许用户通过拖动鼠标绘制曲线。

（1）使用 BufferedImage 作为画布，Graphics2D 用于在这个画布上绘制。

（2）使用 Path2D.Double 来追踪绘制的曲线路径，确保曲线平滑。

```
public class DrawingPanel {
    // 构造函数中初始化绘图面板，设置背景，监听鼠标事件等
}
```

4. 绘制曲线

在鼠标拖动时，连续地将当前鼠标位置添加到 Path2D.Double 路径中，并在 paintComponent 方法中绘制这个路径，实现曲线的绘制。

5. 保存图像

通过一个按钮单击事件触发保存操作，提示用户输入文件名，并将绘制的内容保存到项目的 images 子目录中。如果该目录不存在，则创建它。如果文件已存在，提示用户是否覆盖。使用 ImageIO.write 方法将 BufferedImage 保存为文件。

6. 获取当前工程目录

保存图像时，通过 Project 对象获取当前工程的基路径，然后在该路径下的 images 目录中保存图像文件。如果目录不存在，则先创建它。

```
String projectBasePath = project.getBasePath();
Path imagesDir = Paths.get(projectBasePath, "images");
```

绘图插件通过上述步骤实现，从添加一个新的视图开始，然后实现一个支持绘图的面板，并最终允许用户将绘制的图像保存到项目特定的目录中。插件的实现主要涉及 Swing 组件的使用、图形绘制 API 的应用，以及文件 I/O 操作。通过 ToolWindowFactory 和 plugin.xml 的配置，插件被集成到 IntelliJ IDEA 中，为用户提供了一个交互式的绘图工具。

下面的例子完整地演示了如何编写这个绘图插件。绘制的效果和保存的图像如图 10-9 所示。

图 10-9　绘图组件的效果

工程目录：src/plugins/Drawing

DrawingToolWindowFactory 类：

```
package com.unitymarvel.drawing;
import com.intellij.openapi.project.Project;
import com.intellij.openapi.wm.ToolWindow;
import com.intellij.openapi.wm.ToolWindowFactory;
import com.intellij.ui.content.Content;
import com.intellij.ui.content.ContentFactory;
import org.jetbrains.annotations.NotNull;
public class DrawingToolWindowFactory implements ToolWindowFactory {
    @Override
    public void createToolWindowContent(@NotNull Project project, @NotNull
ToolWindow toolWindow) {
        DrawingPanel drawingPanel = new DrawingPanel(project);        // 修改这里
        ContentFactory contentFactory = ContentFactory.SERVICE.getInstance();
        Content content = contentFactory.createContent(drawingPanel.
getMainPanel(), "", false);
        toolWindow.getContentManager().addContent(content);
    }
}
```

DrawingPanel 类：

```
package com.unitymarvel.drawing;
import com.intellij.openapi.project.Project;
import com.intellij.openapi.vfs.VirtualFileManager;
import javax.imageio.ImageIO;
import javax.swing.*;
import java.awt.*;
import java.awt.event.MouseAdapter;
import java.awt.event.MouseEvent;
import java.awt.geom.Path2D;
import java.awt.image.BufferedImage;
import java.io.File;
import java.nio.file.Files;
import java.nio.file.Path;
import java.nio.file.Paths;
public class DrawingPanel {
    private final JPanel mainPanel = new JPanel(new BorderLayout());
    private BufferedImage image;
    private Graphics2D g2d;
    private Path2D.Double path = new Path2D.Double();
    // 使用 Path2D.Double 来支持浮点坐标
    private Point lastPoint = null;                     // 用于记录上一个点的位置
    private Project project;                            // 添加对 Project 的引用
```

```
    public DrawingPanel(Project project) {      // 修改构造函数以接收 Project 对象
        this.project = project;
        image = new BufferedImage(800, 600, BufferedImage.TYPE_INT_ARGB);
        g2d = image.createGraphics();
        g2d.setColor(Color.WHITE);                  // 设置背景为白色
        g2d.fillRect(0, 0, image.getWidth(), image.getHeight());    // 填充背景色
        setupGraphics2D(g2d);
        JPanel drawingPanel = new JPanel() {
            @Override
            protected void paintComponent(Graphics g) {
                super.paintComponent(g);
                g.drawImage(image, 0, 0, this);
            }
        };
        drawingPanel.setPreferredSize(new Dimension(800, 600));
        initDrawingListeners(drawingPanel);
        JButton saveButton = new JButton("Save");
        saveButton.addActionListener(e -> saveDrawing());
        mainPanel.add(saveButton, BorderLayout.NORTH);
        mainPanel.add(drawingPanel, BorderLayout.CENTER);
    }
    private void setupGraphics2D(Graphics2D g2d) {
        g2d.setRenderingHint(RenderingHints.KEY_ANTIALIASING, RenderingHints.
VALUE_ANTIALIAS_ON);                                  // 抗锯齿
        g2d.setStroke(new BasicStroke(2));            // 设置线条宽度为 2
        g2d.setColor(Color.RED);                      // 设置线条颜色为红色
    }
    private void initDrawingListeners(JPanel drawingPanel) {
        drawingPanel.addMouseListener(new MouseAdapter() {
            @Override
            public void mousePressed(MouseEvent e) {
                path.moveTo(e.getX(), e.getY());
                lastPoint = e.getPoint();
            }
            @Override
            public void mouseReleased(MouseEvent e) {
                lastPoint = null;                    // 鼠标释放时重置上一个点的位置
            }
        });
        drawingPanel.addMouseMotionListener(new MouseAdapter() {
            @Override
            public void mouseDragged(MouseEvent e) {
                if (lastPoint != null) {
                    path.lineTo(e.getX(), e.getY());
                    g2d.draw(path);
                    path.moveTo(e.getX(), e.getY());
                    drawingPanel.repaint();
```

```
                }
            }
        });
    }
    private void saveDrawing() {
        String fileName = JOptionPane.showInputDialog(mainPanel, "Enter file name:");
        if (fileName != null && !fileName.trim().isEmpty()) {
            try {
                // 使用 Project 对象获取当前工程目录
                String projectBasePath = project.getBasePath();
                Path imagesDir = Paths.get(projectBasePath, "images");
                if (!Files.exists(imagesDir)) {
                    Files.createDirectories(imagesDir);
                }
                File file = imagesDir.resolve(fileName + ".png").toFile();
                if (file.exists()) {
                    int result = JOptionPane.showConfirmDialog(mainPanel,
"File exists. Overwrite?", "File Exists", JOptionPane.YES_NO_OPTION);
                    if (result != JOptionPane.YES_OPTION) {
                        return;
                    }
                }
                ImageIO.write(image, "PNG", file);
                JOptionPane.showMessageDialog(mainPanel, "Saved: " +
file.getAbsolutePath(), "Save Successful", JOptionPane.INFORMATION_MESSAGE);
                // 刷新项目视图，使新文件显示
                VirtualFileManager.getInstance().asyncRefresh(null);

            } catch (Exception ex) {
                JOptionPane.showMessageDialog(mainPanel, "Error saving image: "
+ ex.getMessage(), "Error", JOptionPane.ERROR_MESSAGE);
            }
        }
    }
    public JPanel getMainPanel() {
        return mainPanel;
    }
}
```

plugin.xml 的配置：

```
<extensions defaultExtensionNs="com.intellij">
    <toolWindow id="My Drawing" anchor="right"
factoryClass="com.unitymarvel.drawing.DrawingToolWindowFactory" />
</extensions>
```

10.8 小结

本章详细介绍了 IntelliJ IDEA 插件开发的全过程，涵盖了从环境准备到插件发布的各个关键步骤。在技术层面，我们探讨了如何利用 IntelliJ Platform 的 API 和扩展点来实现插件的核心功能。通过实际的插件开发案例，如统计 Java 文件代码行数、图像文件转换为圆形头像、视频转换为 GIF 动画，以及绘图工具等，我们展示了插件开发的多样性和实用性。这些案例不仅展示了插件开发的技术细节，还强调了插件在提升开发效率和用户体验方面的巨大潜力。

总的来说，本章为有意开发 IntelliJ IDEA 插件的开发者提供了宝贵的知识和实践指导，无论是为了个人使用还是为社区贡献，都将成为不可或缺的资源。通过本章的学习，读者将能够掌握插件开发的核心技能，为 IDE 增添无限可能。

第 11 章

代码魔法：释放 ChatGPT 的神力

阿拉丁神灯的故事相信大家都知道，尤其是那盏神灯，简直不要太爽。不过现在有了 ChatGPT，这让每个人都可能成为阿拉丁，而 ChatGPT 就是那盏神灯。你需要什么，只要告诉 ChatGPT，ChatGPT 就会满足你的要求。当然，阿拉丁的神灯只能满足 3 个要求，而 ChatGPT 这盏现代版的 AI 神灯，可以满足你无限多的要求。本章将会向读者展示 ChatGPT 的魔法之一：生成代码。就光这一项技能，足以让世人震惊，当然，如果你完全融入 ChatGPT 的世界，那么魔法将会伴随你终生！

11.1 走进 ChatGPT

本节主要介绍 ChatGPT 的相关内容，包括 ChatGPT 的历史背景，涉及的技术，以及与 ChatGPT 相似的产品等。通过本节的内容，读者可以大体了解什么是 ChatGPT，以及 ChatGPT 与程序员的关系到底是怎样的。

11.1.1 AIGC 概述

AIGC（AI-Generated Content，AI 生成内容）是指基于生成对抗网络（Generative Adversarial Network，GAN）、大型预训练模型等人工智能技术的方法，通过对已有数据进行学习和模式识别，以适当的泛化能力生成相关内容的技术。 AIGC 技术可以用于创造各种类型的内容，如文字、图像、音频等。

AIGC 的目标是超越传统的弱人工智能系统，这些系统通常专注于解决特定任务或处理特定领域的问题。相反，AIGC 旨在实现一种通用智能，能够在多个领域中学习、推理和执行任务，类似于人类的能力。

AIGC 技术的应用领域非常广泛，包括游戏开发、数据分析、计算机图形学、自动控制等多个领域。列举如下。

（1）自动编写代码：是指利用人工智能技术，根据用户的需求或描述，自动生成相应的代码的过程。自动编写代码可以帮助程序员提高编码效率和质量，减少错误和重复工作，以及学习新的编程技能。自动编写代码的主要方法是使用生成式 AI 技术，如生成对抗网络、变分自编码器（Variational Auto Encoder，VAE）和自回归模型（Autoregressive Model）等。这些技术可以在大型代码库上进行训练，并使用机器学习算法生成与训练数据相似的新代码。与自动编写代码相关的特性包括代码生成、代码优化、代码注释、代码转换等。

（2）游戏开发：AIGC 技术可以用于生成游戏中的角色、场景、任务、剧情等内容，提高游戏的丰富性和可玩性。例如，Unity Machine Learning Agents 是一个人工智能工具包，可以用于开发具有智能性的游戏和虚拟环境。

（3）数据分析：AIGC 技术可以用于生成数据集、数据报告、数据可视化等内容，提高数据的质量和价值。例如，OpenAI Codex 是一个可以根据自然语言描述生成代码的程序，可以用于数据分析和处理。

（4）计算机图形学：AIGC 技术可以用于生成图像、视频、动画等内容，提高图形的美观和真实性。例如，Stable Diffusion 是一个可以根据文字提示和风格类型生成图像的平台。

（5）自动控制：AIGC 技术可以用于生成控制策略、控制信号、控制系统等内容，提高控制的效率和性能。例如，AlphaGo 是一个可以下围棋的人工智能程序，使用了深度学习和强化学习等 AIGC 技术。

AIGC 技术具有一些明显的优势和不足，这些可以从以下方面进行概述。

❑ 优势

（1）自动化和效率：AIGC 技术能够自动地生成大量的内容，从而提高生产效率。相对于传统的人工创作方式，AIGC 可以在短时间内生成大量内容，节省了人力资源和时间成本。

（2）创新和灵感：AIGC 技术能够创造出新颖和有趣的内容，从而提高创新性。AIGC 可以根据不同的条件或指导生成与之相关的内容，为用户提供更多选择、更多灵感和更多可能性。

（3）个性化和定制化：AIGC 技术能够根据用户的喜好和需求生成个性化和定制化的内容，从而提高满意度。AIGC 可以根据用户输入的关键词、描述或样本生成与之相匹配的内容，并根据用户对中间结果的反馈进行调整。

❑ 不足

（1）数据质量和隐私：AIGC 技术依赖于大量的数据进行训练和生成，这可能导致数据质量和隐私方面的问题。数据质量方面，如果训练数据存在噪声、偏差或不完整等问题，则可能影响生成内容的质量和准确性。数据隐私方面，如果训练数据涉及个人或机构的敏感信息，则可能存在泄露或滥用的风险。

（2）原创性和创新性：AIGC 技术基于现有的数据信息进行内容生产，因此输出的内容容易缺乏原创性和创新性。AIGC 技术很难创造出超越现有数据范围的新颖和有趣的内容，也很难反映出人类的情感和个性。

（3）版权和伦理：AIGC 技术生成的内容可能涉及版权和伦理方面的问题。版权方面，如果 AIGC 技术生成的内容侵犯了他人的版权，或者与他人的作品相似，则可能引起法律纠纷。伦理方面，如果 AIGC 技术生成的内容违反了社会道德或公序良俗，则可能引起道德争议或社会不安。

11.1.2　AIGC 的落地案例

随着 ChatGPT 的问世，在短时间内，国内外涌现了大量的 AIGC 落地案例。

（1）ChatGPT：ChatGPT 是由 OpenAI 开发的一款人工智能聊天机器人，能够与人类进行自然对话，并根据聊天的上下文进行互动。ChatGPT 的英文缩写是 Chat Generative Pre-trained Transformer，意思是聊天生成预训练变换器。ChatGPT 的优点是它能够生成流畅和有逻辑的对话，回答跟进问题，承认错误，挑战错误的前提，拒绝不恰当的请求等。ChatGPT 还能够根据自然语言描述生成代码、邮件、视频脚本、文案、翻译、论文等内容。

（2）New Bing：微软推出的一款新型搜索引擎，它可以让用户直接输入自然语言的问题，并得到完整的答案。New Bing 不仅可以提供网页搜索结果，还可以提供引用、聊天和创作等功能。

（3）Claude：由 Anthropic 开发的一款人工智能平台，它可以执行各种对话和文本处理任务，同时保持高度的可靠性和可预测性。Claude 可以根据用户输入的关键词或主题，自动生成相关的文章或段落。Claude 还可以根据用户反馈进行自我学习和优化，提高生成内容的质量和适应性。

（4）Bard：由谷歌开发的一款实验性的人工智能服务，它可以让用户与生成式 AI 进行协作。Bard 可以帮助用户提高生产力、加速想法和激发好奇心。Bard 可以根据用户输入的文字提示或样本，自动生成相关的代码、邮件、视频脚本、文案、翻译、论文等内容。Bard 还可以根据用户反馈进行调整和优化，提高生成内容的质量和满意度。

（5）文心一言：由百度开发的一款人工智能写作平台，能够根据用户输入的关键词或主题，自动生成相关的文章或段落。百度文心一言的英文缩写是 Baidu Wenxin Yiyuan，意思是 Baidu Heart of Writing One Sentence。百度文心一言的优点是它能够快速地生成各种类型和风格的文本内容，如新闻、故事、诗歌、广告等，并且支持多种语言和领域。

（6）Kimi Chat：月之暗面科技推出的智能聊天机器人。它拥有大容量知识库，可以进行智能闲聊、解答问题、提供生活助手服务等，具有极高的智能交互能力。

11.1.3　ChatGPT 概述

ChatGPT 的创始人是 OpenAI 的 CEO 和联合创始人 Sam Altman。Sam Altman 是一位知名的技术企业家和投资者，曾经担任 Y Combinator 的总裁，并参与了多个知名的科技项目，如 Airbnb、Dropbox、Stripe 等。Sam Altman 于 2015 年与 Elon Musk、Peter Thiel、Ilya Sutskever 等共同创立了 OpenAI，一个致力于推进人类利益的人工智能研究和部署的公司。

ChatGPT 的发展历程可以追溯到 OpenAI 早期的语言模型项目，如 GPT、GPT-2 和 GPT-3 等。这些语言模型都是基于大规模的文本数据进行训练，能够生成与训练数据相似的新文本。ChatGPT 是在 GPT-3.5 和 GPT-4 的基础上进行了微调，以适应对话应用的需求。ChatGPT 使用了监督学习和强化学习等技术，利用人类反馈来提高模型的性能和安全性。

ChatGPT 于 2022 年 11 月 30 日正式发布，没有进行任何宣传，但很快在社交媒体上引起了广泛关注和讨论。在发布后的 5 天内，ChatGPT 就吸引了超过一百万的用户，并展示了它在各领域的应用价值，如写作、编程、学习等。2023 年 1 月，OpenAI 与微软扩大了长期合作关系，并宣布了数十亿美元的投资计划，以加速全球范围内的人工智能突破。2023 年 2 月 1 日，OpenAI 推出了 ChatGPT Plus，一个付费订阅计划，为用户提供更快速、更安全、更有用的回复。

ChatGPT Plus 是基于 GPT-4 模型的服务，相比于免费版的 ChatGPT（基于 GPT-3.5 模型），它具有以下几个优势。

（1）更高的准确性：由于 GPT-4 模型具有更广泛的常识和问题解决能力，它可以更准确地回答复杂和困难的问题。

（2）更高的创造性：由于 GPT-4 模型具有更强大的生成能力和协作能力，它可以更好地生成、编辑和迭代各种创意和技术写作任务，如创作歌曲、编写剧本或学习用户的写作风格。

（3）更高的安全性：由于 OpenAI 花费了 6 个月时间来提高 GPT-4 模型的安全性和对齐性，它比 GPT-3.5 模型更不容易产生不良或不合适的内容，并且能够生成更符合事实的回复。

11.1.4 ChatGPT vs New Bing

微软的 New Bing 也是基于 GPT-4 模型开发的，但 New Bing 并不是 ChatGPT Plus，它们的主要差异如下。

（1）功能和定位：ChatGPT 主要是一个人工智能聊天机器人，它专注于提供基于 GPT-4.0 的智能聊天体验。ChatGPT 可以与用户进行自然对话，并根据聊天的上下文进行互动。ChatGPT 还可以根据用户输入的文字提示或样本，自动生成相关的代码、邮件、视频脚本、文案、翻译、论文等内容。New Bing 则是一个新型的搜索引擎，它可以让用户直接输入自然语言的问题，并得到完整的答案。New Bing 不仅可以提供网页搜索结果，还可以提供引用、聊天和创作等功能。New Bing 还可以根据用户输入的关键词或主题，自动生成相关的文章或段落。

（2）版本和性能：ChatGPT 使用了 OpenAI 最新发布的 GPT-4.0 模型，这是目前最先进的语言模型之一。GPT-4.0 模型具有更广泛的常识和问题解决能力，能够生成更准确和更具创造性的内容。New Bing 则使用了一个测试版本的 GPT-4.0 模型，这是一个尚未正式发布的版本。测试版本的 GPT-4.0 模型可能存在一些不稳定或不完善的地方，导致生成内容的质量和准确性有所下降。经过测试，单从代码生成来看，New Bing 在生成复杂代码时，的确错误比较多，甚至还不如 ChatGPT 免费版，不过多生成几次，总有一次适合你。

（3）数据和安全：ChatGPT 目前还没有实现实时地从网络上获取数据的功能[①]，它只能依赖于模型中已经存储的数据进行生成。这使得 ChatGPT 在实时性较强的场景下具有劣势。而 New Bing 则可以实时地从网络上获取数据，并为生成内容提供来源和引用。这使得 New Bing 在实时性较强的场景下具有优势。另一方面，ChatGPT 花费了 6 个月时间来提高模型的安全性和对齐性，使得它比以前的版本更不容易产生不良或不合适的内容，并且能够生成更符合事实的回复。而 New Bing 则没有明确说明它对模型安全性和对齐性方面做了哪些改进或措施。

总之，ChatGPT 和 New Bing 都是基于生成式 AI 技术的产品或服务，它们都可以根据用户输入的自然语言，生成相关的回复或内容。但是它们在功能和定位、版本和性能、数据和安全等方面有着明显的差异。

11.1.5　ChatGPT Plus，史上最强 AI

ChatGPT Plus 是 ChatGPT 的收费版本，也被称为目前地球上最强的 AI，据说 ChatGPT Plus 解决问题的能力已经达到了博士水准，而且用于全人类已经公开的知识，所以 ChatGPT Plus 的知识渊博程度绝对碾压一切专家，强烈推荐大家尝试 ChatGPT Plus，因为它可以 24 小时不间断提供服务，无论是在生活中，还是在学习中，或是在工作中，甚至无聊想找人聊天，ChatGPT Plus 都可以提供全方位、无死角的服务。对于那些不使用 ChatGPT Plus 的人来说，完全就是降维打击。那么 ChatGPT Plus 主要有哪些功能呢？ChatGPT Plus 的功能主要包括以下几方面。

（1）编程：ChatGPT Plus 可以根据用户输入的自然语言描述或样本，自动生成相关的代码，如 Python、Java、C++ 等。ChatGPT Plus 还可以根据用户反馈进行调整和优化，提高生成代码的质量和性能。ChatGPT Plus 可以帮助用户提高编程效率和质量，减少错误和重复工作，以及学习新的编程技能。经过大量使用 ChatGPT Plus，还可以解锁更多新技能，例如，有一段 Python 的代码，要转换成功能完全相同的 Java 和 Go 代码，或者给出一段 Rust 语言的代码，给出这段代码可能存在什么缺陷，ChatGPT Plus 在大多数时候，都会给出满意的答案。

（2）写作：曾几何时，作家都是少数人的专利，写小说，写散文并不容易，也只有少数人可以做到，不过有了 ChatGPT Plus，一切都将改变，使人人成为作家变为可能。ChatGPT Plus 可以根据用户输入的关键词或主题，自动生成相关的文章或段落，如新闻、

---

① 在 2023 年 5 月份，OpenAI 为 ChatGPT Plus 版本推出了插件功能，包括可以联网的插件，使得 ChatGPT Plus 可以获得最新的数据，但这仍然属于补丁形式的解决方案。在未来，ChatGPT Plus 应该会像 New Bing 一样，可以实时从网络获取最新的数据。

故事、诗歌、广告等。ChatGPT Plus 还可以根据用户反馈进行调整和优化，提高生成文本的质量和适应性。ChatGPT Plus 可以帮助用户提高写作效率和创造力，节省时间和精力，以及激发灵感。甚至还可以用 ChatGPT Plus 改写文章，以及审核文章，看看哪里有错别字或者不合适的描述。而且 ChatGPT Plus 不是简单机械地根据文字本身去审核，而是理解语义甚至是语境后的审核，所以效果比普通的 AI 审核的效果更好。

（3）音乐：ChatGPT Plus 可以根据用户输入的音乐风格或样本，自动生成相关的音乐作品，如流行、摇滚、古典等。ChatGPT Plus 还可以根据用户反馈进行调整和优化，提高生成音乐的质量和满意度。ChatGPT Plus 可以帮助用户创造出新颖和有趣的音乐，并为音乐产业的发展带来新的机遇和挑战。

（4）视频：ChatGPT Plus 可以根据用户输入的视频类型或样本，自动生成相关的视频内容，如电影、动画、纪录片等。ChatGPT Plus 还可以根据用户反馈进行调整和优化，提高生成视频的质量和真实性。ChatGPT Plus 可以帮助用户创造出新颖和有趣的视频，并为视频产业的发展带来新的机遇和挑战。

（5）图像：ChatGPT Plus 可以根据用户输入的图像类型或样本，自动生成相关的图像内容，如人物、风景、动物等。ChatGPT Plus 还可以根据用户反馈进行调整和优化，提高生成图像的质量和美观性。ChatGPT Plus 可以帮助用户创造出新颖和有趣的图像，并为图像产业的发展带来新的机遇和挑战。

（6）支持插件：ChatGPT Plus 支持与其他平台或服务进行集成，如微软 Office、谷歌 Docs、Slack 等。这些插件可以让用户在使用这些平台或服务时，方便地调用 ChatGPT Plus 来生成或编辑内容。这些插件可以提高用户在各个领域的效率和生产力。

另外，ChatGPT 可以在任何浏览器中使用，而 New Bing 只能在微软的 Edge 浏览器中使用。

11.1.6 有了 ChatGPT，程序员真的会失业吗

自从 ChatGPT 发布以来，关于 ChatGPT 会导致大量失业的声音就一直不绝于耳[①]。不过说实话，ChatGPT 在大多数领域，的确是很酷啊。就拿编程来说，如果描述得当，ChatGPT 会为我们准确生成各种语言的源代码，的确可以加快开发的速度，但这些代码仍然需要人工审核，以及人工调优，所以用 ChatGPT 编程的人本身就要是编程高手，我的结论是某些能力很强的程序员并不会因为 ChatGPT 而失业，反而会让自己更强大。

不过 ChatGPT 的出现肯定会导致另外一个问题，那就是仍然处于初级阶段和刚大学毕业的程序员可能会找不到工作。在没有 ChatGPT 的时候，可能一个项目组是一两个高手，然后带一些初级和中级的程序员。一些难度不大，但比较费时的代码的工作，通常是由这

① 自从 OpenAI 在 2024 年 2 月宣布推出 Sora（可以根据文本生成 1 分钟高清视频的 AIGC），不仅仅程序员面临被 AIGC 抢饭碗的窘境，就连短视频作者、导演、演员在不远的未来都有可能面临失业的风险，至少行业会重新洗牌。会利用 AIGC 的从业者将远远超越那些还在使用传统方法拍摄短视频、拍电影的从业者。

些初级和中级程序员来做的，而那些高手通常编写比较有难度的代码，以及审核这些初级和中级程序员写的代码。那么现在问题来了，有了 ChatGPT，那些基础的代码，完全可以交给 ChatGPT 来做，而高级程序员还是同样做着审核代码，以及编写更复杂代码的工作。在这个场景中，就没初级和中级程序员什么事了。所以，在未来，有可能高级程序员会利用 ChatGPT 以及其他 AIGC 产品大幅度提升自己的工作效率，而那些初级和中级程序员的存在就显得没那么必要了，至少不再需要那么多人了。

总之，ChatGPT 会让一部分人失业，但也同时会让一部分人变得更强、更成功。就像金钱一样，钱越多，就会让你更加有钱，从一次成功走向另一次成功，从一场胜利走向另一场胜利。而 ChatGPT 也同样会让强者更强，会让强者成为强者 .plus。而那些排除 ChatGPT，或者不懂得使用 ChatGPT 的人，未来可能并不会美妙。

不过话又说回来，ChatGPT 也同样会产生大量新的工作岗位，如 ChatGPT 擅长代码自动补全、代码格式化等辅助性工作，对复杂的软件工程师来说难以完全替代。程序设计、系统架构、问题诊断等工作仍然需要程序员完成。ChatGPT 生成的代码质量无法保证，还无法处理复杂的业务逻辑，所以生成的代码通常需要程序员核查与修订，然后才能使用。程序员的工作会从创造性转向验证性，成为 AI 的监督者。新的 AI 技术也会不断产生新的工作岗位，如 AI 架构师、数据科学家、AI 安全工程师等，这需要相关的程序员与专家来担任。

尽管 ChatGPT 会创造更多的就业岗位，但这些岗位基本上都要依赖 ChatGPT 来完成，就像现在的计算机一样，不管是做什么工作，都要使用计算机。如果现在的某个人，不会使用计算机，那基本上 99% 的工作都不适合，所以尽快掌握 ChatGPT 才能让自己的未来更加幸运。

11.2　注册和登录 ChatGPT

第一次使用 ChatGPT，需要打开网址 **https://chat.openai.com**，并注册 ChatGPT 账户。进到该页面后，会显示如图 11-1 所示的内容。

单击 Sign up 按钮进入注册页面，如图 11-2 所示。在文本框中输入 Email 地址，或使用 Gmail、微软账户或苹果账户进行注册，推荐使用 Gmail。

创建账户后，单击 Continue 按钮，会显示如图 11-3 所示的页面，要求输入姓名和生日。

单击 Continue 按钮，进入下一页面，如图 11-4 所示。在该页面输入一个接收验证码的手机号，输入完后单击 Send code 按钮进入下一页面。

图 11-1　ChatGPT 的欢迎页面

如果手机成功接收到短信，那么在如图 11-5 所示的页面中输入 6 位验证码。

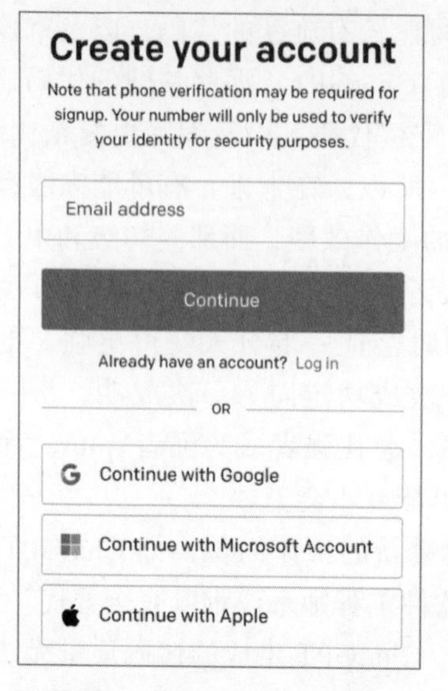

图 11-2 ChatGPT 注册页面

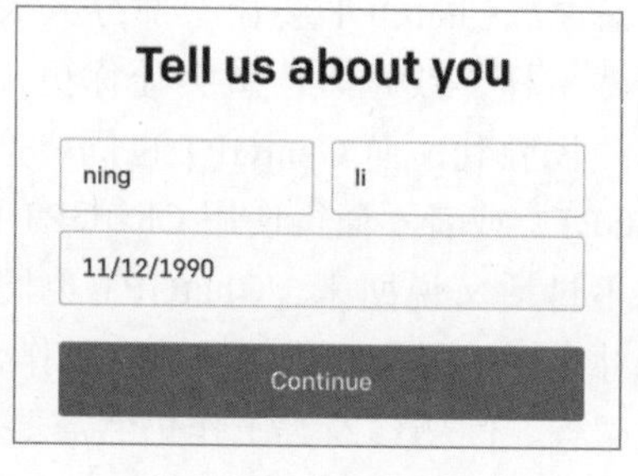

图 11-3 个人信息页

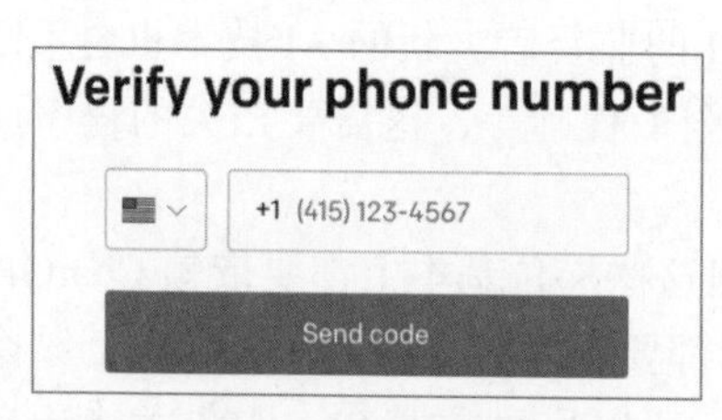

图 11-4 输入手机号

图 11-5 输入验证码

如果验证码通过，就会直接进入 ChatGPT 的聊天首页，如图 11-6 所示，现在可以和 ChatGPT 打个招呼了。

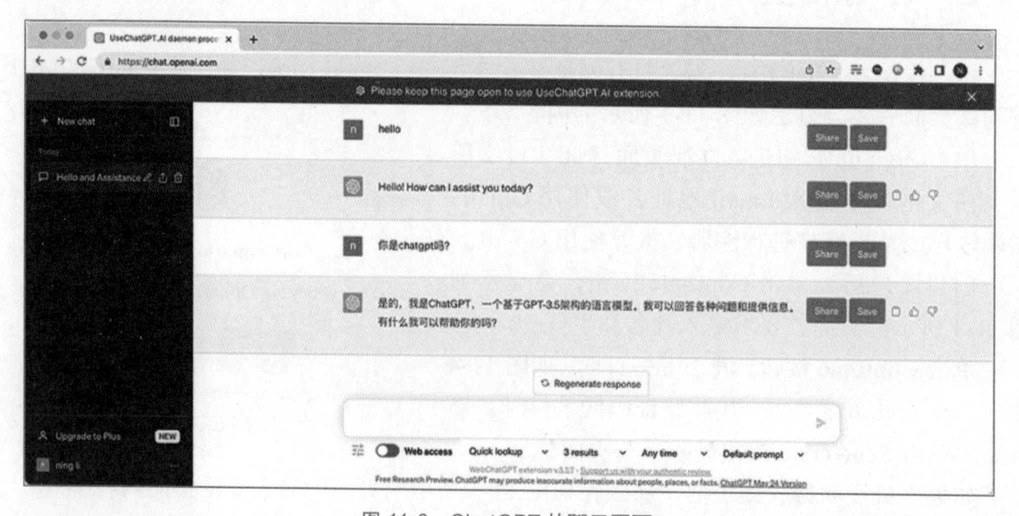

图 11-6 ChatGPT 的聊天页面

到现在为止，我们已经完成了 ChatGPT 的注册，下回再使用 ChatGPT，除非清空浏览器的 Cookie，或退出 ChatGPT 账户，否则会直接进入图 11-6 所示的聊天页面。在注册和使用 ChatGPT 的过程中，可能会涉及 IP、电话号、信用卡等问题。

11.3　升级为 ChatGPT Plus 账户

ChatGPT 的普通账号只能使用 ChatGPT-3.5。这个版本的 ChatGPT，通常只用于完成一般的任务，也就是对结果的精确程度没有过高要求的任务，例如，规划一次旅行，给出红烧肉的 5 种做法等。如果对结果的精度要求非常高，如编程、分析数据，强烈建议升级为 ChatGPT Plus 账号，因为 ChatGPT Plus 用户可以使用 ChatGPT-4.0，这是目前最强大的 AIGC。可以完成比较复杂的编程以及其他各种任务。

升级 ChatGPT Plus 账号比较麻烦，需要的步骤比较多。具体操作流程可以看下面我写的两篇文章：

```
https://www.bilibili.com/read/cv25932153
https://www.bilibili.com/read/cv25952744
```

或者扫描如图 11-7 和图 11-8 所示二维码浏览文章。

图 11-7　注册和使用 ChatGPT Plus

图 11-8　OpenAI API 申请

11.4　让 ChatGPT 帮你写程序

对于程序员来说，最想尝试的就是让 ChatGPT 写程序，那么现在就开始吧。在 ChatGPT 下方的文本框中输入下面的内容：

用 Java 编写一个堆排序程序，并给出测试案例

等 1 ～ 2 秒钟，ChatGPT 就会给我们答案，如图 11-9 所示。

单击代码区域右上角的 Copy code，就可以将生成的 Java 代码复制到剪贴板中，然后可以将剪贴板中的代码复制到任何 Java IDE，然后再执行这些代码。对于比较简单的编程要求，ChatGPT-4 生成的代码基本上是准确的，可以得到我们想要的完美结果。

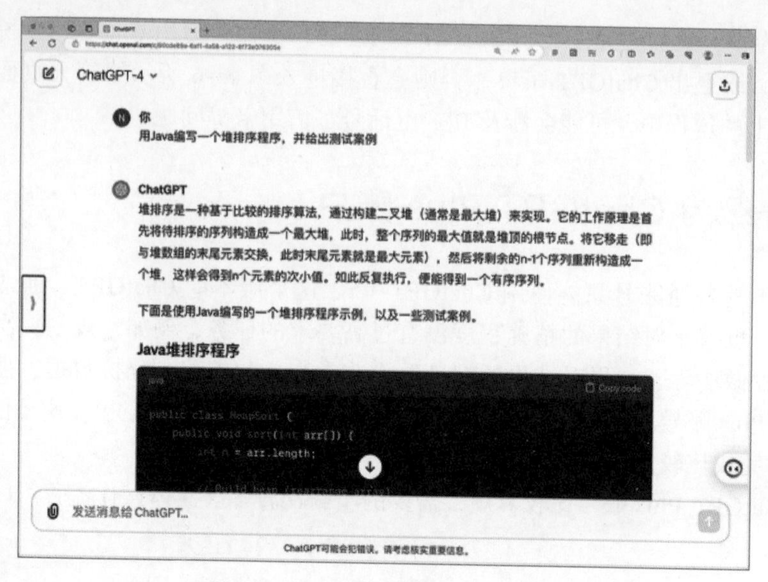

图 11-9　ChatGPT 编写的 Java 代码

11.5　将 ChatGPT 嵌入自己的程序：OpenAI API

OpenAI API 是由 OpenAI 提供的一组编程接口，旨在使开发者、研究人员和企业能够轻松地集成和利用 OpenAI 开发的先进人工智能模型，如 GPT、DALL•E（用于生成图像的 AI）、Codex（面向编程的 AI）等。这些 API 提供了访问 OpenAI 强大的语言模型和其他 AI 工具的能力，允许用户创建、测试和部署各种基于 AI 的应用和服务。

OpenAI API 的主要功能如下。

（1）文本生成和处理：通过 GPT 模型，OpenAI API 能够生成连贯、相关且多样化的文本。这可应用于自动写作、聊天机器人、内容创作、自动化客户服务、语言翻译、摘要提取等领域。

（2）编码辅助和自动化：Codex 模型专为编程任务设计，可以理解自然语言并生成代码。它可以帮助开发者自动化编程任务，包括编写、解释代码，生成编程语言之间的转换等。

（3）图像生成：DALL•E 模型能够基于文本描述生成高质量、创意性的图像。这对于艺术创作、游戏设计、广告、产品原型设计等领域特别有用。

（4）语音处理：虽然 OpenAI API 最初主要关注文本和图像，但 OpenAI 也在探索语音相关的应用，例如自动语音识别（Automatic Speech Recognition，ASR）和语音生成。

（5）自然语言理解：OpenAI API 能够理解和解释自然语言，支持情感分析、文本分类、问答系统等功能。这有助于分析客户反馈、社交媒体监控、自动问答服务等。

（6）教育和研究：提供强大的工具来辅助教育内容的创作、学术研究、数据分析等，特别是在编程、语言学习、数学等领域。

OpenAI API 的使用场景和优势如下。

（1）创造力和生产力工具：通过自动化内容创作和编程任务，OpenAI API 帮助提高个人和企业的创造力和生产力。

（2）客户支持自动化：创建高度互动和个性化的自动客户支持体验，降低人力成本，提高效率。

（3）教育技术：提供个性化学习经验，辅助编程教育和语言学习。

（4）内容和媒体：自动化新闻报道、文章写作、图像创作等，提高内容生成的速度和多样性。

（5）研究和分析：在学术研究、市场分析、数据处理等领域提供深度分析和见解提取的能力。

11.6　用 Java 访问 OpenAI API

OpenAI 官方只提供了 Python 版的用于访问 OpenAI API 的模块，其他编程语言并没有官方提供的库。所以使用 Java 语言访问 OpenAI API 通常有如下 2 种方式。

（1）使用第三方开发的 Java 库。

（2）直接通过 URL 访问 OpenAI API。

本章采用第（2）种方式，因为这种访问 OpenAI API 的方式比较通用，只要是支持 HTTP/HTTPS 的编程语言，都可以访问 OpenAI API。

不同的功能，OpenAI API 有不同的 URL，对于聊天服务，可以使用下面的 URL：

```
https://api.openai.com/v1/chat/completions
```

然后需要使用 Content-Type 和 Authorization 设置请求头：

（1）Content-Type：用于设置请求数据的格式，这里是 application/json。

（2）Authorization：用于指定 API Key，如 "Bearer " + apiKey。其中 Bearer 是 OpenAI 要求的关键字，后面有一个空格。apiKey 是保存 API Key 的变量。

这里的关键就是申请 API Key。将信用卡与 OpenAI API 绑定后（在前面提供的文章中已经描述如何绑定），可以到如下页面申请 API Key。

```
https://platform.openai.com/api-keys
```

进入该页面，单击 Create new secret key 按钮即可申请新的 API Key，如图 11-10 所示。

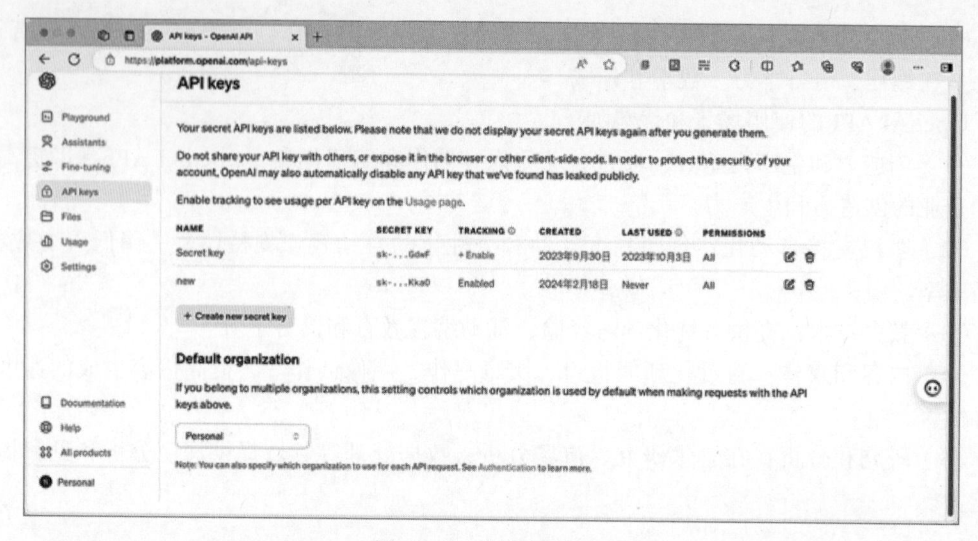

图 11-10　申请 API Key

下面的例子演示了 Java 如何通过 URL 访问 OpenAI API，发送和接收聊天信息。本例使用了 gpt-3.5-turbo 模型。

代码位置：src/main/java/chatgpt/OpenAIChatExample.java

```
package chatgpt;
import java.io.IOException;
import java.net.URI;
import java.net.http.HttpClient;
import java.net.http.HttpRequest;
import java.net.http.HttpResponse;
import java.net.http.HttpRequest.BodyPublishers;
import java.net.http.HttpResponse.BodyHandlers;
public class OpenAIChatExample {
    public static void main(String[] args) throws IOException, InterruptedException {
        String apiKey = "在这里指定 API Key";
        String text = "hello";
        HttpClient httpClient = HttpClient.newHttpClient();
        HttpRequest request = HttpRequest.newBuilder()
                .uri(URI.create("https://api.openai.com/v1/chat/completions"))
                .header("Content-Type", "application/json")
                .header("Authorization", "Bearer " + apiKey)
                .POST(BodyPublishers.ofString("""
                        {
                          "model": "gpt-3.5-turbo",
                          "messages": [
                            {"role": "user", "content": "%s"}
                          ]
                        }
                        """.formatted(text)))
```

```
                .build();
        HttpResponse<String> response = httpClient.send(request,
BodyHandlers.ofString());
        System.out.println(response.body());
    }
}
```

在这段代码中，通过 POST 方法向 OpenAI 服务端提交 JSON 格式的数据。如果是提交问题，需要设置 role 为 user，而 content 则是具体提交的问题。

运行程序，会输出如下信息：


```
{
  "id": "chatcmpl-8u9Q0OLxLwgnovzAJNPRpXTDxRqqm",
  "object": "chat.completion",
  "created": 1708393264,
  "model": "gpt-3.5-turbo-0125",
  "choices": [
    {
      "index": 0,
      "message": {
        "role": "assistant",
        "content": "Hello! How can I assist you today?"
      },
      "logprobs": null,
      "finish_reason": "stop"
    }
  ],
  "usage": {
    "prompt_tokens": 8,
    "completion_tokens": 9,
    "total_tokens": 17
  },
  "system_fingerprint": "fp_69420321d0"
}
```


其中 message 的 content 就是 OpenAI API 返回的信息，在处理返回信息时，可以直接提取 content 属性的值。

11.7　编程魔匣

本节的例子利用 OpenAI API 实现了一个可以用文字让 OpenAI 编程的应用：编程魔匣。在窗口下方输入要编写什么程序（本例输入了“用 Java 编写冒泡排序算法，并给出测试用例”），单击“写程序”按钮，过一会儿，就会在窗口的上方显示 OpenAI API 返回的代码和相关描述，如图 11-11 所示。

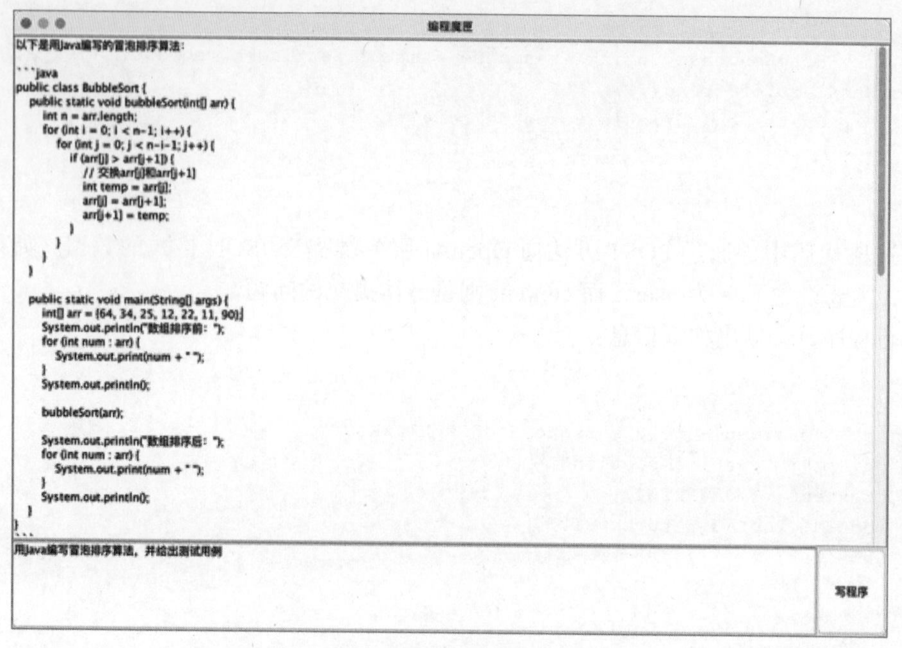

图 11-11　编程魔匣

本例使用 Java Swing 实现 UI。Swing 是 JDK 内置的 GUI 图形库，不需要单独安装，使用起来比较简单。本例的界面也非常简单。单击“写程序”按钮，执行的代码与 11.6 节的例子类似，同时使用 HTTP POST 方法向 OpenAI 服务端提交 JSON 格式的请求，并返回 JSON 格式的数据。不过本例有一个关键的 extractContentFromResponse 方法，是从 JSON 格式的返回数据中提取出 content 属性值，这样就可以直接显示返回结果了。

```
private String extractContentFromResponse(String jsonResponse) {
    try {
        ObjectMapper mapper = new ObjectMapper();
        var rootNode = mapper.readTree(jsonResponse);
        var choicesNode = rootNode.path("choices");
        if (!choicesNode.isEmpty()) {
            var firstChoice = choicesNode.get(0);
            var messageNode = firstChoice.path("message");
            if (!messageNode.isMissingNode()) {
                return messageNode.path("content").asText();
            }
        }
    } catch (Exception e) {
        e.printStackTrace();
    }
    return "解析响应失败";
}
```

11.8　聊天机器人

在 11.7 节实现了一个可以自动编程的应用，这一节会实现一个更复杂的聊天机器人，效果如图 11-12 所示。

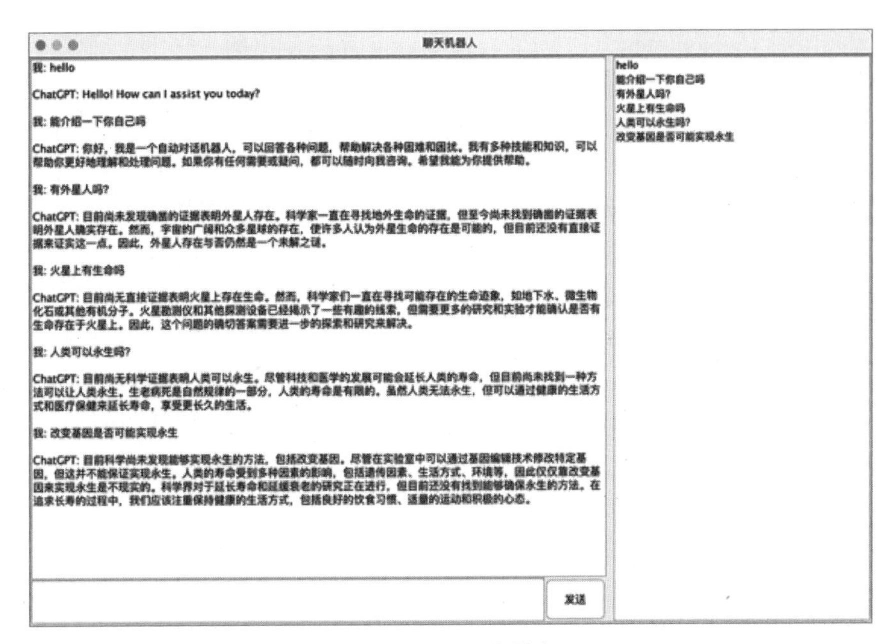

图 11-12　聊天机器人

界面的左侧是聊天区域，在下方的聊天输入框输入要交流的内容，单击“发送”按钮，过一会儿，就会在上方的文本输入框中按顺序显示 ChatGPT 返回的内容。在界面的右侧显示了聊天历史（我说的内容），单击某一条历史记录，会将记录的文本重新显示在聊天输入框中。

这个例子总体的实现与编程魔匣差不多，只是在发送和接收数据时采用了异步的方式，这样有效避免了由于 OpenAI API 的调用延迟导致的程序假死现象。

在本例中有一个核心方法 sendMessage，用于获取输入的聊天内容，将聊天内容添加到历史列表，最后用线程的方式向 OpenAI 服务端发送消息。

```
private void sendMessage(ActionEvent e) {
    String userInput = inputTextArea.getText().trim();
    if (!userInput.isEmpty()) {
        listModel.addElement(userInput);
        displayMessage("我：" + userInput);
        inputTextArea.setText("");
        new Thread(() -> {
            String response = sendMsg(userInput);
            SwingUtilities.invokeLater(() -> displayMessage("ChatGPT: "
```

```
+ response));
            }).start();
        }
}
```

这里主要解释一下 SwingUtilities.invokeLater 方法。

SwingUtilities.invokeLater 方法是在 Java Swing 库中使用的，它用于确保一个特定的任务（Runnable 的实现）在事件分发线程（Event Dispatching Thread，EDT）上被执行。EDT 负责处理图形用户界面（GUI）事件，如用户的单击和键盘输入，以及更新 GUI 的显示。

当需要从非事件分发线程修改 GUI 组件时（比如从一个工作线程中），应该使用 SwingUtilities.invokeLater 来确保这些更改在 EDT 上执行，这样可以避免潜在的线程安全问题。如果直接从非 EDT 线程修改 GUI 组件，可能会导致不可预测的行为和显示问题，因为 Swing 组件并不是线程安全的。

使用 SwingUtilities.invokeLater 方法，可以将一个 Runnable 对象传递给它，Runnable 对象的 run 方法中包含了需要在 EDT 执行的代码。这样，无论 invokeLater 被调用时当前线程是什么，都可以保证 Runnable 对象的 run 方法在事件分发线程上异步执行，从而安全地更新 GUI 组件。

简而言之，SwingUtilities.invokeLater() 方法的主要目的如下。

（1）保证 GUI 的线程安全性。

（2）允许从非 GUI 线程安全地更新 GUI 组件。

（3）确保 GUI 更新和事件处理的顺序性和正确性。

11.9 小结

尽管本章只是向大家展示了 ChatGPT 和 OpenAI API 的极小的一部分功能，不过单从这一点，就足以勾起大家的好奇心，ChatGPT 到底能为我们带来什么呢？ ChatGPT 会将人类的科技带向何方呢？当然，这些问题，ChatGPT 和我都无法准确回答，只有靠广大读者自己在深入使用 ChatGPT 后，得到令自己满意的答案了。不管答案是什么，可以肯定的是，有了以 ChatGPT 为首的各种 AIGC 应用，未来的世界一定与现在有很大差异，至于是怎样的差异，那就仁者见仁智者见智了。

第 12 章

读写 Office 文档

微软的 Office 是目前最流行的办公软件，在 Windows 和 macOS 平台都可以使用。尽管 Office 的每一个成员（如 Excel、Word、PowerPoint 等）都支持使用 VBA 实现办公自动化，但 VBA 并非现代编程语言，而且功能有限，与其他技术结合比较费劲。而 Java 通过与众多第三方模块的结合，可以毫无压力地读写 Excel、Word、PowerPoint 等 Office 文档。所以本章将会向广大读者展示如何使用第三方 Java 库读写 Office 文档。

12.1 Apache POI 简介

Apache POI 是一个非常强大的 Java 库，用于读写 Microsoft Office 格式的文件，包括 Excel、Word 和 PowerPoint 文档。Apache POI 提供了不同的组件来处理不同类型的文件。

（1）HSSF 和 XSSF：用于读写 Excel 文件（.xls 和 .xlsx）。HSSF 是用于处理 Excel 97-2003 文件的，而 XSSF 是用于处理 Excel 2007 及以上版本文件的。对于 Excel 文件，还有一个简化的 API 叫作 SXSSF，它对于大数据量的写操作进行了优化。

（2）HWPF 和 XWPF：用于处理 Word 文件（.doc 和 .docx）。HWPF 是用于处理 Word 97-2003 文件的，而 XWPF 是用于处理 Word 2007 及以上版本文件的。

（3）HSLF 和 XSLF：用于处理 PowerPoint 文件（.ppt 和 .pptx）。HSLF 是用于处理 PowerPoint 97-2003 文件的，而 XSLF 是用于处理 PowerPoint 2007 及以上版本文件的。

Apache POI 提供了丰富的 API 来创建、修改和显示这些文件类型。使用 Apache POI，开发者可以在 Java 程序中轻松地处理 Office 文档，不需要 Office 软件的支持。这使得 Apache POI 成为在 Java 环境中处理 Office 文件的首选库。

12.2 下载和引用 Apache POI 库

读者可以从下面的页面下载 Apache POI 的最新版：

```
https://archive.apache.org/dist/poi/release/bin
```

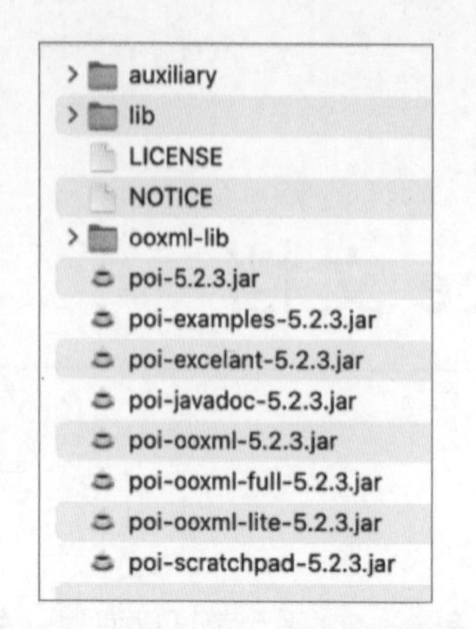

图 12-1 Apache POI 压缩包的目录结构

下载后，直接解压，会看到如图 12-1 所示的文件和目录。

在 Apache POI 压缩包中有多个 JAR 文件，这些 JAR 文件的解释如下。

（1）poi-5.2.3.jar：这是主要的库，它包含了处理老版本的 Excel (HSSF)、Word (HWPF)、PowerPoint (HSLF) 等文件的 API。如果只需要处理这些老版本的文件，引用这个 JAR 文件就足够了。

（2）poi-ooxml-5.2.3.jar：这个库用于处理 Office Open XML (OOXML) 格式的文件，即现代版本的 Office 文件，如 Excel (.xlsx)、Word (.docx)、PowerPoint (.pptx) 等。

（3）poi-ooxml-full-5.2.3.jar：包含了 poi-ooxml 的所有依赖库，如果想确保所有 OOXML 相关的功能都能用，并且不想自己管理所有的依赖，可以使用这个。

（4）poi-ooxml-lite-5.2.3.jar：这是一个较轻版本的 poi-ooxml 库，它提供了对 OOXML 文件的基本支持，但可能不包括一些高级功能。

（5）poi-excelant-5.2.3.jar：这个库使 Apache POI 功能可以通过 Apache Ant 任务使用。

（6）poi-examples-5.2.3.jar：包含了使用 POI 库的示例代码。

（7）poi-javadoc-5.2.3.jar：包含了 API 的 Javadoc 文档，对于理解和使用 API 很有帮助。

（8）poi-scratchpad-5.2.3.jar：这个库包含了对老版本的 Word (HWPF) 和 PowerPoint (HSLF) 的额外支持，以及对其他格式如 Visio 和 Publisher 文件的支持。

在一般情况下，可以引用 poi-5.2.3.jar 和 poi-ooxml-5.2.3.jar。以及 lib 和 ooxml 目录中的 JAR 文件（这些文件是 Apache POI 的一些依赖）。

12.3 读写 Excel 文档

本节将展示如何使用第三方 Java 库读写 Excel 文档。

12.3.1 生成 Excel 表格

本节会使用 Apache POI 库创建一个表格，需要使用如下一些操作。

1. 添加一行数据到工作表

要将一行数据（以列表形式）添加到工作表中，首先需要创建一个 Row 对象。这可以通过调用 Sheet 对象的 createRow(int) 方法来实现，其中参数是行的索引（从 0 开始）。一旦创建了 Row，就可以通过调用 Row 的 createCell(int) 方法来创建单元格 Cell，并通过 setCellValue 方法为其赋值。

```
Row row = sheet.createRow(rowNum);                    // rowNum 是行号
for (int i = 0; i < dataList.size(); i++) {
    Cell cell = row.createCell(i);
    cell.setCellValue(dataList.get(i));               // dataList 是一个包含数据的列表
}
```

2. 设置数字单元格显示千分号

为了设置数字格式，我们需要创建一个 CellStyle 并设置数据格式。CellStyle 对象通过 Workbook 的 createCellStyle 方法创建，并通过 DataFormat 的 getFormat 方法设定一个格式字符串，例如“#,##0”表示千分位。

```
CellStyle style = wb.createCellStyle();
style.setDataFormat(wb.createDataFormat().getFormat("#,##0"));
cell.setCellStyle(style);                                      // 应用样式到单元格
```

3. 设置表格行的背景色

要为行设置背景色，我们同样使用 CellStyle 对象，并设置前景色和填充模式。我们可以使用 setFillForegroundColor 和 setFillPattern 方法来实现。

```
CellStyle fillStyle = wb.createCellStyle();
fillStyle.setFillForegroundColor(IndexedColors.LIGHT_BLUE.index);      // 设置颜色
fillStyle.setFillPattern(FillPatternType.SOLID_FOREGROUND);     // 设置填充模式
row.setRowStyle(fillStyle);                                     // 应用样式到整行
```

4. 隐藏网格线

隐藏工作表的网格线可以通过 Sheet 对象的 setDisplayGridlines 方法实现，传入 false 即可隐藏网格线。

```
sheet.setDisplayGridlines(false);
```

5. 设置单元格的边框

设置单元格边框涉及定义一个 CellStyle 并指定顶部、底部、左侧和右侧的边框样式。可以使用 setBorderTop、setBorderBottom、setBorderLeft 和 setBorderRight 方法。

```
CellStyle borderStyle = wb.createCellStyle();
borderStyle.setBorderTop(BorderStyle.THIN);
borderStyle.setBorderBottom(BorderStyle.THIN);
borderStyle.setBorderLeft(BorderStyle.THIN);
borderStyle.setBorderRight(BorderStyle.THIN);
cell.setCellStyle(borderStyle);                          // 应用边框样式到单元格
```

下面的例子演示了如何用 Apache POI 创建 Excel 表格，并将 Excel 表格保存为 table.xlsx。

代码位置：src/main/java/office/CreateExcelExample.java

```
package office;
import org.apache.poi.ss.usermodel.*;
import org.apache.poi.xssf.usermodel.XSSFWorkbook;
import java.io.FileOutputStream;
import java.io.IOException;
import java.util.Random;
public class CreateExcelExample {
    public static void main(String[] args) throws IOException {
        Workbook wb = new XSSFWorkbook();
        Sheet sheet = wb.createSheet();
        Row headerRow = sheet.createRow(0);
        String[] cities = {"北京", "上海", "广州", "深圳"};
        // 设置表头和列头
        headerRow.createCell(0);
        for (int i = 1; i <= 5; i++) {
            headerRow.createCell(i).setCellValue("第" + i + "年");
        }
        // 填充城市和随机数据
        Random rand = new Random();
        for (int i = 0; i < cities.length; i++) {
            Row row = sheet.createRow(i + 1);
            row.createCell(0).setCellValue(cities[i]);
            for (int j = 1; j <= 5; j++) {
                row.createCell(j).setCellValue(rand.nextInt(9000) + 1000);
            }
        }
        // 设置数字格式和行背景颜色
        CellStyle style = wb.createCellStyle();
        style.setDataFormat(wb.createDataFormat().getFormat("#,##0"));
        for (int i = 1; i <= sheet.getLastRowNum(); i++) {
            Row row = sheet.getRow(i);
            for (int j = 1; j < row.getLastCellNum(); j++) {
                Cell cell = row.getCell(j);
                cell.setCellStyle(style);
            }
            // 设置行背景颜色
            CellStyle fillStyle = wb.createCellStyle();
            fillStyle.cloneStyleFrom(row.getCell(1).getCellStyle());
            fillStyle.setFillForegroundColor(IndexedColors.LIGHT_BLUE.index);
            fillStyle.setFillPattern(FillPatternType.SOLID_FOREGROUND);
            if (i % 2 != 0) {
                for (Cell cell : row) {
                    cell.setCellStyle(fillStyle);
                }
            }
        }
        // 隐藏网格线
        sheet.setDisplayGridlines(false);
        // 设置边框
```

```
        CellStyle borderStyle = wb.createCellStyle();
        borderStyle.setBorderTop(BorderStyle.THICK);
        borderStyle.setBorderBottom(BorderStyle.THICK);
        borderStyle.setBorderLeft(BorderStyle.THICK);
        borderStyle.setBorderRight(BorderStyle.THICK);
        for (Row row : sheet) {
            for (Cell cell : row) {
                CellStyle cellStyle = wb.createCellStyle();
                cellStyle.cloneStyleFrom(cell.getCellStyle());
                cellStyle.setBorderTop(BorderStyle.THIN);
                cellStyle.setBorderBottom(BorderStyle.THIN);
                cellStyle.setBorderLeft(BorderStyle.THIN);
                cellStyle.setBorderRight(BorderStyle.THIN);
                cell.setCellStyle(cellStyle);
            }
        }
        // 保存文件
        try (FileOutputStream fileOut = new FileOutputStream("table.xlsx")) {
            wb.write(fileOut);
        }
        wb.close();
    }
}
```

运行程序，会看到当前目录生成了一个 table.xlsx 文件，文件内容如图 12-2 所示。

	A	B	C	D	E	F
1		第1年	第2年	第3年	第4年	第5年
2	北京	6,478	2,777	3,620	8,151	2,790
3	上海	4,471	9,723	6,952	5,267	4,035
4	广州	6,258	3,710	5,829	7,235	3,287
5	深圳	1,790	3,653	3,129	8,121	4,990

图 12-2　Excel 表格

12.3.2　读取 Excel 表格

本节会读取 12.3.1 节创建的 table.xlsx 文件中的表格，并将表格内容输出到终端。读取表格数据要比创建表格容易得多。首先要打开 table.xlsx 文件，然后获取第 1 个 Sheet。接下来先扫描 Row，再扫描列（双 for 循环），并判断每一个 cell 的类型，最后将 cell 的内容输出到终端。例如，如果 cell 的类型是字符串，那么使用下面的代码获取 cell 的内容，并将内容输出到终端。

```
System.out.print(cell.getRichStringCellValue().getString());
```

下面是本例的完整代码。

代码位置：src/main/java/office/ReadExcelExample.java

```
package office;
import org.apache.poi.ss.usermodel.*;
import org.apache.poi.xssf.usermodel.XSSFWorkbook;
import java.io.FileInputStream;
import java.io.IOException;
public class ReadExcelExample {
    public static void main(String[] args) throws IOException {
        try (FileInputStream fileIn = new FileInputStream("table.xlsx")) {
            Workbook wb = new XSSFWorkbook(fileIn);
            Sheet sheet = wb.getSheetAt(0);
            // 扫描表格的所有行和列
            for (Row row : sheet) {
                for (Cell cell : row) {
                    switch (cell.getCellType()) {
                        case STRING:
                            System.out.print(cell.getRichStringCellValue().
getString());
                            break;
                        case NUMERIC:
                            if (DateUtil.isCellDateFormatted(cell)) {
                                System.out.print(cell.getDateCellValue());
                            } else {
                                System.out.print(cell.getNumericCellValue());
                            }
                            break;
                        case BOOLEAN:
                            System.out.print(cell.getBooleanCellValue());
                            break;
                        case FORMULA:
                            System.out.print(cell.getCellFormula());
                            break;
                        default:
                            System.out.print("");
                    }
                    System.out.print("\t");
                }
                System.out.println();
            }
        }
    }
}
```

运行程序，会在终端输出如图 12-3 所示的内容。

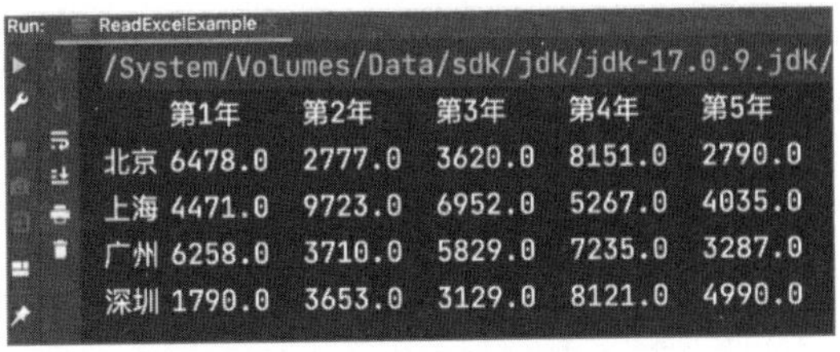

```
Run:  ReadExcelExample
/System/Volumes/Data/sdk/jdk/jdk-17.0.9.jdk/
     第1年    第2年    第3年    第4年    第5年
北京 6478.0  2777.0  3620.0  8151.0  2790.0
上海 4471.0  9723.0  6952.0  5267.0  4035.0
广州 6258.0  3710.0  5829.0  7235.0  3287.0
深圳 1790.0  3653.0  3129.0  8121.0  4990.0
```

图 12-3　将表格内容输出到终端

12.3.3　为 Excel Cell 指定公式

本节的例子会使用 Apache POI 为 Excel 的 cell 指定公式，具体的步骤如下。

（1）创建或获取单元格：首先，需要有一个单元格对象，这通常意味着得先创建一个工作簿（Workbook），然后创建一个工作表（Sheet），接着在工作表中创建或获取一个行（Row）对象，最后在行中创建或获取一个单元格（Cell）对象。

（2）设置公式：使用单元格对象的 setCellFormula(String formula) 方法来设置公式。设置的字符串应该是 Excel 中有效的公式，比如 "SUM(A1:B1)" 或 "A1*2"。

例如，如果想要为第 1 行第 3 列的单元格设置一个公式，用来计算第 1 行第 1 列和第 2 列单元格的和，可以这样做：

```
Cell cell = row.createCell(2);                          // 创建 C1 单元格
cell.setCellFormula("A1+B1");                           // 设置公式
```

（3）计算公式：公式设置后，它并不会立即计算结果。要得到计算结果，需要使用公式计算器（FormulaEvaluator）。可以通过调用工作簿的 getCreationHelper().createFormulaEvaluator() 方法来获取公式计算器。然后，可以使用 evaluateInCell(Cell cell) 方法来计算和更新特定单元格的公式，或者使用 evaluateAll() 方法来计算工作簿中所有单元格的公式。

当使用 evaluateInCell() 方法时，单元格类型会自动更改为计算后的值类型，例如：

```
FormulaEvaluator evaluator = workbook.getCreationHelper().
createFormulaEvaluator();
evaluator.evaluateInCell(cell);                          // 计算 C1 单元格的公式
```

这样，公式就会被计算，并且计算结果会存储在单元格中。如果打开这个工作簿，Excel 也会自动计算这些公式。

注意： 在 Java 程序中计算公式时，Apache POI 使用的是 Excel 的公式引擎，所以计算结果应该和 Excel 中得到的结果一致。然而，在某些情况下，如果公式特别复杂或者使用了 Excel 的某些特定功能，Apache POI 可能无法计算这些公式，这时可能需要依赖 Excel 应用程序本身来计算。

下面的代码完整地演示了如何在 Cell 中插入公式。

代码位置：src/main/java/office/ExcelFormulaExample.java

```
package office;
import org.apache.poi.ss.usermodel.*;
import org.apache.poi.xssf.usermodel.XSSFWorkbook;
import java.io.FileOutputStream;
import java.io.IOException;
public class ExcelFormulaExample {
    public static void main(String[] args) {
        // 创建一个新的空白工作簿
        Workbook workbook = new XSSFWorkbook();
        // 创建一个工作表
        Sheet sheet = workbook.createSheet("Example");
        // 创建行（Row），0 表示第一行
        Row row = sheet.createRow(0);
        // 创建单元格并设置值
        Cell cellA = row.createCell(0);                          // A1 单元格
        cellA.setCellValue(10);
        Cell cellB = row.createCell(1);                          // B1 单元格
        cellB.setCellValue(20);
        // 创建一个单元格并设置公式
        Cell cellC = row.createCell(2);                          // C1 单元格
        // 设置公式计算 A1 和 B1 单元格的和
        cellC.setCellFormula("A1+B1");
        // 计算公式
        FormulaEvaluator evaluator = workbook.getCreationHelper().
createFormulaEvaluator();
        // 对所有单元格公式进行计算更新
        evaluator.evaluateAll();
        // 将工作簿写入文件系统
        try (FileOutputStream out = new FileOutputStream
("FormulaExample.xlsx")) {
            workbook.write(out);
        } catch (IOException e) {
            e.printStackTrace();
        } finally {
            // 最后确保关闭工作簿以释放资源
            try {
                workbook.close();
            } catch (IOException e) {
                e.printStackTrace();
            }
        }
    }
}
```

运行程序，会在当前目录生成一个 FormulaExample.xlsx 文件，打开该文件。双击 C1 单元格，会看到 C1 实际上是通过 A1 + B1 计算得到的，如图 12-4 所示。

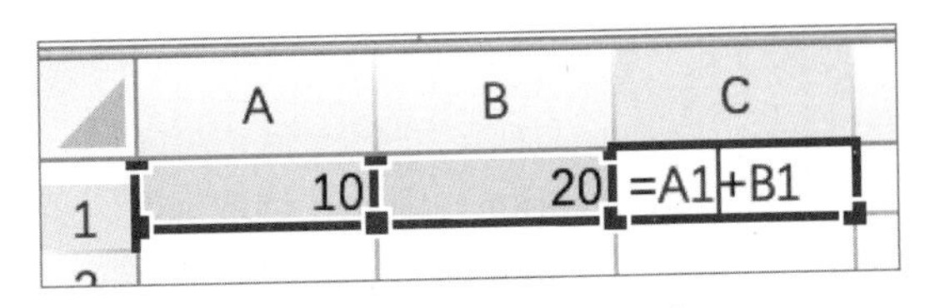

图 12-4　在 Cell 中插入公式

12.4　读写 Word 文档

本节将展示如何使用第三方 Java 库读写 Word 文档。

12.4.1　创建 Word 文档

本节的例子会创建一个 Word 文档，并插入一些文字。具体步骤如下。

1. 初始化文档

首先，需要创建一个 XWPFDocument 类的实例，这将代表 Word 文档。XWPFDocument 是 Apache POI 库中用来处理 Word 文档的主要类。

2. 创建段落

（1）使用 XWPFDocument 对象，可以创建一个或多个段落。段落是文档中的基本结构单元，类似于 Word 文档中按下 Enter 键生成的一个新的行。

（2）通过调用 createParagraph 方法，可以在文档中生成一个新的 XWPFParagraph 对象。

3. 添加文本运行

（1）在段落中，文本以“运行”（Run）的形式出现，每个“运行”可以有自己的样式和属性。在 Word 中，当开始输入并在之后更改样式（如颜色、字体、大小等）时，就创建了一个新的运行。

（2）通过 XWPFParagraph 对象的 createRun 方法，可以生成一个新的 XWPFRun 对象。

（3）使用 setText 方法，可以为这个运行添加文本。

4. 设置样式

（1）可以为每个 XWPFRun 设置样式，例如字体颜色、字体大小、字体类型等。

（2）字体颜色可以通过 setColor 方法设置，该方法接收一个代表颜色的字符串（通常是一个 6 位的十六进制数，与 HTML 颜色编码相同）。

（3）如果需要其他样式（如加粗、斜体等），可以使用 setBold、setItalic 等方法。

5. 保存文档

一旦完成了文档的编辑，就可以使用 write 方法将 XWPFDocument 对象的内容输出到一个文件流中，通常是 FileOutputStream。最后，确保调用 close 方法来关闭文档对象和文件流，以释放资源和锁定的文件。

6. 异常处理

在处理文件时，可能会遇到 IOException，因此在读取和写入文件时应该使用 try-catch 块来妥善处理这些潜在的异常。

在上述过程中，Apache POI 库负责与 Word 文档的底层结构交互，使得开发者可以用类似操作对象的方式来创建和编辑文档，而无须直接处理复杂的文件格式细节。这大大简化了自动化 Word 文档处理的过程。

下面的代码演示了使用 Apache POI 库创建 Word 文档，以及插入文字、设置样式的完整过程。

代码位置：src/main/java/office/CreateWordDocument.java

```
package office;
import org.apache.poi.xwpf.usermodel.*;
import java.io.FileOutputStream;
import java.io.IOException;
public class CreateWordDocument {
    public static void main(String[] args) throws IOException {
        // 创建一个新的空文档
        XWPFDocument document = new XWPFDocument();
        // 添加第一行文本
        XWPFParagraph paragraph = document.createParagraph();
        XWPFRun run = paragraph.createRun();
        run.setText("红酥手");
        // 设置字体颜色为红色
        run.setColor("FF0000");                    // 使用十六进制颜色代码
        run = paragraph.createRun();
        run.setText(",黄藤酒,满城春色宫墙柳。东风恶,欢情薄。一怀愁绪,几年离索。错、错、错！");
        // 添加第二行文本
        paragraph = document.createParagraph();
        run = paragraph.createRun();
        run.setText("春如旧，人空瘦，泪痕红浥鲛绡透。桃花落，闲池阁。山盟虽在，锦书难托。
莫、莫、莫！");
        // 保存文档
        try (FileOutputStream out = new FileOutputStream("basic.docx")) {
            document.write(out);
        }
        document.close();
        System.out.println("Word document created successfully!");
    }
}
```

运行程序，会在当前目录生成一个 basic.docx 文件，效果如图 12-5 所示。

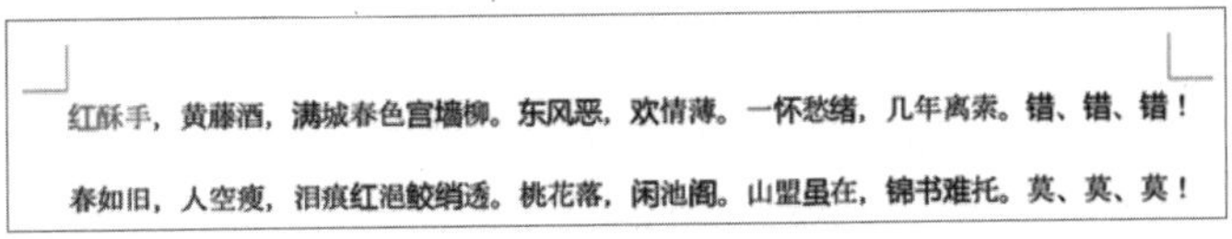

红酥手，黄藤酒，满城春色宫墙柳。东风恶，欢情薄。一怀愁绪，几年离索。错、错、错！

春如旧，人空瘦，泪痕红浥鲛绡透。桃花落，闲池阁。山盟虽在，锦书难托。莫、莫、莫！

图 12-5　向 Word 中插入文本

12.4.2　插入图像

本节的例子会向 Word 文档插入一张图片。Apache POI 库中的 XWPFRun.addPicture 方法是用来将图像插入 Word 文档中的。这个方法属于 XWPFRun 类，XWPFRun 是表示文档中文本的运行部分，可以包含文本、图片或其他元素。

addPicture 方法的原型如下。

```
public void addPicture(InputStream pictureData,
                       int pictureType,
                       String filename,
                       int width,
                       int height) throws InvalidFormatException, IOException
```

参数含义如下。

（1）pictureData：这是一个 InputStream，用于读取要插入的图片文件的数据。需要创建一个指向图片文件的 InputStream，例如使用 FileInputStream。

（2）pictureType：这是一个整型值，指定图片的类型。Apache POI 提供了几种图片类型的常量，如 XWPFDocument.PICTURE_TYPE_JPEG、XWPFDocument.PICTURE_TYPE_PNG 等。我们需要根据图片文件格式选择合适的常量。

（3）filename：这是图片文件的名称。在某些情况下，Word 文档中的图像可能会以文件形式存储，此参数可以用于指定这个文件的名字。

（4）width：这是插入图片的宽度，单位是 EMU（English Metric Unit）。EMU 是 Office Open XML 标准中使用的度量单位，用于确保跨不同分辨率和应用程序的一致性。可以使用 Units.pixelToEMU(int pixels) 方法将像素转换为 EMU。

（5）height：这是插入图片的高度，单位同样是 EMU。

下面的例子演示了如何将一张图片插入 Word 文档中。

代码位置：src/main/java/office/InsertSingleImageToWord.java

```
package office;
import org.apache.poi.xwpf.usermodel.*;
import org.apache.poi.util.Units;
import org.apache.poi.openxml4j.exceptions.InvalidFormatException;
import javax.imageio.ImageIO;
import java.awt.image.BufferedImage;
import java.io.*;
```

```
public class InsertSingleImageToWord {
    public static void main(String[] args) {
        // 创建一个新的 Word 文档
        XWPFDocument document = new XWPFDocument();
        // 设置图片路径
        String imgPath = "images/girl1.png";           // 替换为实际图片路径
        // 插入图片
        try {
            // 使用 ImageIO 读取图片
            BufferedImage image = ImageIO.read(new File(imgPath));
            // 获取图片的原始尺寸
            int originalWidth = image.getWidth();
            int originalHeight = image.getHeight();
            // 转换为 EMU
            int widthInEMU = Units.pixelToEMU(originalWidth);
            int heightInEMU = Units.pixelToEMU(originalHeight);
            // 创建一个段落
            XWPFParagraph paragraph = document.createParagraph();
            XWPFRun run = paragraph.createRun();
            // 添加图片到运行
            run.addPicture(new FileInputStream(imgPath),
                    XWPFDocument.PICTURE_TYPE_PNG,
                    imgPath,
                    widthInEMU,
                    heightInEMU);
        } catch (InvalidFormatException | IOException e) {
            e.printStackTrace();
        }
        // 保存 Word 文档
        try (FileOutputStream out = new FileOutputStream("SingleImageWord.docx")) {
            document.write(out);
            System.out.println("Word document with a single image created
successfully!");
        } catch (IOException e) {
            e.printStackTrace();
        } finally {
            try {
                document.close();
            } catch (IOException e) {
                e.printStackTrace();
            }
        }
    }
}
```

12.4.3 插入页眉、页脚和页码

本节的例子会向 Word 文档中插入页眉、页脚和页码，具体步骤如下。

（1）创建一个 XWPFDocument 对象，代表一个新的 Word 文档。

（2）获取文档的 CTSectPr 对象，用于配置文档的属性，包括页眉和页脚。

（3）创建 XWPFHeaderFooterPolicy 对象，这个对象用于定义文档中页眉和页脚的展现策略。

（4）通过 createHeader 方法添加一个默认的页眉，并在其中创建一个居中对齐的段落，然后添加文本“这是页眉”。

（5）通过 createFooter 方法添加一个默认的页脚，并在其中创建一个居中对齐的段落。

（6）在页脚段落中，首先添加文本“页码”，然后使用 getCTP().addNewFldSimple().setInstr("PAGE * MERGEFORMAT") 添加一个字段代码，这个代码在 Word 中会被转换为当前的页码。

（7）添加文本“ | 这是页脚”，表示页脚中除了页码外还包含其他文本。

（8）使用 FileOutputStream 将文档保存为 .docx 文件。

（9）关闭 XWPFDocument 对象，释放资源。

具体的实现代码如下。

代码位置：src/main/java/office/InsertSingleImageToWord.java

```
package office;
import org.apache.poi.xwpf.model.XWPFHeaderFooterPolicy;
import org.apache.poi.xwpf.usermodel.*;
import org.openxmlformats.schemas.wordprocessingml.x2006.main.CTSectPr;
import java.io.FileOutputStream;
import java.io.IOException;
public class InsertHeaderFooterPageNumber {
    public static void main(String[] args) {
        // 创建一个新的 Word 文档
        XWPFDocument document = new XWPFDocument();
        // 获取文档的 CTSectPr 属性，用于设置页眉和页脚
        CTSectPr sectPr = document.getDocument().getBody().addNewSectPr();
        // 创建页眉和页脚配置策略
        XWPFHeaderFooterPolicy headerFooterPolicy = new XWPFHeaderFooterPolicy
(document, sectPr);
        // 创建页眉
        XWPFHeader header = headerFooterPolicy.createHeader
(XWPFHeaderFooterPolicy.DEFAULT);
        // 向页眉添加文本
        XWPFParagraph headerParagraph = header.createParagraph();
        headerParagraph.setAlignment(ParagraphAlignment.CENTER);
        XWPFRun headerRun = headerParagraph.createRun();
        headerRun.setText("这是页眉");
        // 创建页脚
        XWPFFooter footer = headerFooterPolicy.createFooter
(XWPFHeaderFooterPolicy.DEFAULT);
        // 向页脚添加文本
        XWPFParagraph footerParagraph = footer.createParagraph();
        footerParagraph.setAlignment(ParagraphAlignment.CENTER);
        // 向页脚添加页码
```

```
            XWPFRun footerRun = footerParagraph.createRun();
            footerRun.setText(" 页码 ");
            // 添加字段代码，用于自动生成页码
            footerParagraph.getCTP().addNewFldSimple().setInstr("PAGE \\* MERGEFORMAT");
            footerRun = footerParagraph.createRun();
            footerRun.setText(" | 这是页脚 ");
            // 保存 Word 文档
            try (FileOutputStream out = new FileOutputStream
("HeaderFooterPageNumber.docx")) {
                document.write(out);
            } catch (IOException e) {
                e.printStackTrace();
            }
            // 关闭文档
            try {
                document.close();
            } catch (IOException e) {
                e.printStackTrace();
            }
            System.out.println("Word document with header, footer, and page number
created successfully!");
        }
    }
```

12.4.4 统计 Word 文档生成云图

本节会统计 Word 文档，并根据文档中字符的出现频率，绘制一张云图。这里涉及两个库：HanLP 和 kumo。其中 HanLP 是用来中文分词的，kumo 是用来生成云图的。

读者可以从下面的页面下载 HanLP 最新版的 JAR 文件：

https://mvnrepository.com/artifact/com.hankcs/hanlp

安装 kumo 库，需要在 pom.xml 文件中添加如下依赖：

```
<dependency>
     <groupId>com.kennycason</groupId>
     <artifactId>kumo-tokenizers</artifactId>
     <version>1.28</version>
</dependency>
```

下面分别对这两个库的使用方法进行详细的介绍。

- HanLP 库

HanLP 是一款由一系列模型与算法支持的中文语言处理库，能够处理中文自然语言处理（Natural Language Processing，NLP）的多种任务，如分词、词性标注、命名实体识别等。HanLP 具有高效和准确的特点，广泛用于中文文本分析领域。

使用 HanLP 库分词的具体步骤如下。

（1）导入 HanLP 包：确保项目中已经添加了 HanLP 的依赖。

（2）分词：使用 HanLP.segment(String text) 方法对文本进行分词，这将返回一个 List<Term> 对象，其中 Term 对象包含词语和对应的词性等信息。如果只需要词语，可以进一步处理这个列表。

```
List<String> words = HanLP.segment(text).stream()
                         .map(term -> term.word)
                         .toList();
```

❑ kumo 库

kumo 是一个能够生成词云图的 Java 库。它支持不同形状、颜色、字体和算法的词云生成，使得生成的词云图既美观又富有信息量。

使用 kumo 绘制词云图的步骤如下。

（1）设置词频分析器：使用 FrequencyAnalyzer 类来分析文本的词频。可以设置返回的最大词频数、最小词语长度等。

```
final FrequencyAnalyzer frequencyAnalyzer = new FrequencyAnalyzer();
frequencyAnalyzer.setWordFrequenciesToReturn(300);
frequencyAnalyzer.setMinWordLength(2);
```

（2）生成词频列表：将文本或词列表加载到分析器中，生成词频列表。

```
final List<WordFrequency> wordFrequencies = frequencyAnalyzer.load(wordList);
```

（3）配置并生成词云：使用 WordCloud 类来配置词云图的大小、背景、字体缩放策略和颜色方案。然后，使用 build(List<WordFrequency> wordFrequencies) 方法构建词云，并通过 writeToFile(String outputPath) 方法将词云保存为图片。

```
final Dimension dimension = new Dimension(600, 600);
final WordCloud wordCloud = new WordCloud(dimension, CollisionMode.PIXEL_PERFECT);
// 设置参数 ...
wordCloud.build(wordFrequencies);
wordCloud.writeToFile(outputFilePath);
```

通过这种方式，可以将 HanLP 的强大分词能力和 kumo 的词云生成能力结合起来，对中文文本进行分析并以词云图的形式直观展示分词结果。这种方法特别适用于文本数据的可视化分析，可以帮助理解和探索文本内容中的关键词和主题。

下面的例子演示了生成云图的具体过程。

代码位置：src/main/java/office/WordCloudFromDocx.java

```
package office;
import com.hankcs.hanlp.HanLP;
import com.kennycason.kumo.CollisionMode;
import com.kennycason.kumo.WordCloud;
```

```
import com.kennycason.kumo.WordFrequency;
import com.kennycason.kumo.bg.RectangleBackground;
import com.kennycason.kumo.font.scale.SqrtFontScalar;
import com.kennycason.kumo.nlp.FrequencyAnalyzer;
import com.kennycason.kumo.palette.ColorPalette;
import org.apache.poi.xwpf.usermodel.XWPFDocument;
import org.apache.poi.xwpf.usermodel.XWPFParagraph;
import java.awt.*;
import java.io.FileInputStream;
import java.io.FileNotFoundException;
import java.io.IOException;
import java.util.List;
public class WordCloudFromDocx {
    public static void main(String[] args) throws FileNotFoundException,
IOException {
        // 定义要处理的 Word 文档的文件名和词云图的输出文件名
        String docxFilePath = "word.docx";
        String outputImagePath = "wordcloud.png";
        // 从 Word 文档中提取所有的文本
        String text = extractTextFromDocx(docxFilePath);
        // 对文本进行分词，并返回一个词汇列表
        List<String> words = HanLP.segment(text).stream()
                .map(term -> term.word)
                .toList();
        // 根据词汇列表生成词云图，并保存为 PNG 图像
        generateWordCloud(words, outputImagePath);
    }
    private static String extractTextFromDocx(String docxFilePath) throws
IOException {
        StringBuilder text = new StringBuilder();
        try (FileInputStream fis = new FileInputStream(docxFilePath);
             XWPFDocument document = new XWPFDocument(fis)) {
            for (XWPFParagraph paragraph : document.getParagraphs()) {
                text.append(paragraph.getText());
            }
        }
        return text.toString();
    }
    private static void generateWordCloud(List<String> wordList, String
outputFilePath) {
        final FrequencyAnalyzer frequencyAnalyzer = new FrequencyAnalyzer();
        frequencyAnalyzer.setWordFrequenciesToReturn(300);
        frequencyAnalyzer.setMinWordLength(2);
        // 分析词频
        final List<WordFrequency> wordFrequencies = frequencyAnalyzer.load
(wordList);
        // 设置词云图的大小、背景、字体等参数
        final Dimension dimension = new Dimension(600, 600);
        final WordCloud wordCloud = new WordCloud(dimension, CollisionMode.
PIXEL_PERFECT);
        wordCloud.setPadding(2);
```

```
        wordCloud.setBackground(new RectangleBackground(dimension));
        wordCloud.setFontScalar(new SqrtFontScalar(10, 100));
        wordCloud.setColorPalette(new ColorPalette(new Color(0x4055F1),
new Color(0x408DF1),
                new Color(0x40AAF1), new Color(0x40C5F1), new Color(0xFFFFFF)));
        wordCloud.build(wordFrequencies);
        wordCloud.writeToFile(outputFilePath);
    }
}
```

运行程序，会在当前目录生成一个 wordcloud.png 文件，效果如图 12-6 所示。

图 12-6　云图

12.5　PowerPoint 文档

本节将展示如何使用第三方 Java 库读写 PowerPoint 文档。

12.5.1　创建 PowerPoint 文档

用 Java 程序创建 PPT 文档（使用 Apache POI 库）大致分为以下几个步骤。

1. 初始化 PPT 文档

创建 XMLSlideShow 对象。这是创建 PPT 文档的第一步。XMLSlideShow 是 Apache POI 库中用于代表整个 PPT 文档的类。通过实例化这个类，我们创建了一个空白的 PPT 文档。这一步不需要任何参数，因为我们是在创建一个全新的文档。

```
XMLSlideShow ppt = new XMLSlideShow();
```

2. 获取默认幻灯片布局

在 PPT 中，每个幻灯片都可以有不同的布局，如仅标题、标题加内容等。XSLFSlideMaster 对象代表了幻灯片的母版，它包含了不同的布局设置。通过 ppt.getSlideMasters().get(0) 获取第一个（通常也是唯一的）幻灯片母版。然后，通过 defaultMaster.getLayout(SlideLayout.TITLE) 获取一个包含标题和内容的默认布局。这个步骤对于设置幻灯片的基础布局非常重要。

```
XSLFSlideMaster defaultMaster = ppt.getSlideMasters().get(0);
XSLFSlideLayout titleLayout = defaultMaster.getLayout(SlideLayout.TITLE);
```

3. 创建幻灯片并添加内容

（1）循环创建幻灯片：接下来使用一个循环来添加三个幻灯片。这里展示了如何重复使用同一个布局来创建多个幻灯片。

（2）创建幻灯片：对于每个幻灯片，我们调用 ppt.createSlide(titleLayout) 方法。这个方法接收一个 XSLFSlideLayout 对象作为参数，根据提供的布局创建一个新的幻灯片。

（3）添加标题和内容：每个幻灯片创建后，我们通过调用 slide.getPlaceholder(0) 和 slide.getPlaceholder(1) 获取标题和内容的占位符。然后，分别为这些占位符设置文本，代表幻灯片的标题和主体内容。这里的数字 0 和 1 分别代表布局中的标题占位符和内容占位符的索引。

```
for (int i = 1; i <= 3; i++) {
    XSLFSlide slide = ppt.createSlide(titleLayout);
    XSLFTextShape title = slide.getPlaceholder(0);
    title.setText("第" + i + "页");
    XSLFTextShape body = slide.getPlaceholder(1);
    body.setText("这是第" + i + "页的内容");
}
```

4. 保存 PPT 文档到文件

最后一步是将创建的 PPT 文档保存到文件系统中。这通过创建一个 FileOutputStream 对象来实现，指定要写入的文件名。然后，使用 ppt.write(out) 方法将 PPT 文档的内容写入文件中。完成写入后，确保通过 out.close() 方法关闭文件输出流，释放资源。在本示例中，使用 try-with-resources 语句自动管理资源，因此不需要显式调用 close() 方法。

```
try (FileOutputStream out = new FileOutputStream("basic.pptx")) {
    ppt.write(out);
}
```

下面的代码完整地演示了如何使用 Apache POI 创建 PPT 文档。

代码位置：src/main/java/office/CreatePPT.java

```
package office;
import org.apache.poi.xslf.usermodel.*;
```

```java
import java.io.FileOutputStream;
import java.io.IOException;
public class CreatePPT {
    public static void main(String[] args) {
        // 创建一个空白的 PPT 文档
        XMLSlideShow ppt = new XMLSlideShow();
        // 获取默认的幻灯片布局
        XSLFSlideMaster defaultMaster = ppt.getSlideMasters().get(0);
        XSLFSlideLayout titleLayout = defaultMaster.getLayout(SlideLayout.TITLE);
        // 添加三个幻灯片
        for (int i = 1; i <= 3; i++) {
            // 根据默认布局添加新的幻灯片
            XSLFSlide slide = ppt.createSlide(titleLayout);
            // 在幻灯片中添加标题
            XSLFTextShape title = slide.getPlaceholder(0);      // 获取标题占位符
            title.setText("第" + i + "页");
            // 在幻灯片中添加文本
            XSLFTextShape body = slide.getPlaceholder(1);       // 获取正文占位符
            body.setText("这是第" + i + "页的内容");
        }
        // 保存 PPT 文档到文件
        try (FileOutputStream out = new FileOutputStream("basic.pptx")) {
            ppt.write(out);
        } catch (IOException e) {
            e.printStackTrace();
        }
    }
}
```

运行程序，在当前目录会生成一个 basic.pptx 文件，效果如图 12-7 所示。

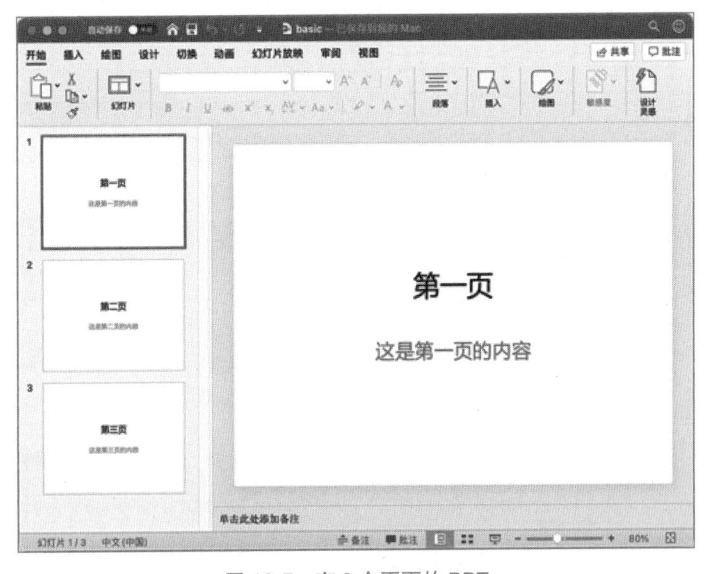

图 12-7　有 3 个页面的 PPT

12.5.2 将图片批量插入 PPT 中

在 Apache POI 库中，addPicture 方法是用于向 PPT 文档中添加图片的关键方法。这个方法通常属于 XMLSlideShow 类或其父类。插入图片到幻灯片时，首先需要将图片文件读入到一个字节数组中，然后使用这个数组作为 addPicture 方法的参数来添加图片。

addPicture 方法的原型如下。

```
XSLFPictureData addPicture(byte[] pictureData, PictureData.PictureType
pictureType) throws IOException;
```

参数含义如下。

（1）pictureData：这是一个字节数组，包含了要添加到 PPT 中的图片的原始字节数据。我们可以通过读取图片文件来获得这些数据，如使用 java.nio.file.Files.readAllBytes(Path path) 方法。

（2）pictureType：这个参数指定了添加图片的格式类型。PictureData.PictureType 是一个枚举类型，它定义了支持的图片格式，如 JPEG、PNG、GIF 等。正确指定图片格式是确保图片能正确添加和显示的关键。

下面的代码演示了如何将图片批量插入到 PPT 中，每页一张图片。

代码位置：src/main/java/office/InsertImagesToPPT.java

```
package office;
import org.apache.poi.xslf.usermodel.XMLSlideShow;
import org.apache.poi.xslf.usermodel.XSLFSlide;
import org.apache.poi.xslf.usermodel.XSLFPictureData;
import org.apache.poi.xslf.usermodel.XSLFPictureShape;
import org.apache.poi.sl.usermodel.PictureData;
import javax.imageio.ImageIO;
import java.awt.Dimension;
import java.awt.image.BufferedImage;
import java.io.File;
import java.io.FileInputStream;
import java.io.FileOutputStream;
import java.io.IOException;
import java.nio.file.Files;
import java.nio.file.Paths;
public class InsertImagesToPPT {
    public static void main(String[] args) {
        // 创建一个空白的 PPT 文件对象
        XMLSlideShow ppt = new XMLSlideShow();
        // 获取 images 目录中所有的 PNG 格式文件名
        File imgDir = new File("images");
        File[] images = imgDir.listFiles((dir, name) -> name.endsWith(".png"));
        if (images != null) {
            for (File image : images) {
```

```
                try {
                    // 创建一个新的幻灯片
                    XSLFSlide slide = ppt.createSlide();
                    // 使用 ImageIO 读取图片文件，获取图片对象
                    BufferedImage img = ImageIO.read(new FileInputStream(image));
                    // 获取图片的原始宽度和高度
                    double imgWidth = img.getWidth();
                    double imgHeight = img.getHeight();
                    // 计算图片的原始宽高比
                    double imgRatio = imgWidth / imgHeight;
                    // 获取幻灯片的宽度和高度（单位：像素）
                    Dimension pgsize = ppt.getPageSize();
                    double slideWidth = pgsize.getWidth();
                    double slideHeight = pgsize.getHeight();
                    // 计算幻灯片的宽高比
                    double slideRatio = slideWidth / slideHeight;
                    // 根据宽高比调整图片大小
                    double scaledWidth, scaledHeight;
                    if (imgRatio > slideRatio) {
                        scaledWidth = slideWidth;
                        scaledHeight = slideWidth / imgRatio;
                    } else {
                        scaledHeight = slideHeight;
                        scaledWidth = slideHeight * imgRatio;
                    }
                    // 读取图片文件到 byte 数组
                    byte[] pictureData = Files.readAllBytes(Paths.get(image.
getAbsolutePath()));
                    // 将图片添加到幻灯片
                    XSLFPictureData pd = ppt.addPicture(pictureData,
                            PictureData.PictureType.JPEG);
                    XSLFPictureShape picture = slide.createPicture(pd);
                    picture.setAnchor(new java.awt.Rectangle(0, 0, (int)
scaledWidth, (int) scaledHeight));
                } catch (IOException e) {
                    e.printStackTrace();
                }
            }
        }
        // 保存 PPT 文件
        try (FileOutputStream out = new FileOutputStream("images.pptx")) {
            ppt.write(out);
        } catch (IOException e) {
            e.printStackTrace();
        }
    }
}
```

运行程序，在当前目录会生成一个 images.pptx 文件，效果如图 12-8 所示。

图 12-8 插入图片的 PPT

12.6 小结

本章主要介绍了如何使用 Java 编程语言结合 Apache POI 库来读写 Microsoft Office 文档，包括 Excel、Word 和 PowerPoint。Apache POI 是一个强大的 Java 库，它提供了丰富的 API 来处理 Office 文档，使得开发者能够在 Java 程序中轻松创建、修改和读取这些文档，而无须依赖于 Microsoft Office 软件。

在 Excel 文档处理方面，本章详细介绍了如何使用 Apache POI 的 HSSF 和 XSSF 组件来读写 Excel 文件（.xls 和 .xlsx）。通过示例代码，展示了如何创建表格、设置单元格样式、隐藏网格线、设置单元格边框以及为单元格指定公式。此外，还介绍了如何读取 Excel 表格数据，包括处理不同类型的单元格内容。

对于 Word 文档，本章介绍了如何使用 XWPFDocument 类来创建和编辑 Word 文档。示例代码展示了如何添加文本、设置样式、插入图片以及如何插入页眉、页脚和页码。此外，还介绍了如何使用 HanLP 库进行中文分词和 kumo 库生成词云图，以可视化文档内容。

在 PowerPoint 文档处理方面，本章讲解了如何使用 XMLSlideShow 类创建 PPT 文档，包括获取默认幻灯片布局、创建幻灯片、添加标题和内容。还介绍了如何批量插入图片到 PPT 中，以及如何调整图片大小以适应幻灯片的布局。

随着办公自动化和数据分析需求的增长，对 Office 文档处理的需求将持续上升。Apache POI 库作为处理 Office 文档的强大工具，其功能和性能有望继续优化，以支持更复杂的文档操作和自动化任务。同时，随着云计算和人工智能技术的发展，未来可能会有更多集成云服务和 AI 分析功能的 Office 文档处理工具出现，以进一步提高办公效率和数据分析能力。此外，随着跨平台办公软件的兴起，如 LibreOffice 和 Google Workspace，对这些软件文档格式的支持也将成为 Apache POI 等库未来发展的方向。

第 13 章

读写 PDF 文档

本章使用 Java 和 Apache PDFBox 库读写 PDF 文档，PDF 文档尽管没有 Word 和 Excel 灵活，但更方便阅读，所以 PDF 通常是发布文档的不二文件格式，几乎所有的现代浏览器都支持阅读 PDF 文档。像 Word、Excel 一样，使用 Java 同样可以向 PDF 文档中插入基本信息，如图像、表格等，还可以对 PDF 文档加密和解密。如果与更多的第三方模块结合，甚至可以在 PDF 文档上绘制复杂的图表。

13.1 Apache PDFBox 简介

Apache PDFBox 是一个开源的 Java 库，用于处理 PDF 文档。它提供了一系列广泛的功能来创建、渲染、打印、分割、合并、更改、校验和提取文本和元数据等。Apache PDFBox 是 Apache 软件基金会下的一个项目，它的主要目标是提供一个简单而强大的接口来操作 PDF 文件。下面是对 Apache PDFBox 的一些主要功能和应用场景的详细介绍。

❑ 主要功能

（1）PDF 创建和编辑：PDFBox 允许开发者从头开始创建新的 PDF 文件，或修改现有的 PDF 文件。这包括添加文本、图像、表格等。

（2）PDF 文本抽取：PDFBox 可以从 PDF 文件中抽取文本，这对于内容分析、搜索引擎索引或数据迁移等场景非常有用。

（3）PDF 文档分割与合并：可以将单个 PDF 文档分割成多个文档，或将多个 PDF 文档合并为一个。

（4）PDF 加密和解密：PDFBox 支持 PDF 文档的加密和解密，提供了文档保护的功能。

（5）表单填充和提交：支持在 PDF 表单中填充数据，以及从填充的表单中提取数据。

（6）PDF 打印：可以将 PDF 文档直接发送到打印机进行打印。

（7）PDF 转换：支持将 PDF 转换为其他格式，如图片。

（8）PDF 渲染：提供了一个 PDF 渲染器，可以将 PDF 页面渲染为图像，以在 Web 页面或其他媒体中显示。

（9）文档校验：提供了验证 PDF/A 文档合规性的工具。

❑ 应用场景

（1）报告生成：在企业和科研机构中，PDFBox 被广泛用于自动化报告生成，如财务报表、研究结果等。

（2）内容管理系统：在内容管理系统（Content Management System，CMS）中，PDFBox 用于文档的导入导出功能，以及内容的抽取和索引。

（3）文档归档与备份：用于将各种格式的文档转换为 PDF 进行长期存储。

（4）数据提取：从 PDF 文档中提取文本和数据，用于数据分析、机器学习模型训练等。

（5）自动化文档处理：用于自动化的文档审查、合并、拆分和加密等任务。

（6）电子表单处理：自动填充和提交 PDF 表单，以及从已填充的表单中提取数据。

（7）文档转换服务：作为服务器端的服务，提供文档格式的转换，如将 PDF 转换为图片格式以便在网页上显示。

Apache PDFBox 的这些功能和应用场景展示了其作为 PDF 处理工具的强大能力和灵活性。无论是需要处理文档的企业、开发者还是研究人员，PDFBox 都提供了一个可靠和高效的解决方案。

读者可以从下面的地址下载 Apache PDFBox 的最新版：

```
https://pdfbox.apache.org/download.html
```

13.2 生成简单的 PDF 文档

本节使用 Apache POI 生成一个简单的 PDF 文档，其中涉及如下技术。

1. 初始化 PDF 文档和页面

首先，创建一个 PDDocument 对象，这代表整个 PDF 文档。接着，创建一个 PDPage 对象，表示单个页面，并将其添加到文档中。这是创建任何 PDF 内容的基础。

```
PDDocument document = new PDDocument();
PDPage page = new PDPage();
document.addPage(page);
```

2. 创建内容流

PDPageContentStream 对象用于向页面添加内容。它是与页面交互的主要方式，允许用户绘制文本、图形等。

```
PDPageContentStream contentStream = new PDPageContentStream(document, page);
```

3. 绘制文字

使用 contentStream.beginText() 和 contentStream.endText() 标记文本块的开始和结束。

contentStream.setFont() 用于设置文本的字体和大小。contentStream.setNonStrokingColor() 设置文本颜色。contentStream.newLineAtOffset() 定位文本的开始位置。contentStream.showText() 用于显示文本内容。

```
contentStream.beginText();
contentStream.setFont(new PDType1Font(Standard14Fonts.FontName.HELVETICA), 24);
contentStream.setNonStrokingColor(1, 0, 0);                          // 红色
contentStream.newLineAtOffset(72, page.getMediaBox().getHeight() - 72);
contentStream.showText("Hello, world!");
contentStream.endText();
```

4. 设置字体和颜色

PDFBox 允许通过 PDType1Font 和 Standard14Fonts.FontName 枚举轻松设置标准字体。setNonStrokingColor() 方法用于定义字体颜色，接收 RGB 值作为参数。

5. 绘制背景色和矩形

背景色的设置通过 setNonStrokingColor() 方法完成，定义了之后的填充颜色。addRect() 定义了矩形的位置和尺寸，fill() 方法用于填充该矩形。这可以用于设置背景色或突出显示区域。

```
contentStream.setNonStrokingColor(0.9F, 0.9f, 0.9f);            // 浅灰色
contentStream.addRect(72, page.getMediaBox().getHeight() - 288, 172, 36);
contentStream.fill();
```

6. 保存和关闭文档

最后，使用 document.save("basic.pdf") 方法保存文档到文件系统。创建和编辑完成后，务必关闭 contentStream 和 document 以释放资源。

```
document.save("basic.pdf");
document.close();
```

下面的代码完整地演示了如何用 Apache POI 创建一个 PDF 文档。

代码位置：src/main/java/pdf/CreatePDFExample.java

```
package pdf;
import org.apache.pdfbox.pdmodel.PDDocument;
import org.apache.pdfbox.pdmodel.PDPage;
import org.apache.pdfbox.pdmodel.PDPageContentStream;
import org.apache.pdfbox.pdmodel.font.PDType1Font;
import org.apache.pdfbox.pdmodel.font.Standard14Fonts;
import java.io.IOException;
public class CreatePDFExample {
    public static void main(String[] args) {
        try (PDDocument document = new PDDocument()) {
            PDPage page = new PDPage();
```

```
            document.addPage(page);
            // 创建内容流以向 PDF 页面添加内容
            try (PDPageContentStream contentStream = new PDPageContentStream
(document, page))
{
                // 设置文档标题
                document.getDocumentInformation().setTitle("Basic PDF");
                // 设置字体和颜色并写入文字
                contentStream.beginText();
                contentStream.setFont(new PDType1Font(Standard14Fonts.FontName.
HELVETICA), 24);
                contentStream.setNonStrokingColor(1, 0, 0);            // 红色
                contentStream.newLineAtOffset(72, page.getMediaBox().getHeight() -
72); // 1 inch from top
                contentStream.showText("Hello, world!");
                contentStream.endText();
                // 改变字体和颜色并写入文字
                contentStream.beginText();
                contentStream.setFont(new PDType1Font(Standard14Fonts.FontName.
TIMES_ROMAN), 18);
                contentStream.setNonStrokingColor(0, 1, 0);            // 绿色
                contentStream.newLineAtOffset(72, page.getMediaBox().getHeight() -
144);
                                                                       // 下移
                contentStream.showText("This is an example PDF.");
                contentStream.endText();
                // 再次改变字体和颜色并写入文字
                contentStream.beginText();
                contentStream.setFont(new PDType1Font(Standard14Fonts.FontName.
COURIER), 12);
                contentStream.setNonStrokingColor(0, 0, 1);            // 蓝色
                contentStream.newLineAtOffset(72, page.getMediaBox().getHeight() -
 216);
                                                                       // 下移
                contentStream.showText("Created by Bing with Java and Apache
PDFBox.");
                contentStream.endText();
                // 设置背景色并画矩形
                contentStream.setNonStrokingColor(0.9F, 0.9f, 0.9f); // 浅灰色
                contentStream.addRect(72, page.getMediaBox().getHeight() - 288,
172, 36);
                contentStream.fill();
                // 改变字体和颜色并写入文字
                contentStream.beginText();
                contentStream.setFont(new PDType1Font(Standard14Fonts.FontName.
COURIER_BOLD), 14);
                contentStream.setNonStrokingColor(0, 0, 0);            // 黑色
                contentStream.newLineAtOffset(72, page.getMediaBox().getHeight() -
 280);
                contentStream.showText("This is a rectangle with background
color.");
                contentStream.endText();
            }
```

```
            // 保存 PDF 文档
            document.save("basic.pdf");
        } catch (IOException e) {
            e.printStackTrace();
        }
    }
}
```

运行程序，会在当前目录生成一个 basic.pdf 文件，打开该文件，会看到如图 13-1 所示的效果。

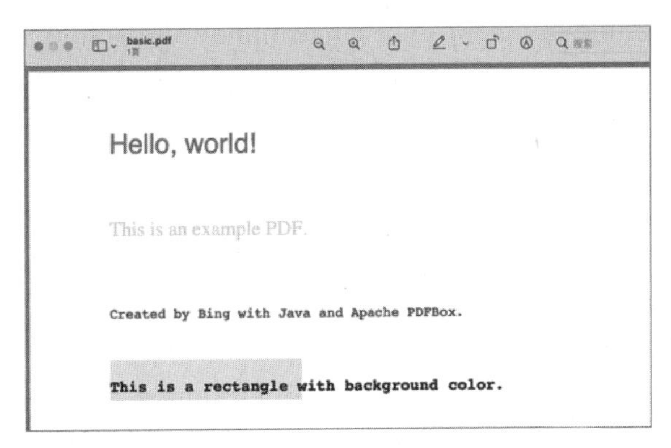

图 13-1　basic.pdf 的效果

13.3　在 PDF 文档中插入图像和表格

本例会向 PDF 文档插入图片和表格，实现原理如下。

1. 插入图像

要在 PDF 中插入图像，需要使用 PDImageXObject 类。通过调用 createFromFile 方法，可以从文件系统加载图像。然后，使用 PDPageContentStream 对象的 drawImage 方法将图像绘制到页面上的指定位置。图像位置和尺寸的单位是点（1/72 英寸），所以如果使用厘米作为单位，需要转换成点。

```
PDImageXObject image = PDImageXObject.createFromFile("images/girl1.png", document);
contentStream.drawImage(image, 5 * 28.35f, 25 * 28.35f, 2 * 28.35f, 2 * 28.35f);
```

2. 绘制表格

绘制表格稍微复杂一些，因为需要手动绘制表格的线条和填充内容。表格的绘制分为以下几个步骤。

（1）计算表格尺寸：基于页面的宽度和预定的边距计算表格的宽度。表格的每一行高度固定，从而可以计算整个表格的高度。

（2）绘制行和列：使用 contentStream.moveTo() 方法和 contentStream.lineTo() 方法绘制表格的行和列。对于每一行和每一列，分别绘制一个从起点到终点的直线。

（3）填充内容：计算文本的位置，并使用 contentStream.beginText、contentStream.setFont、contentStream.newLineAtOffset、contentStream.showText 和 contentStream.endText 方法在表格的每个单元格内填充文本。

通过这种方式，可以在 PDF 文档中手动绘制复杂的布局，如表格。虽然这种方法比较烦琐，但它提供了灵活性，可以根据需要自定义表格的外观和内容。

下面的代码完整地演示了如何向 PDF 文档中插入图像和表格。

代码位置：src/main/java/pdf/CreatePDFWithImageAndTable.java

```
package pdf;
import org.apache.pdfbox.pdmodel.PDDocument;
import org.apache.pdfbox.pdmodel.PDPage;
import org.apache.pdfbox.pdmodel.PDPageContentStream;
import org.apache.pdfbox.pdmodel.common.PDRectangle;
import org.apache.pdfbox.pdmodel.font.PDType1Font;
import org.apache.pdfbox.pdmodel.font.Standard14Fonts;
import org.apache.pdfbox.pdmodel.graphics.image.PDImageXObject;
import java.io.IOException;
public class CreatePDFWithImageAndTable {
    public static void main(String[] args) {
        try (PDDocument document = new PDDocument()) {
            PDPage page = new PDPage(PDRectangle.A4);
            document.addPage(page);
            PDPageContentStream contentStream = new PDPageContentStream
(document, page);
            // 插入图像
            PDImageXObject image = PDImageXObject.createFromFile("images/
girl1.png", document);
            // 在 PDF 页面上以厘米为单位绘制图像，图像左下角位于（5,25）处，大小为 2×2。
            // PDF 单位为 1/72 英寸，1 厘米约等于 28.35 单位。
            contentStream.drawImage(image, 5 * 28.35f, 25 * 28.35f, 2 * 28.35f,
2 * 28.35f);
            // 开始绘制表格
            String[][] content = {
                    {"Name", "Age", "Gender"},
                    {"Bill", "20", "Female"},
                    {"Bob", "22", "Male"},
                    {"Charlie", "24", "Male"}
            };
            float margin = 50;
            float tableTopY = 700;
            drawTable(page, contentStream, tableTopY, margin, content);

            contentStream.close();
            document.save("image_table.pdf");
        } catch (IOException e) {
```

```
            e.printStackTrace();
        }
    }

    private static void drawTable(PDPage page, PDPageContentStream 
contentStream,
                                  float y, float margin, String[][] content)
throws IOException {
        final int rows = content.length;
        final int cols = content[0].length;
        final float rowHeight = 20f;
        final float tableWidth = page.getMediaBox().getWidth() - (2 * margin);
        final float tableHeight = rowHeight * rows;
        final float colWidth = tableWidth / (float) cols;
        // 绘制行
        float nexty = y ;
        for (int i = 0; i <= rows; i++) {
            contentStream.moveTo(margin, nexty);
            contentStream.lineTo(margin + tableWidth, nexty);
            contentStream.stroke();
            nexty -= rowHeight;
        }
        // 绘制列
        float nextx = margin;
        for (int i = 0; i <= cols; i++) {
            contentStream.moveTo(nextx, y);
            contentStream.lineTo(nextx, y - tableHeight);
            contentStream.stroke();
            nextx += colWidth;
        }
        // 写内容
        float textx = margin + 2;                            // 从左边留出一点空白
        float texty = y - 15;                                // 从顶部行向下移动
        for (String[] row : content) {
            textx = margin + 2;                              // 每行结束后重置
            for (String cell : row) {
                contentStream.beginText();
                contentStream.setFont(new PDType1Font(Standard14Fonts.FontName.
HELVETICA), 12);
                contentStream.newLineAtOffset(textx, texty);
                contentStream.showText(cell);
                contentStream.endText();
                textx += colWidth;
            }
            texty -= rowHeight;
        }
    }
}
```

运行程序，会在当前目录生成一个 image_table.pdf 文件，效果如图 13-2 所示。

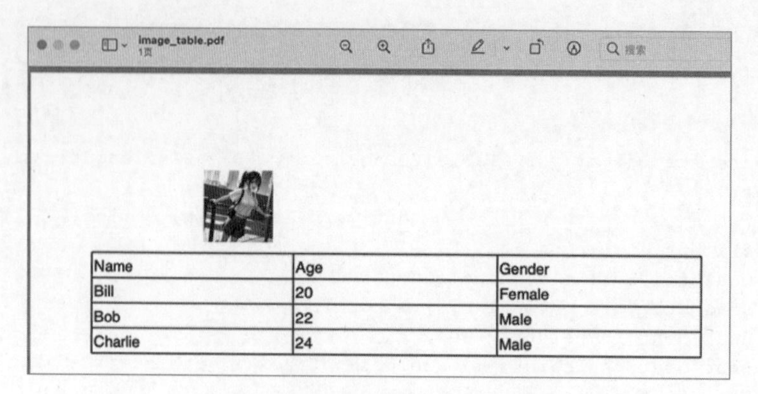

Name	Age	Gender
Bill	20	Female
Bob	22	Male
Charlie	24	Male

图 13-2　插入图像和表格的 PDF 文档

13.4　加密 PDF 文档

本节的例子会对 PDF 文档使用 Apache PDFBox 进行加密并保存。在使用 Apache PDFBox 进行 PDF 文件加密和保存时，整个过程可以细分为以下详细步骤。

1. 创建加密策略

加密策略由 StandardProtectionPolicy 类实现，这个类需要 3 个主要参数来初始化。

（1）用户密码（userPassword）：这是打开加密 PDF 文件所需的密码。如果设置了用户密码，任何尝试打开加密 PDF 文件的用户都需要输入这个密码。

（2）所有者密码（ownerPassword）：这个密码允许用户访问文档的完整权限，包括更改加密设置和编辑文档的权限。拥有所有者密码的用户可以更改或撤销用户密码。

（3）访问权限（AccessPermission）：这个对象定义了哪些操作是允许的，如打印、编辑内容、复制内容等。通过设置这个对象，可以精细控制用户在输入正确的用户密码后能对文档执行哪些操作。

2. 应用加密策略

一旦创建了 StandardProtectionPolicy 对象，需要将它应用到 PDDocument 对象上。这是通过调用 PDDocument 的 protect 方法完成的。这个方法接收一个 StandardProtectionPolicy 实例作为参数，应用这个策略来加密 PDF 文档。

3. 保存加密的 PDF 文档

加密策略应用后是保存文档。这通过调用 PDDocument 的 save 方法完成，需要指定保存加密 PDF 文件的路径。在调用 save 方法时，PDFBox 会根据之前应用的加密策略对文档内容进行加密。完成保存操作后，生成的 PDF 文件将被加密，且只能通过提供正确的密码（用户密码或所有者密码）来访问。

4. 关闭文档

操作完成后，应该调用 PDDocument 的 close 方法来关闭文档并释放系统资源。这是一个好习惯，特别是在处理大量或复杂的 PDF 文件时，可以帮助避免内存泄漏和其他资源管理问题。

下面的例子演示了如何使用 Apache PDFBox 对 PDF 文档加密的完整过程。

代码位置：src/main/java/pdf/PDFEncryptionExample.java

```
package pdf;
import org.apache.pdfbox.pdmodel.PDDocument;
import org.apache.pdfbox.pdmodel.PDPage;
import org.apache.pdfbox.pdmodel.PDPageContentStream;
import org.apache.pdfbox.pdmodel.encryption.AccessPermission;
import org.apache.pdfbox.pdmodel.encryption.StandardProtectionPolicy;
import org.apache.pdfbox.pdmodel.font.PDType1Font;
import org.apache.pdfbox.pdmodel.font.Standard14Fonts;
import java.io.IOException;
public class PDFEncryptionExample {
    public static void main(String[] args) {
        String encryptedPDF = "encrypted.pdf";          // 加密后的 PDF 文件路径
        try (PDDocument document = new PDDocument()) {
            PDPage page = new PDPage();
            document.addPage(page);
            // 在 PDF 中添加一些内容
            try (PDPageContentStream contentStream = new PDPageContentStream
(document, page))
{
                contentStream.beginText();
                contentStream.setFont(new PDType1Font(Standard14Fonts.FontName.
HELVETICA), 12);
                contentStream.newLineAtOffset(100, 700);
                contentStream.showText("Hello, PDFBox - Encrypted PDF");
                contentStream.endText();
            }
            // 设置访问权限
            AccessPermission accessPermission = new AccessPermission();
            // 创建加密策略
            StandardProtectionPolicy spp = new StandardProtectionPolicy("1234",
"4321", accessPermission);
            spp.setEncryptionKeyLength(128);          // 设置加密密钥长度
            // 应用加密策略
            document.protect(spp);
            // 保存加密的 PDF 文件
            document.save(encryptedPDF);
        } catch (IOException e) {
            e.printStackTrace();
        }
    }
}
```

运行程序，会看到当前目录生成了一个 encrypted.pdf 文件，打开该文件，会要求输入密码。如图 13-3 所示。输入用户密码和所有者密码都可以，只是输入用户密码，只能查看，输入所有者密码，不仅可以查看，还可以修改。

图 13-3 输入 PDF 密码

13.5 在 PDF 文档上绘制图表

本节的例子会使用 JFreeChart 绘制一个图表，然后放到 PDF 文档上。读者可以从下面的地址下载 JFreeChart 的最新版（JAR 文件），然后引用这个 JAR 文件。

```
https://repo1.maven.org/maven2/org/jfree/jfreechart/1.5.3
```

1. JFreeChat 简介

JFreeChart 库允许开发者在 Java 应用程序中创建多种类型的图表。使用 JFreeChart 创建图表的过程涉及以下几个基本概念。

（1）数据集（Dataset）：图表的基础，用于存储所有要显示的数据。根据不同的图表类型（如折线图、饼图等），选择合适的数据集类型来存储数据。

（2）图表（Chart）：数据的视觉表示。JFreeChart 提供了多种图表类型，如 JFreeChart 对象，每种图表类型都有相应的工厂方法来创建。

（3）图表区域（Plot）：图表的核心区域，负责绘制图表的主体，包括数据系列、坐标轴等。不同类型的图表有不同的 Plot 对象。

（4）渲染器（Renderer）：控制图表的具体绘制方式，如颜色、线型等。每种类型的 Plot 都有相应的 Renderer。

2. 创建数据集

图表的创建始于数据集的构建。根据图表类型（折线图、饼图、柱状图等），选择合适的数据集类，并填充数据。

（1）对于折线图或散点图，通常使用 XYSeriesCollection，它包含一个或多个 XYSeries，每个 XYSeries 代表一组数据点。

（2）对于饼图，使用 DefaultPieDataset，它存储了各个部分的名称和值。

3. 创建图表

有了数据集后，使用 JFreeChart 提供的 ChartFactory 类来创建图表。ChartFactory 类为不同类型的图表提供了静态方法，只需传入适当的参数（如图表标题、数据集等）即可。

4. 定制图表

创建图表后，可能需要根据需求对图表的外观进行定制，包括但不限于：

（1）设置图表的背景颜色、字体、边框等。

（2）定制图表的 Plot，如设置背景颜色、网格线颜色等。

（3）使用不同的 Renderer 定制数据系列的外观，如线条颜色、点的形状等。

5. 展示或保存图表

图表创建并定制完成后，可以将其展示在 Java 应用程序的 GUI 中，或者保存为图片文件。JFreeChart 支持将图表保存为多种格式的图片，如 PNG、JPEG 等。

（1）要在 GUI 中展示图表，可以使用 ChartPanel，它是一个轻量级的 Swing 组件。

（2）要将图表保存为图片，可以使用 ChartUtilities.saveChartAsPNG 或 saveChartAsJPEG 方法，这些方法允许指定文件路径、图表对象以及图片的宽度和高度。

通过这些步骤，可以使用 JFreeChart 库创建、定制和展示多种类型的图表。JFreeChart 的强大和灵活性使它成为 Java 中最受欢迎的图表库之一。

下面的例子完整地演示了如何用 JFreeChart 绘制图表，并用 Apache PDFBox 将图表嵌入 PDF 文档中。

代码位置：src/main/java/pdf/ChartToPDF.java

```
package pdf;
import org.apache.pdfbox.pdmodel.PDDocument;
import org.apache.pdfbox.pdmodel.PDPage;
import org.apache.pdfbox.pdmodel.PDPageContentStream;
import org.apache.pdfbox.pdmodel.common.PDRectangle;
import org.apache.pdfbox.pdmodel.font.PDType1Font;
import org.apache.pdfbox.pdmodel.font.Standard14Fonts;
import org.apache.pdfbox.pdmodel.graphics.image.PDImageXObject;
import org.jfree.chart.ChartFactory;
import org.jfree.chart.ChartUtils;
import org.jfree.chart.JFreeChart;
import org.jfree.data.xy.XYSeries;
import org.jfree.data.xy.XYSeriesCollection;
import java.io.File;
import java.io.IOException;
```

```
public class ChartToPDF {
    public static void main(String[] args) throws IOException {
        // 创建一个 XY 数据序列
        XYSeries series = new XYSeries("XYGraph");
        series.add(1, 1);
        series.add(2, 4);
        series.add(3, 9);
        series.add(4, 16);
        // 将数据序列添加到数据集合中
        XYSeriesCollection dataset = new XYSeriesCollection();
        dataset.addSeries(series);
        // 创建一个图表
        JFreeChart chart = ChartFactory.createXYLineChart("Line Chart", "x",
"y", dataset);
        // 将图表保存为 PNG 图片
        ChartUtils.saveChartAsPNG(new File("chart.png"), chart, 400, 300);
        // 创建 PDF 文档，并添加一个页面
        try (PDDocument document = new PDDocument()) {
            PDPage page = new PDPage(PDRectangle.A4);
            document.addPage(page);
            // 将 PNG 图片插入 PDF 文档中
            PDImageXObject pdImage = PDImageXObject.createFromFile("chart.png",
document);
            try (PDPageContentStream contentStream = new PDPageContentStream
(document, page)) {
                contentStream.drawImage(pdImage, 70, 500, 400, 300);
                // 图片的位置和大小
                // 在 PDF 文档中添加一些文字说明
                contentStream.beginText();
                contentStream.setFont(new PDType1Font(Standard14Fonts.FontName.
HELVETICA), 12);
                contentStream.newLineAtOffset(70, 480);
                contentStream.showText("This is a line chart created by
JFreeChart.");
                contentStream.endText();
            }
            // 保存并关闭 PDF 文档
            document.save("chart.pdf");
        }
    }
}
```

运行程序，会在当前目录生成一个 chart.pdf 文件，打开 chart.pdf 文件，效果如图 13-4 所示。

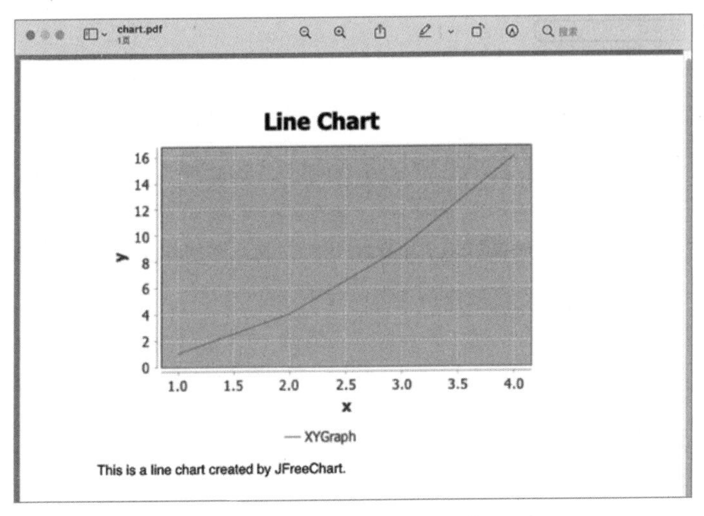

图 13-4　带图表的 PDF 文档

13.6　小结

本章主要使用了 Apache PDFBox 库读写 PDF 文档，这个库的主要功能在 13.1 节已经介绍了。如果只是使用 Apache PDFBox 库读写 PDF 文档，那么功能很有限。如果想让 PDF 文档丰富多彩，可以使用 JFreeChart 或其他第三方库与这些读写 PDF 文档的库结合，例如，通过 JFreeChart 可以绘制非常复杂的图表，然后将这些图表保存为图像，最后将这些图像插入 PDF 文档中。

第 14 章

控制软件和设备

尽管 Java 的功能强大，但要想更强大，最直接的方式就是与其他软件和设备交互或融合，通过 Java 强大的第三方模块，可以轻松达到我们的目的。例如，Java 可以与浏览器交互、可以与剪贴板交互数据，甚至控制键盘和鼠标、录制键盘和鼠标动作等。本章将深入介绍如何利用第三方库来完成这些炫酷的功能。

14.1 浏览器

Java 可以通过 Selenium 与各种浏览器交互。读者可以从下面的地址下载 Selenium Java 最新版的 JAR 文件。

```
https://mvnrepository.com/artifact/org.seleniumhq.selenium/selenium-java
```

Selenium 是一个用于 Web 应用程序测试的工具，它可以模拟用户的操作，如打开网页、输入、单击等，来实现对网页的自动化控制。Selenium 需要通过 WebDriver 控制浏览器，所以想要在 Java 中使用 Selenium，除了要下载 Selenium Java JAR 文件外，还需要下载对应浏览器的 WebDriver。Selenium 是跨平台的，支持 Windows、macOS 和 Linux。Selenium 支持多种浏览器，如 IE、Firefox、Chrome、Safari、Opera、Edge 等。

WebDriver 是自动化测试工具 Selenium 的主要组件之一。它是一个 W3C 标准，用于与实际浏览器进行通信并控制其行为。

WebDriver 允许通过编程方式来控制浏览器，例如：

（1）启动和关闭浏览器。

（2）浏览网页。

（3）获取当前页面的 URL 和标题。

（4）找到元素并进行交互（单击、输入文本等）。

（5）执行 JavaScript。

（6）处理 Cookies 和其他浏览器设置，主流浏览器都提供了对应的 WebDriver 实现。

WebDriver 是一个协议，它定义了一组命令和响应，以及一些事件和错误。

WebDriver 的实现包括两部分：客户端和服务器。客户端是一个库，它提供了一个编程接口，让我们可以发送命令给服务器。服务器是一个程序，它运行在与浏览器相同或者不同的机器上，它接收客户端的命令，并将其转换为浏览器能够理解的操作。

WebDriver 支持多种编程语言，如 Java、Python、Ruby、C# 等。读者可以根据喜好和需求选择合适的语言来编写测试脚本。

WebDriver 还支持一些高级功能，如隐式等待、显式等待、页面对象模式、截图、远程执行等。这些功能可以让我们更方便地处理一些复杂或者特殊的情况。

大多数浏览器都有对应的 WebDriver，例如：

（1）ChromeDriver 用于 Chrome 浏览器。

（2）GeckoDriver 用于 Firefox 浏览器。

（3）EdgeDriver 用于 Edge 浏览器。

Selenium 可以自动化执行动作并做验证，这样可以大大提高 Web 应用测试的效率，而无须人工进行大量重复操作。WebDriver 针对不同的浏览器提供不同的 Driver 实现，并通过 W3C 标准定义的通用接口与这些 Driver 进行交互。这样可以实现跨浏览器的自动化测试，我们可以编写一次测试脚本，然后对不同浏览器进行执行。总之，WebDriver 为 Web 应用自动化测试提供了极其强大的功能和支持。它简化了测试流程，使测试维护和扩展变得更简单和更安全。

在 Java 中使用 Selenium，最重要的是下载对应浏览器的 WebDriver，例如，本例使用了 Microsoft Edge 浏览器，该浏览器的 WebDriver 可以在微软官网下载，URL 如下：

```
https://developer.microsoft.com/en-us/microsoft-edge/tools/webdriver/?form=
MA13LH#downloads
```

Edge 为大多数 OS 都提供了对应的版本，下载时要考虑浏览器的版本号是否与 WebDriver 版本号一致，大版本号必须相同。通常浏览器与 WebDriver 都保持最新版本即可。WebDriver 本身是一个可执行程序，本例将其复制到当前目录的 webdrivers 子目录中，文件名是 msedgedriver，这是 macOS 版本的 WebDriver，如果是 Windows 版本，则是 msedgedriver.exe。

下面的例子使用 Selenium 控制 Edge 浏览器访问百度，并将“Python 从菜鸟到高手 第二版”输入搜索框，然后单击右侧的按钮进行搜索。

代码位置：src/main/java/controller/SeleniumExample.java

```
package controller;
import org.openqa.selenium.By;
import org.openqa.selenium.Keys;
import org.openqa.selenium.WebDriver;
import org.openqa.selenium.WebElement;
import org.openqa.selenium.edge.EdgeDriver;
```

```
import org.openqa.selenium.support.ui.ExpectedConditions;
import org.openqa.selenium.support.ui.WebDriverWait;
import java.time.Duration;
public class SeleniumExample {
    public static void main(String[] args) {
        // 设置 WebDriver 的路径
        System.setProperty("webdriver.edge.driver", "webdrivers/msedgedriver");
        // 创建一个 Edge 浏览器实例
        WebDriver driver = new EdgeDriver();
        // 打开百度首页
        driver.get("https://www.baidu.com");
        // 找到搜索框元素
        WebElement input = driver.findElement(By.id("kw"));
        // 在搜索框中输入 "Python 从菜鸟到高手  第二版 " 并按下 Enter 键
        input.sendKeys("Python 从菜鸟到高手  第二版 ", Keys.ENTER);
        // 添加显式等待，等待搜索结果页面加载完成
        new WebDriverWait(driver, Duration.ofSeconds(5)).until
(ExpectedConditions.presenceOfElementLocated(By.id("content_left")));
    }
}
```

运行程序，会看到自动启动了 Microsoft Edge 浏览器，并自动打开百度，然后输入相应的字符串，最后会进行搜索，结果如图 14-1 所示。

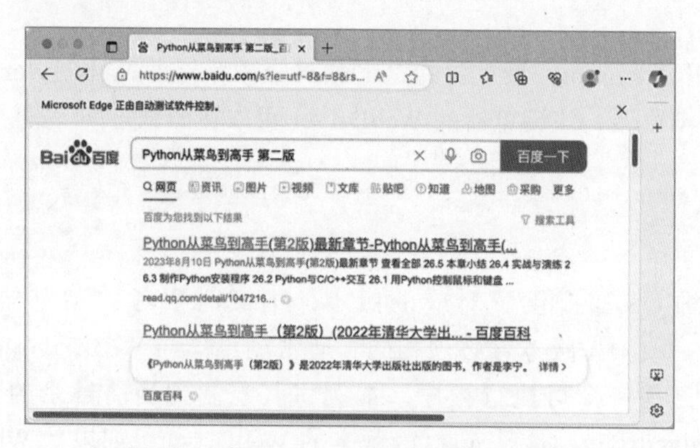

图 14-1　用 Selenium 控制浏览器进行搜索

14.2　鼠标和键盘

本节会介绍如何使用 JNativeHook 库录制鼠标和键盘动作，并利用 java.awt.Robot 回放录制的动作。

14.2.1　录制键盘和鼠标的动作

JNativeHook 是第三方库，读者可以从下面的页面下载最新版 JNativeHook 的 JAR 文件。

```
https://mvnrepository.com/artifact/com.github.kwhat/jnativehook
```

在本例中还使用了 javax.json 库，这是一个处理 JSON 的第三方库，需要到如下的两个页面下载两个 JAR 文件。

```
https://repo1.maven.org/maven2/javax/json/javax.json-api
https://repo1.maven.org/maven2/org/glassfish/javax.json
```

这两个页面分别下载 javax.json-api-1.1.4.jar 和 javax.json-1.1.4.jar 文件，也就是说，运行本节的例子，一共需要引用三个 JAR 文件。

录制键盘和鼠标动作的原理基于几个关键步骤和概念，涉及监听系统级的键盘和鼠标事件，将这些事件转换为可理解的数据格式，并将这些数据保存到文件中。这些主要依赖 JNativeHook 和 javax.json 实现。下面是详细步骤。

1. 初始化和注册全局监听器

（1）引入 JNativeHook 库：这个库允许 Java 应用监听全局的键盘和鼠标事件，即使应用没有获得焦点。首先，需要将 JNativeHook 库引入项目中，还有另外两个 javax.json 库也需要引用。

（2）注册全局监听器：在 main 方法中，使用 GlobalScreen.registerNativeHook 方法注册一个全局的监听器。这使得应用能够开始监听全局的键盘和鼠标事件。

（3）关闭日志输出：JNativeHook 库默认会产生一些日志输出。通过获取库的 Logger 并将其日志级别设置为 Level.OFF，可以关闭这些不必要的日志输出，使得应用运行时更加清爽。

2. 监听和处理键盘事件

（1）实现 NativeKeyListener 接口：通过实现 NativeKeyListener 接口，定义当键盘事件发生时应该执行的操作。主要关注的方法有 nativeKeyPressed 和 nativeKeyReleased。

（2）记录键盘按下动作（nativeKeyPressed）：当任何键被按下时，此方法被触发。可以在这里记录按键动作，包括按下的键的标志。

（3）记录键盘释放动作（nativeKeyReleased）：当任何键被释放时，此方法被触发。同样地，记录释放键的动作。

（4）退出和保存（可选）：在某些条件下（如按下 Esc 键），执行保存动作，并注销全局监听器，停止监听事件。

3. 监听和处理鼠标事件

（1）实现 NativeMouseInputListener 接口：通过实现 NativeMouseInputListener 接口，定义当鼠标事件发生时应该执行的操作。主要关注的方法有 nativeMouseClicked 和 nativeMouseMoved。

（2）记录鼠标单击动作（nativeMouseClicked）：当鼠标单击事件发生时，此方法被触发。记录单击的鼠标按钮和单击位置。

（3）记录鼠标移动动作（nativeMouseMoved）：当鼠标移动事件发生时，此方法被触发。记录鼠标的新位置。

4. 数据记录和保存

（1）使用JSON格式记录事件：使用javax.json库构建JSON对象，将每个事件的类型、相关数据（如按键标志、鼠标位置等）保存到一个JsonArrayBuilder中。这种格式易于理解和后续处理。

（2）将动作保存到JSON文件：定义一个saveActionsToFile方法，将累积的事件数据转换为JSON字符串，并写入一个文件中。这个文件可以在之后被用于回放这些动作。

下面的例子演示了如何用JNativeHook库记录键盘和鼠标动作的完整过程，按Esc键停止录制，并在当前目录生成用于记录动作的actions.json文件。

代码位置：src/main/java/controller/EventRecorder.java

```
package controller;
import com.github.kwhat.jnativehook.GlobalScreen;
import com.github.kwhat.jnativehook.NativeHookException;
import com.github.kwhat.jnativehook.keyboard.NativeKeyEvent;
import com.github.kwhat.jnativehook.keyboard.NativeKeyListener;
import com.github.kwhat.jnativehook.mouse.NativeMouseEvent;
import com.github.kwhat.jnativehook.mouse.NativeMouseInputListener;
import javax.json.Json;
import javax.json.JsonArrayBuilder;
import java.io.FileWriter;
import java.io.IOException;
import java.util.logging.Level;
import java.util.logging.Logger;
public class EventRecorder implements NativeKeyListener, NativeMouseInputListener {
    // 使用JsonArrayBuilder记录所有动作事件
    private JsonArrayBuilder actions = Json.createArrayBuilder();
    public static void main(String[] args) {
        try {
            // 关闭JNativeHook库的日志输出，以避免在控制台中显示不必要的信息
            Logger logger = Logger.getLogger(GlobalScreen.class.getPackage().
getName());
            logger.setLevel(Level.OFF);
            // 注册全局事件监听器，开始监听键盘和鼠标事件
            GlobalScreen.registerNativeHook();
            EventRecorder recorder = new EventRecorder();
            GlobalScreen.addNativeKeyListener(recorder);
            GlobalScreen.addNativeMouseListener(recorder);
            GlobalScreen.addNativeMouseMotionListener(recorder);

        } catch (NativeHookException e) {
            e.printStackTrace();
        }
    }
    // 当键盘按键被按下时触发
```

```
    @Override
    public void nativeKeyPressed(NativeKeyEvent e) {
        // 记录按键按下事件，包括按键的文本表示
        actions.add(Json.createObjectBuilder()
                .add("type", "press")
                .add("key", NativeKeyEvent.getKeyText(e.getKeyCode())));
    }
    // 当键盘按键被释放时触发
    @Override
    public void nativeKeyReleased(NativeKeyEvent e) {
        // 记录按键释放事件
        actions.add(Json.createObjectBuilder()
                .add("type", "release")
                .add("key", NativeKeyEvent.getKeyText(e.getKeyCode())));
        // 如果按下的是 Esc 键，则保存事件到文件并停止监听
        if (e.getKeyCode() == NativeKeyEvent.VC_ESCAPE) {
            saveActionsToFile();
            try {
                GlobalScreen.unregisterNativeHook();
            } catch (NativeHookException ex) {
                ex.printStackTrace();
            }
        }
    }
    // 当鼠标被单击时触发
    @Override
    public void nativeMouseClicked(NativeMouseEvent e) {
        // 记录鼠标单击事件，包括单击的按钮和位置
        actions.add(Json.createObjectBuilder()
                .add("type", "click")
                .add("button", e.getButton())  // 记录哪个鼠标按钮被单击
                .add("x", e.getX())            // 记录单击时鼠标的 X 坐标
                .add("y", e.getY()));          // 记录单击时鼠标的 Y 坐标
    }
    // 当鼠标移动时触发
    @Override
    public void nativeMouseMoved(NativeMouseEvent e) {
        // 记录鼠标移动事件，包括新的位置
        actions.add(Json.createObjectBuilder()
                .add("type", "move")
                .add("x", e.getX())            // 记录鼠标的 X 坐标
                .add("y", e.getY()));          // 记录鼠标的 Y 坐标
    }
    // 将记录的动作事件保存到文件中
    private void saveActionsToFile() {
        try (FileWriter file = new FileWriter("actions.json")) {
            // 将动作事件转换为 JSON 字符串并写入文件
            file.write(actions.build().toString());
        } catch (IOException e) {
            e.printStackTrace();
        }
```

```
    }
}
```

运行程序，然后控制鼠标的移动和单击动作，以及键盘的按键，最后按 Esc 键，会在当前目录生成一个 actions.json 文件。在下一节的例子中会读取这个文件，回放记录的鼠标和键盘动作。

14.2.2　回放键盘和鼠标的动作

在这一节会读取 actions.json 文件，并使用 java.awt.Robot 回放 actions.json 文件中的动作。

代码位置：src/main/java/controller/EventRecorder.java

```
package controller;
import javax.json.Json;
import javax.json.JsonArray;
import javax.json.JsonObject;
import javax.json.JsonReader;
import java.awt.Robot;
import java.awt.event.InputEvent;
import java.awt.event.KeyEvent;
import java.io.FileReader;
import java.util.HashMap;
import java.util.Map;

public class EventPlayer {
    private static final Map<String, Integer> keyMap = new HashMap<>();

    static {
        // 初始化键映射
        keyMap.put("A", KeyEvent.VK_A);
        keyMap.put("Esc", KeyEvent.VK_ESCAPE);
        // 这里可以尽可能添加按键的对应关系，本例只添加了 A 和 Esc
    }
    public static void main(String[] args) {
        try {
            Robot robot = new Robot();
            robot.setAutoDelay(40);                     // 设置操作之间的延迟
            try (FileReader fileReader = new FileReader("actions.json");
                 JsonReader jsonReader = Json.createReader(fileReader)) {
                JsonArray actions = jsonReader.readArray();
                for (JsonObject action : actions.getValuesAs(JsonObject.class)) {
                    String type = action.getString("type");
                    switch (type) {
                        case "move":
                            robot.mouseMove(action.getInt("x"), action.getInt("y"));
```

```
                    break;
                case "click":
                    int buttonCode = action.getInt("button");
                    // 使用 getInt 获取 button 值
                    // 假设数字 1、2、3 分别代表鼠标左键、中键、右键
                    int inputEventMask = 0;
                    switch (buttonCode) {
                        case 1:
                            inputEventMask = InputEvent.BUTTON1_DOWN_MASK;
                            break;
                        case 2:
                            inputEventMask = InputEvent.BUTTON2_DOWN_MASK;
                            break;
                        case 3:
                            inputEventMask = InputEvent.BUTTON3_DOWN_MASK;
                            break;
                    }
                    if (inputEventMask != 0) {
                        robot.mousePress(inputEventMask);
                        robot.mouseRelease(inputEventMask);
                    }
                    break;

                case "press":
                case "release":
                    String key = action.getString("key");
                    if (keyMap.containsKey(key)) {
                        int keyCode = keyMap.get(key);

                        if (type.equals("press")) {
                            robot.keyPress(keyCode);
                            System.out.println(keyCode);
                        } else {
                            robot.keyRelease(keyCode);
                        }
                    }
                    break;
                // 可以根据需要添加更多的事件处理逻辑
            }
        }
    }
    } catch (Exception e) {
        e.printStackTrace();
    }
  }
}
```

运行程序，如果 actions.json 文件存在，就会按录制的顺序播放记录的鼠标和键盘动作。

14.3 剪贴板

通过 java.awt.Toolkit 和 java.awt.datatransfer 包中相关的类，可以复制和粘贴剪贴板中的文本数据，代码如下。

代码位置：src/main/java/controller/ ClipboardExample.java

```
package controller;
import java.awt.Toolkit;
import java.awt.datatransfer.Clipboard;
import java.awt.datatransfer.StringSelection;
import java.awt.datatransfer.DataFlavor;
import java.awt.datatransfer.Transferable;

public class ClipboardExample {
    public static void main(String[] args) {
        try {
            // 获取系统剪贴板
            Clipboard clipboard = Toolkit.getDefaultToolkit().getSystemClipboard();

            // 将文本复制到剪贴板
            String textToCopy = "Hello World!";
            StringSelection stringSelection = new StringSelection(textToCopy);
            clipboard.setContents(stringSelection, null);

            // 从剪贴板中粘贴文本
            Transferable contents = clipboard.getContents(null);
            String pastedText = (String) contents.getTransferData
(DataFlavor.stringFlavor);

            // 打印剪贴板中的文本
            System.out.println(pastedText);
        } catch (Exception e) {
            e.printStackTrace();
        }
    }
}
```

14.4 小结

本章介绍了多个库，包括 Selenium、JNativeHook、java.awt.Robot、java.awt.Toolkit、java.awt.datatransfer 等，这些库可以实现对浏览器、键盘和鼠标、剪贴板等外部软件或设备的控制。尤其是 JNativeHook 和 java.awt.Robot，可以录制鼠标和键盘的动作，并可以回放，这给控制各种软件和设备提供了极大的便利，也大大增强了 Java 的功能。

第 15 章

加密与解密

本章会使用 Java 的内置 API 实现对数据的编码、解码、加密和解密。主要包括 MD5 摘要算法、SHA 摘要算法、Base64 编码和解码、DES 加密和解密、AES 加密和解密、RSA 加密和解密。

15.1 MD5 摘要算法

MD5 摘要算法是一种被广泛使用的密码散列函数，可以产生出一个 128 位（16 字节）的散列值（hash value），用于确保信息传输完整一致。MD5 摘要算法的原理可简要地叙述为：MD5 码以 512 位分组来处理输入的信息，且每一分组又被划分为 16 个 32 位子分组，经过一系列的处理后，算法的输出由 4 个 32 位分组组成，将这 4 个 32 位分组级联后将生成一个 128 位散列值 2。

Java 中可以使用内置的 hashlib 模块来实现 MD5 摘要算法，hashlib 模块提供了常见的摘要算法，如 MD5，SHA1 等。使用 hashlib 模块实现 MD5 摘要算法的步骤如下。

（1）导入 hashlib 模块：import hashlib。

（2）创建一个 MD5 对象：md = hashlib.md5()。

（3）使用 update 方法传入要加密的内容，注意要先编码为字节串：md.update(data.encode('utf-8'))。

（4）使用 hexdigest 方法获取加密后的十六进制字符串：md.hexdigest()。

MD5 摘要算法的应用领域和场景主要有以下几种。

（1）文件校验：MD5 可以对任何文件（不管其大小、格式、数量）产生一个同样独一无二的 MD5“数字指纹”，如果任何人对文件做了任何改动，其 MD5 也就是对应的“数字指纹”都会发生变化。这样就可以用来检验文件的完整性和一致性，防止文件被篡改或损坏。

（2）密码加密：MD5 可以对用户的密码进行加密，存储在数据库中，这样即使数据库被泄露，也不会暴露用户的明文密码。同时，MD5 是不可逆的，即不能从散列值还原出原始数据，这样就增加了破解的难度。

（3）数字签名：MD5 可以用来生成数字签名，用于验证信息的来源和完整性。数字签名是将信息的摘要（如 MD5）和发送者的私钥进行加密，然后附在信息上发送给接收者。接收者收到信息后，可以用发送者的公钥解密数字签名，得到信息的摘要，并与自己计算出的摘要进行比较，如果一致，则说明信息没有被篡改，并且确实来自发送者。

（4）其他领域：MD5 还可以用于生成全局唯一标识符（GUID）、数据去重、数据分片等领域。

下面的例子完整地演示了如何使用 Java 实现 MD5 摘要算法。

代码位置：src/main/java/crypto/MD5Example.java

```
package crypto;
import java.security.MessageDigest;
import java.security.NoSuchAlgorithmException;
public class MD5Example {
    public static void main(String[] args) {
        // 要加密的内容
        String data = "世界，你好，Hello world！";
        try {
            // 创建一个 MessageDigest 实例，并指定 MD5 摘要算法
            MessageDigest md = MessageDigest.getInstance("MD5");
            // 使用 update 方法传入要加密的内容
            md.update(data.getBytes("UTF-8"));
            // 获取加密后的字节数组
            byte[] digest = md.digest();
            // 将字节数组转换为十六进制的字符串
            StringBuilder result = new StringBuilder();
            for (byte b : digest) {
                result.append(String.format("%02x", b));
            }
            // 打印结果
            System.out.println(result.toString());
        } catch (NoSuchAlgorithmException | java.io.UnsupportedEncodingException e) {
            e.printStackTrace();
        }
    }
}
```

运行程序，会输出如下内容：

```
3a83d8702bf538a64682d1031b6a8e17
```

15.2 SHA 摘要算法

SHA 摘要算法是一种密码散列函数家族，是 FIPS 所认证的安全散列算法。能计算出一个数字消息所对应到的，长度固定的字符串（又称消息摘要）的算法。且若输入的消息

不同，它们对应到不同字符串的概率很高。SHA 摘要算法的原理可简要地叙述为：SHA 码以 512 位分组来处理输入的信息，且每一分组又被划分为 16 个 32 位子分组，经过了一系列的处理后，算法的输出由四个 32 位分组组成，将这四个 32 位分组级联后将生成一个 128 位散列值。

SHA 摘要算法有三大类，分别是 SHA-1、SHA-2 和 SHA-3。其中，SHA-1 已经被破解，不再安全；SHA-3 是最新的标准，还没有广泛应用；目前最常用和相对安全的是 SHA-2 算法。SHA-2 算法又包括了 SHA-224、SHA-256、SHA-384、SHA-512、SHA-512/224、SHA-512/256 等变体，它们的区别主要在于输出长度和运算速度。

MD5 和 SHA 是两种常见的密码散列函数，它们都可以将任意长度的信息转换为固定长度的散列值，也称为消息摘要。MD5 和 SHA 的区别主要有以下几点。

（1）MD5 的输出长度是 128 位，SHA 的输出长度可以是 160 位、224 位、256 位、384 位或 512 位，根据不同的算法而定。

（2）MD5 的运行速度比 SHA 快，但是 MD5 的安全性比 SHA 低，因为 MD5 已经被发现存在多个碰撞，即不同的输入可以得到相同的输出。

（3）MD5 和 SHA 都是单向的，即不能从散列值还原出原始数据，但是 MD5 更容易受到暴力破解或彩虹表攻击，因为其输出空间较小。

（4）MD5 和 SHA 都可以用于文件校验、密码加密、数字签名等领域，但是一般推荐使用 SHA 系列算法，因为它们更安全。

下面的例子完整地演示了如何使用 Java 实现 SHA-256 加密算法。

代码位置：src/main/java/crypto/SHA256Example.java

```
package crypto;
import java.security.MessageDigest;
import java.security.NoSuchAlgorithmException;
public class SHA256Example {
    public static void main(String[] args) {
        // 要加密的内容
        String data = "世界，你好，Hello world！";
        try {
            // 创建一个 MessageDigest 实例，并指定 SHA-256 加密
            MessageDigest sha = MessageDigest.getInstance("SHA-256");
            // 使用 update 方法传入要加密的内容
            sha.update(data.getBytes("UTF-8"));
            // 获取加密后的字节数组
            byte[] digest = sha.digest();
            // 将字节数组转换为十六进制的字符串
            StringBuilder result = new StringBuilder();
            for (byte b : digest) {
                result.append(String.format("%02x", b));
            }
            // 打印结果
```

```
            System.out.println(result.toString());
        } catch (NoSuchAlgorithmException | java.io.UnsupportedEncodingException e) {
            e.printStackTrace();
        }
    }
}
```

运行程序，会输出如下内容：

```
f51e95d5e12184ebac0995187fc216eb1d1247950e46fb452cddcbc65fb3305a
```

15.3 Base64 编码和解码

Base64 编码是一种将任意二进制数据转换为可打印字符的编码方式，它可以用于在文本协议中传输二进制数据，如电子邮件、网页、XML 等。Base64 编码的原理是将每 3 字节的二进制数据拆分为 4 个 6 位的二进制数据，然后用一个 64 个字符的编码表将它们映射为 4 个可打印字符。如果原始数据不是 3 的倍数，那么在最后一组数据后面补足 0，并用 = 号表示补充的字节数 1。

Base64 编码的应用场景主要有以下几种。

（1）电子邮件：电子邮件协议只支持 ASCII 字符，所以如果要发送图片、音频、视频等二进制文件，就需要用 Base64 编码将它们转换为文本格式。

（2）网页：网页中可以使用 data URI scheme 来嵌入图片、字体等资源，这样可以减少 HTTP 请求的次数，提高网页加载速度。data URI scheme 的格式是 data:[<media type>][;base64],<data>，其中 <data> 部分就是 Base64 编码的二进制数据。

（3）XML：XML 是一种常用的数据交换格式，它也只支持文本数据，所以如果要在 XML 中包含二进制数据，就需要用 Base64 编码将它们转换为文本格式。

下面的例子完整地演示了如何使用 Java 实现对字符串的 Base64 编码和解码。

代码位置：src/main/java/crypto/Base64Example.java

```
package crypto;
import java.nio.charset.StandardCharsets;
import java.util.Base64;
public class Base64Example {
    public static void main(String[] args) {
        // 要编码和解码的内容
        String data = "世界你好，Hello, world!";
        // 使用 Base64 编码
        String ciphertext =
            Base64.getEncoder().encodeToString(data.getBytes(StandardCharsets.UTF_8));
        // 打印编码后的结果
        System.out.println(ciphertext);
```

```
        // 使用Base64 解码
        String plaintext = new String(Base64.getDecoder().decode(ciphertext),
StandardCharsets.UTF_8);
        // 打印解码后的结果
        System.out.println(plaintext);
    }
}
```

运行程序，会输出如下内容：

```
5LiW55WM5L2g5aW977yMSGVsbG8sIHdvcmxkIQ==
世界你好，Hello, world!
```

15.4　DES 加密和解密

DES 加密算法是一种对称加密算法，它以 64 位为分组对数据加密①，使用 56 位的密钥（实际上是 64 位的密钥，但每 8 位中有 1 位用于奇偶校验）。DES 算法的基本原理是：首先对明文进行初始置换，然后将置换后的结果分为左右两个 32 位的分组，接着进行 16 轮迭代，每轮迭代中，右分组经过扩展置换、与子密钥异或、S 盒替换、P 盒置换等操作，得到一个新的 32 位的值，再与左分组异或，作为下一轮的右分组；左分组则直接作为下一轮的左分组。最后一轮迭代后，不交换左右分组，而是直接进行逆初始置换，得到 64 位的密文。解密过程与加密过程相同，只是子密钥的顺序相反。

Java 进行 DES 加密和解密的过程涉及几个关键步骤，这些步骤遵循加密 / 解密的一般流程，但在实现上具体依赖于 Java 加密架构（Java Cryptography Architecture，JCA）。下面是详细的实现步骤。

1. 导入必要的加密类库

需要导入 Java 加密相关的类库，主要包括 javax.crypto.Cipher、javax.crypto.SecretKey、javax.crypto.spec.SecretKeySpec 等。这些类库提供了加密和解密所需的功能和操作。

2. 生成密钥

DES 加密算法需要一个固定长度（通常是 64 位，实际使用 56 位，剩下 8 位作为奇偶校验位）的密钥。在 Java 中，可以直接指定一个字节数组作为密钥，然后使用 SecretKeySpec 类来根据这个字节数组创建一个 SecretKey 实例。这个密钥将用于后续的加密和解密过程。

---

① 对称加密算法是一种使用相同的密钥进行加密和解密的加密算法，也称为共享密钥加密算法。对称加密算法的优点是加密速度快，适合对大量数据进行加密；缺点是密钥的传输和管理比较困难，容易被破解。

3. 创建 Cipher 实例

使用 Cipher 类的 getInstance 方法创建一个 Cipher 实例，这个方法需要一个字符串参数来指定所使用的加密算法、工作模式和填充方式。对于 DES 加密，常见的参数是 DES/ECB/PKCS5Padding，其中 DES 指定了加密算法，ECB 指定了加密的工作模式（电子密码本模式），PKCS5Padding 指定了填充模式。

4. 初始化 Cipher 实例

通过调用 Cipher 实例的 init 方法来初始化这个实例，这个方法需要两个参数：操作模式（加密或解密）和密钥。操作模式通过 Cipher.ENCRYPT_MODE 或 Cipher.DECRYPT_MODE 来指定。

5. 加密或解密数据

（1）加密：调用 Cipher 实例的 doFinal 方法，并传入待加密的字节数组，该方法会返回加密后的字节数组。如果待加密的数据是字符串，需要先将其转换为字节数组（通常使用 UTF-8 编码）。

（2）解密：同样调用 Cipher 实例的 doFinal 方法，并传入待解密的字节数组（通常是加密后的数据），该方法会返回解密后的原始字节数组。解密后，如果原始数据是字符串，还需要将字节数组转换回字符串。

6. 使用 Base64 编码和解码

由于加密后的数据通常包含无法直接打印的字节，因此在实际应用中，常常需要使用 Base64 编码将加密后的字节数组转换为字符串形式，便于存储和传输。同样，解密前需要将 Base64 编码的字符串解码回原始的字节数组形式。

下面的例子完整地演示了如何使用 Java 实现 DES 加密和解密的过程。

代码位置：src/main/java/crypto/DESExample.java

```
package crypto;
import javax.crypto.Cipher;
import javax.crypto.SecretKey;
import javax.crypto.spec.SecretKeySpec;
import java.nio.charset.StandardCharsets;
import java.util.Base64;
public class DESExample {
    public static void main(String[] args) {
        try {
            // 要加密和解密的内容
            String data = "世界，你好，Hello, world!";
            // 设置加密密钥，DES 密钥长度必须是 8 字节
            String key = "abcdefgh";
            SecretKey secretKey = new SecretKeySpec(key.getBytes
(StandardCharsets.UTF_8), "DES");
            // 创建 Cipher 对象并指定其支持的 DES 算法
            Cipher cipher = Cipher.getInstance("DES/ECB/PKCS5Padding");
            // 初始化 Cipher 对象为加密模式
```

```
            cipher.init(Cipher.ENCRYPT_MODE, secretKey);
            byte[] encryptedData = cipher.doFinal(data.getBytes
(StandardCharsets.UTF_8));
            // 使用 Base64 编码加密后的数据，以便打印和观察
            String encodedData = Base64.getEncoder().encodeToString
(encryptedData);
            System.out.println("加密后的数据（Base64 编码）: " + encodedData);
            // 初始化 Cipher 对象为解密模式
            cipher.init(Cipher.DECRYPT_MODE, secretKey);
            byte[] decryptedData = cipher.doFinal(encryptedData);
            // 将解密后的数据转换为字符串
            String decryptedString = new String(decryptedData,
StandardCharsets.UTF_8);
            System.out.println("解密后的数据: " + decryptedString);
        } catch (Exception e) {
            e.printStackTrace();
        }
    }
}
```

运行程序，会输出如下内容：

```
加密后的数据（Base64 编码）: JkrR4pJWWieWgagwuHjdxl3UqEbwne8cXymzFbt1DWQ=
解密后的数据: 世界，你好，Hello, world!
```

15.5 AES 加密和解密

AES 加密算法是一种对称加密算法，也称为高级加密标准（Advanced Encryption Standard）。它是美国国家标准技术研究院（NIST）于 2001 年发布的一种分组密码标准，用来替代原先的 DES 算法。AES 算法使用的分组长度固定为 128 位，密钥长度可以是 128 位、192 位或 256 位。AES 算法的加密和解密过程都是在一个 4×4 的字节矩阵上进行，这个矩阵又称为状态（state）。

使用 Java 实现 AES 加密和解密的步骤如下：

1. 导入必要的 Java 加密类库

Java 加密与解密操作需要用到 javax.crypto 包中的类，例如 Cipher、SecretKeySpec 等，以及 java.util.Base64 用于 Base64 编码和解码。

2. 定义 AES 加密和解密的方法

在 Java 中定义两个方法：encrypt 和 decrypt，分别用于加密和解密操作。

3. 实现加密方法

（1）使用 SecretKeySpec 类来根据给定的字节数组（AES 密钥）创建一个 AES 密钥。

（2）使用 Cipher.getInstance("AES/ECB/PKCS5Padding") 获取 Cipher 实例，指定使用

AES 加密、ECB 模式和 PKCS5Padding 填充模式。

（3）初始化 Cipher 实例为加密模式，并使用前面创建的 AES 密钥。

（4）对输入的文本数据进行加密，并使用 Base64 编码处理加密后的字节数据，以便于存储和传输。

4．实现解密方法

（1）与加密过程类似，使用相同的 AES 密钥和 Cipher 配置。

（2）初始化 Cipher 实例为解密模式。

（3）对输入的 Base64 编码的加密字符串进行解码，然后使用 Cipher 实例对解码后的字节数据进行解密。

（4）将解密后的字节数据转换为字符串。

下面的例子完整地演示了如何使用 Java 实现 AES 加密和解密的过程。

代码位置：src/main/java/crypto/AESExample.java

```
package crypto;

import javax.crypto.Cipher;
import javax.crypto.spec.SecretKeySpec;
import java.util.Base64;
public class AESExample {
    // AES 加密方法
    public static String encrypt(String text, String key) throws Exception {
        // 使用 AES 密钥初始化一个 SecretKeySpec 对象
        SecretKeySpec secretKeySpec = new SecretKeySpec(key.getBytes("UTF-8"),
"AES");
        // 创建 Cipher 实例，并指定其算法为 AES/ECB/PKCS5Padding
        Cipher cipher = Cipher.getInstance("AES/ECB/PKCS5Padding");
        // 初始化 Cipher 为加密模式，并指定密钥
        cipher.init(Cipher.ENCRYPT_MODE, secretKeySpec);
        // 对输入文本进行加密
        byte[] encryptedBytes = cipher.doFinal(text.getBytes("UTF-8"));
        // 使用 Base64 编码加密后的字节数据
        return Base64.getEncoder().encodeToString(encryptedBytes);
    }
    // AES 解密方法
    public static String decrypt(String encryptedText, String key) throws
Exception {
        // 使用 AES 密钥初始化一个 SecretKeySpec 对象
        SecretKeySpec secretKeySpec = new SecretKeySpec(key.getBytes("UTF-8"),
"AES");
        // 创建 Cipher 实例，并指定其算法为 AES/ECB/PKCS5Padding
        Cipher cipher = Cipher.getInstance("AES/ECB/PKCS5Padding");
        // 初始化 Cipher 为解密模式，并指定密钥
        cipher.init(Cipher.DECRYPT_MODE, secretKeySpec);
        // 将 Base64 编码的字符串解码成字节数据
```

```
        byte[] encryptedBytes = Base64.getDecoder().decode(encryptedText);
        // 对加密后的字节数据进行解密
        byte[] decryptedBytes = cipher.doFinal(encryptedBytes);
        // 将解密后的字节数据转换为字符串
        return new String(decryptedBytes, "UTF-8");
    }
    // 测试加密和解密
    public static void main(String[] args) {
        try {
            String originalText = "世界你好，Hello World";
            String key = "1234567812345678";
            // 加密
            String encryptedText = encrypt(originalText, key);
            System.out.println("加密后的结果：" + encryptedText);
            // 解密
            String decryptedText = decrypt(encryptedText, key);
            System.out.println("解密后的结果：" + decryptedText);
        } catch (Exception e) {
            e.printStackTrace();
        }
    }
}
```

运行程序，会输出如下内容：

```
加密后的结果：Xe/dGk4EYdATuNHJwUr+acxOImFqDsfSsVijaHuSCx0=
解密后的结果：世界你好，Hello World
```

15.6　RSA 加密和解密

RSA 加密算法是一种非对称加密算法，它使用一对密钥，即公钥和私钥。公钥是可公开的，用于加密数据；私钥则是保密的，用于解密数据。RSA 算法的安全性基于大数分解的困难性，即将一个大数分解成两个较小的质数的难度。

RSA 算法的具体描述如下：

（1）任意选取两个不同的大素数 p 和 q 计算乘积 n=pq。

（2）任意选取一个大整数 e，满足 gcd(e,(p-1)(q-1))=1，整数 e 用作加密钥。

（3）计算 d 使得 ed ≡ 1(mod (p-1)(q-1))，整数 d 用作解密钥。

（4）公钥为 (n,e)，私钥为 (n,d)。

（5）在加密时，明文 m 被转换为整数 M，使得 0 ≤ M<n。密文 c=M^e(mod n)。在解密时，明文 m=M^d(mod n)。

下面的例子完整地演示了如何使用 Java 实现 RSA 加密和解密的过程。

代码位置：src/main/java/crypto/RSAExample.java

```
package crypto;
import javax.crypto.Cipher;
import java.security.*;
import java.util.Base64;
public class RSAExample {
    // 生成 RSA 公钥和私钥
    public static KeyPair generateKeyPair() throws Exception {
        KeyPairGenerator generator = KeyPairGenerator.getInstance("RSA");
        generator.initialize(2048, new SecureRandom());
        return generator.generateKeyPair();
    }
    // 使用公钥加密数据
    public static String encrypt(String plainText, PublicKey publicKey) throws
Exception {
        Cipher cipher = Cipher.getInstance("RSA/ECB/OAEPWithSHA-
256AndMGF1Padding");
        cipher.init(Cipher.ENCRYPT_MODE, publicKey);
        byte[] encryptedBytes = cipher.doFinal(plainText.getBytes("UTF-8"));
        return Base64.getEncoder().encodeToString(encryptedBytes);
    }
    // 使用私钥解密数据
    public static String decrypt(String encryptedText, PrivateKey privateKey)
throws Exception {
        byte[] encryptedBytes = Base64.getDecoder().decode(encryptedText);
        Cipher cipher = Cipher.getInstance("RSA/ECB/OAEPWithSHA-
256AndMGF1Padding");
        cipher.init(Cipher.DECRYPT_MODE, privateKey);
        byte[] decryptedBytes = cipher.doFinal(encryptedBytes);
        return new String(decryptedBytes, "UTF-8");
    }
    public static void main(String[] args) {
        try {
            // 生成公钥和私钥
            KeyPair keyPair = generateKeyPair();
            PublicKey publicKey = keyPair.getPublic();
            PrivateKey privateKey = keyPair.getPrivate();
            // 加密数据
            String encryptedText = encrypt("世界，你好，Hello World!", publicKey);
            System.out.println("密文：" + encryptedText);
            // 解密数据
            String decryptedText = decrypt(encryptedText, privateKey);
            System.out.println("明文：" + decryptedText);
        } catch (Exception e) {
            e.printStackTrace();
        }
    }
}
```

运行程序，会显示如下内容：

```
密文：
YMmnKL7SENHP8F3cMS+wnwT+REhWRk2Svc/xo3JStixxhONmUf9ZM953h7Difz0CHlVBfSfdkzDGel
BhJExWLSsxHQBEkzplZKomHwo0jWbht5R7Aj+3JY6z8/HhCuIW4QVgROzTpWBQveTLpkLFP5aomwK+
Ecrg3qgxL/nOt5pKnnXOdT8L04YYUQo3LKwdSMoTUOcQul1lmMeJa0tizqcgSzhqcfdNHoE2kVY/sd0
7UwKRxZgAx+tg62x4lQCdG5PJUP1VCTYwbmAknw1ux9nPoPiro8kDhXSI7ig9YRJmyTeATgL50V5fIf
sGwoB7KFGxU/f5Zk/ba9u5R20CBw==
明文：世界，你好，Hello World!
```

15.7 小结

本章主要介绍了使用了 Java 内建的 API 对数据进行编码、加密和解密。数据加密和解密的应用场景有很多。例如，数据库中的数据加密和解密处理，以防止数据被他人窃取；透明数据加密技术适用于对数据库中的数据执行实时加解密的应用场景，尤其是在对数据加密透明化有要求，以及对数据加密后数据库性能有较高要求的场景中；加密使用网络安全来防御恶意软件和勒索软件等暴力破解和网络攻击。读者可以在不同的应用场景中选择合适的加密和解密技术。

第 16 章

文件压缩与解压

本章会介绍如何使用 Java 以及 java.util.zip 包、Apache Commons Compress 库将文件和目录压缩成 zip 和 7z 文件，以及如何解压这两种压缩格式的文件。

16.1 zip 格式

本节介绍如何使用 java.util.zip 包中相关的 API 将文件和目录压缩成 zip 格式的文件，以及解压 zip 文件。

16.1.1 压缩成 zip 文件

使用 Java 压缩文件或目录到 zip 格式需要使用 java.util.zip 包中的 ZipOutputStream 类，具体步骤如下。

1. 创建 ZipOutputStream 实例

首先，创建一个 ZipOutputStream 实例。这个类负责将数据写入 zip 文件。可以通过传递一个 FileOutputStream 实例给 ZipOutputStream 的构造器来指定要创建的 zip 文件的位置。

```
FileOutputStream fos = new FileOutputStream("output.zip");
ZipOutputStream zipOut = new ZipOutputStream(fos);
```

2. 添加文件或目录到 zip 文件

接下来，根据需要压缩的内容是单个文件还是整个目录，可以采取不同的策略。

（1）压缩单个文件：需要创建一个 ZipEntry 对象，该对象代表 zip 文件中的一个条目。然后，可以读取源文件的内容，并写入到 ZipOutputStream 中。

（2）压缩整个目录：需要递归地遍历目录中的所有文件和子目录。对于目录中的每个文件，可以使用与压缩单个文件相同的方法来处理。对于子目录，可以创建以斜杠（/）结尾的 ZipEntry 来表示它们。

3. 关闭 ZipOutputStream 实例

完成所有文件和目录的添加后，不要忘记关闭 ZipOutputStream 实例。这一步很重要，因为它会确保所有数据都被写入 zip 文件中，并且 zip 文件被正确地关闭。

下面的代码完整地演示了如何使用 Java 压缩 zip 文件。

代码位置：src/main/java/zip/ZipUtil.java

```
package zip;
import java.io.File;
import java.io.FileInputStream;
import java.io.FileOutputStream;
import java.io.IOException;
import java.nio.file.Files;
import java.nio.file.Path;
import java.nio.file.Paths;
import java.util.zip.ZipEntry;
import java.util.zip.ZipOutputStream;
public class ZipUtil {
    /**
     * 压缩目录到 zip 文件中
     * @param sourceDirPath 要压缩的目录路径
     * @param zipFilePath zip 文件的路径
     * @throws IOException 如果发生 I/O 错误
     */
    public static void zipDirectory(String sourceDirPath, String zipFilePath)
throws IOException {
        Path zipPath = Files.createFile(Paths.get(zipFilePath));
        try (ZipOutputStream zs = new ZipOutputStream(Files.newOutputStream
(zipPath))) {
            Path sourcePath = Paths.get(sourceDirPath);
            Files.walk(sourcePath)
                    .filter(path -> !Files.isDirectory(path))
                    .forEach(path -> {
                        ZipEntry zipEntry = new ZipEntry(sourcePath.relativize
(path).toString());
                        try {
                            zs.putNextEntry(zipEntry);
                            Files.copy(path, zs);
                            zs.closeEntry();
                        } catch (IOException e) {
                            System.err.println(e);
                        }
                    });
        }
    }
    /**
     * 将文件或目录压缩到 zip 文件中
     * @param sourcePath 要压缩的文件或目录的路径
     * @param zipFilePath zip 文件的路径
     * @throws IOException 如果发生 I/O 错误
```

```
    */
    public static void zipFileOrDirectory(String sourcePath, String zipFilePath)
throws IOException {
        File sourceFile = new File(sourcePath);
        try (ZipOutputStream zipOut = new ZipOutputStream(new FileOutputStream
(zipFilePath))) {
            if (sourceFile.isFile()) {
                zipFile(sourceFile, sourceFile.getName(), zipOut);
            } else {
                File[] fileList = sourceFile.listFiles();
                if (fileList != null) {
                    for (File file : fileList) {
                        zipFile(file, file.getName(), zipOut);
                    }
                }
            }
        }
    }

    /**
     * 将单个文件添加到 zip 文件中
     * @param file 要压缩的文件
     * @param fileName 文件在 zip 中的条目名
     * @param zipOut zip 输出流
     * @throws IOException 如果发生 I/O 错误
     */
    private static void zipFile(File file, String fileName, ZipOutputStream zipOut)
throws IOException {
        if (file.isHidden()) {
            return;
        }
        if (file.isDirectory()) {
            if (fileName.endsWith("/")) {
                zipOut.putNextEntry(new ZipEntry(fileName));
                zipOut.closeEntry();
            } else {
                zipOut.putNextEntry(new ZipEntry(fileName + "/"));
                zipOut.closeEntry();
            }
            File[] children = file.listFiles();
            for (File childFile : children) {
                zipFile(childFile, fileName + "/" + childFile.getName(), zipOut);
            }
            return;
        }
        FileInputStream fis = new FileInputStream(file);
        ZipEntry zipEntry = new ZipEntry(fileName);
        zipOut.putNextEntry(zipEntry);
        byte[] bytes = new byte[1024];
        int length;
        while ((length = fis.read(bytes)) >= 0) {
```

```
            zipOut.write(bytes, 0, length);
        }
        fis.close();
    }
    public static void main(String[] args) {
        try {
            // 创建 zip 文件并添加文件或目录
            zipFileOrDirectory("images", "images.zip");
        } catch (IOException e) {
            e.printStackTrace();
        }
    }
}
```

运行程序之前，要保证当前目录为 images 目录，运行程序后，会在当前目录生成一个 images.zip 文件。

16.1.2　解压 zip 文件

本节的例子会解压 zip 文件，具体的实现步骤如下。

（1）读取 zip 文件：使用 ZipInputStream 读取 zip 文件。ZipInputStream 提供了一个方便的接口来读取 zip 文件条目。

（2）遍历 zip 条目：通过循环调用 getNextEntry 方法遍历 zip 文件中的每个条目。对于每个条目，根据其是否是目录来决定是创建目录还是提取文件。

（3）提取文件：如果条目是文件，则调用 extractFile 方法将文件内容写入目标路径。这个方法使用 BufferedOutputStream 来写文件。

（4）关闭流：在提取完所有文件后，记得关闭 ZipInputStream 和在 extractFile 方法中打开的 BufferedOutputStream。

具体实现代码如下。

代码位置：src/main/java/zip/UnzipUtility.java

```
package zip;
import java.io.*;
import java.util.zip.ZipEntry;
import java.util.zip.ZipInputStream;
public class UnzipUtility {
    /**
     * 解压 zip 文件到指定的目录
     * @param zipFilePath zip 文件的路径
     * @param destDirectory 解压到的目标目录
     * @throws IOException 如果发生 I/O 错误
     */
    public static void unzip(String zipFilePath, String destDirectory) throws
IOException {
        // 确保目标目录存在，如果不存在，则创建它
```

```
        File destDir = new File(destDirectory);
        if (!destDir.exists()) {
            destDir.mkdir();
        }
        // 创建 ZipInputStream 来读取 zip 文件
        FileInputStream fis = new FileInputStream(zipFilePath);
        ZipInputStream zipIn = new ZipInputStream(fis);
        ZipEntry entry = zipIn.getNextEntry();
        // 遍历 zip 文件中的每个条目
        while (entry != null) {
            String filePath = destDirectory + File.separator + entry.getName();
            if (!entry.isDirectory()) {
                // 如果条目不是目录，则提取文件
                extractFile(zipIn, filePath);
            } else {
                // 如果条目是目录，则创建目录
                File dir = new File(filePath);
                dir.mkdir();
            }
            zipIn.closeEntry();
            entry = zipIn.getNextEntry();
        }
        zipIn.close();
    }
    /**
     * 从 zip 文件中提取单个文件
     * @param zipIn zip 输入流
     * @param filePath 提取文件的目标文件路径
     * @throws IOException 如果发生 I/O 错误
     */
    private static void extractFile(ZipInputStream zipIn, String filePath)
throws IOException {
        // 创建一个输出流来写入提取的文件
        BufferedOutputStream bos = new BufferedOutputStream(new FileOutputStream
(filePath));
        byte[] bytesIn = new byte[4096];
        int read = 0;
        while ((read = zipIn.read(bytesIn)) != -1) {
            bos.write(bytesIn, 0, read);
        }
        bos.close();
    }
    public static void main(String[] args) {
        // 指定 zip 文件和目标解压目录
        String zipFilePath = "images.zip";       // 待解压的文件
        String destDirectory = "unzip";          // 会将压缩文件的内容解压到 unzip 目录
        try {
            unzip(zipFilePath, destDirectory);
            System.out.println("zip 文件解压完成 .");
        } catch (IOException e) {
            e.printStackTrace();
```

```
            }
        }
    }
```

16.2　7z 格式

本节介绍使用 Apache Commons Compress 库将文件和目录压缩成 7z 格式的文件，并解压 7z 文件。读者可以从下面的页面下载 Apache Commons Compress 库最新版的 jar 文件。

```
https://mvnrepository.com/artifact/org.apache.commons/commons-compress
```

Apache Commons Compress 需要一个 XZ 库。XZ 是一个用于数据压缩的库，Apache Commons Compress 在处理某些压缩格式（包括 7z）时会使用到它。读者可以从下面的页面下载 XZ 库最新版的 jar 文件。

```
https://mvnrepository.com/artifact/org.tukaani/xz
```

16.2.1　压缩成 7z 格式

本节的例子会使用 Apache Commons Compress 库将文件或目录压缩成 7z 格式，步骤如下。

1. 创建 SevenZOutputFile 实例

（1）通过调用 SevenZOutputFile 的构造函数创建一个指向 7z 压缩文件的输出流。这个输出流代表了最终将要生成的 7z 压缩文件。

（2）构造函数接收一个 File 对象，指定压缩文件的保存位置。

2. 判断压缩目标是文件还是目录

（1）compress7z 方法接收两个参数：源路径（可以是文件也可以是目录）和目标 7z 文件的路径。

（2）通过检查源路径指向的是文件还是目录，该方法决定是直接压缩单个文件，还是遍历目录下所有内容进行压缩。

3. 递归遍历目录

（1）如果源路径是目录，addDirectoryToArchive() 方法被调用，递归地遍历目录中的所有文件和子目录。

（2）对于每个找到的文件，它调用 addFileToArchive() 方法将文件添加到 7z 压缩文件中。

（3）对于子目录，方法将为目录中的每个文件和子目录递归地重复这个过程。

4. 压缩文件

（1）addFileToArchive() 方法为每个文件创建一个 SevenZArchiveEntry，这个条目在压缩文件中代表了一个文件。

（2）方法读取源文件的内容，并将其写入 7z 文件中。这一步是通过读取文件内容到一个缓冲区，然后将缓冲区的内容写入 SevenZOutputFile 完成的。

5. 关闭资源

在所有文件和目录都被添加到压缩文件后，SevenZOutputFile 资源被关闭，完成压缩过程。

SevenZCompression 类通过组合 Apache Commons Compress 库提供的功能，实现了文件和目录的 7z 格式压缩。该过程涉及创建压缩文件条目、递归遍历目录结构、读取文件数据并写入压缩文件中。通过这种方式，无论是单个文件还是复杂的目录结构，都可以被有效地压缩成单个 7z 文件，从而节省空间并方便存储和传输。

下面的代码使用 Apache Commons Compress 库压缩 images 目录，并生成了 images.7z 文件。

代码位置：src/main/java/zip/SevenZCompression.java

```
package zip;
import org.apache.commons.compress.archivers.sevenz.SevenZOutputFile;
import org.apache.commons.compress.archivers.sevenz.SevenZArchiveEntry;
import java.io.File;
import java.io.FileInputStream;
import java.io.IOException;
public class SevenZCompression {

    /**
     * 压缩文件或目录为 7z 格式
     * @param fileOrDirPath 要压缩的文件或目录的路径
     * @param archivePath 压缩后的 7z 文件的存储路径
     * @throws IOException 如果发生 I/O 错误
     */
    public static void compress7z(String fileOrDirPath, String archivePath)
throws IOException {
        // 创建源文件或目录的 File 对象
        File source = new File(fileOrDirPath);
        // 创建 7z 压缩文件的输出流
        try (SevenZOutputFile sevenZOutput = new SevenZOutputFile(new File
(archivePath))) {
            // 判断源是目录还是文件，并相应地处理
            if (source.isDirectory()) {
                // 如果是目录，则递归添加目录内容到压缩文件中
                addDirectoryToArchive(sevenZOutput, source, source.getName());
            } else {
                // 如果是文件，则直接添加文件到压缩文件中
                addFileToArchive(sevenZOutput, source, source.getName());
```

```
            }
        }
    }
    /**
     * 将单个文件添加到 7z 压缩文件中
     * @param sevenZOutput 7z 压缩文件的输出流
     * @param file 要添加的文件
     * @param baseName 文件在压缩文件中的条目名称
     * @throws IOException 如果发生 I/O 错误
     */
    private static void addFileToArchive(SevenZOutputFile sevenZOutput, File
file, String baseName) throws IOException {
        // 创建压缩文件条目并添加到压缩文件
        SevenZArchiveEntry entry = sevenZOutput.createArchiveEntry(file, baseName);
        sevenZOutput.putArchiveEntry(entry);
        // 读取文件内容并写入压缩文件中
        try (FileInputStream fis = new FileInputStream(file)) {
            byte[] buffer = new byte[1024];
            int length;
            while ((length = fis.read(buffer)) > 0) {
                sevenZOutput.write(buffer, 0, length);
            }
        }
        // 关闭当前条目，完成文件的添加
        sevenZOutput.closeArchiveEntry();
    }

    /**
     * 递归地将目录及其内容添加到 7z 压缩文件中
     * @param sevenZOutput 7z 压缩文件的输出流
     * @param directory 要添加的目录
     * @param baseName 目录在压缩文件中的基础名称
     * @throws IOException 如果发生 I/O 错误
     */
    private static void addDirectoryToArchive(SevenZOutputFile sevenZOutput,
File directory, String baseName) throws IOException {
        // 获取目录下的所有文件和子目录
        File[] files = directory.listFiles();
        if (files != null) {
            for (File file : files) {
                // 对于每个文件或子目录，递归调用相应的方法处理
                if (file.isDirectory()) {
                    // 如果是子目录，则递归处理
                    addDirectoryToArchive(sevenZOutput, file, baseName + "/"
+ file.getName());
                } else {
                    // 如果是文件，则添加到压缩文件中
                    addFileToArchive(sevenZOutput, file, baseName + "/" + file.
getName());
                }
            }
```

```
        }
    }
    public static void main(String[] args) {
        // 调用 compress7z 方法压缩指定的文件或目录
        try {
            compress7z("images", "images.7z");
            System.out.println("Compression completed successfully");
        } catch (IOException e) {
            // 处理可能发生的异常
            e.printStackTrace();
        }
    }
}
```

执行代码之前，要保证当前目录存在 images 子目录。运行代码后，会在当前目录生成 images.7z 文件。

16.2.2 解压 7z 文件

本节的例子会使用 Apache Commons Compress 库将 images.7z 文件解压到指定的目录，代码如下。

代码位置：src/main/java/zip/SevenZDecompression.java

```
package zip;
import org.apache.commons.compress.archivers.sevenz.SevenZFile;
import org.apache.commons.compress.archivers.sevenz.SevenZArchiveEntry;
import java.io.File;
import java.io.FileOutputStream;
import java.io.IOException;
import java.io.OutputStream;
public class SevenZDecompression {
    /**
     * 解压 7z 格式的压缩文件到指定目录
     * @param archivePath 7z 压缩文件的路径
     * @param destDirectory 解压到的目标目录路径
     * @throws IOException 如果发生 I/O 错误
     */
    public static void decompress7z(String archivePath, String destDirectory)
throws IOException {
        // 创建指向压缩文件的 File 对象
        File archiveFile = new File(archivePath);
        // 创建指向目标解压目录的 File 对象。如果目录不存在，则创建它
        File destDir = new File(destDirectory);
        if (!destDir.exists()) {
            destDir.mkdirs();
        }
        // 使用 SevenZFile 对象打开 7z 压缩文件
        try (SevenZFile sevenZFile = new SevenZFile(archiveFile)) {
```

```
            SevenZArchiveEntry entry;
            // 遍历压缩文件中的所有条目（文件和目录）
            while ((entry = sevenZFile.getNextEntry()) != null) {
                // 为当前条目构造在解压目录中的完整路径
                File outFile = new File(destDir, entry.getName());
                // 如果条目是目录，则创建目录
                if (entry.isDirectory()) {
                    outFile.mkdirs();
                } else {
                    // 确保条目的父目录存在
                    outFile.getParentFile().mkdirs();
                    // 解压当前条目（文件）
                    extractFile(sevenZFile, outFile);
                }
            }
        }
    }
    /**
     * 从 7z 压缩文件中提取单个文件
     * @param sevenZFile 7z 压缩文件对象
     * @param outFile 解压后的文件应该存储的路径
     * @throws IOException 如果发生 I/O 错误
     */
    private static void extractFile(SevenZFile sevenZFile, File outFile) throws
IOException {
        try (OutputStream out = new FileOutputStream(outFile)) {
            byte[] buffer = new byte[4096];
            int bytesRead;
            // 从压缩文件中读取数据到缓冲区，然后写入输出文件中
            while ((bytesRead = sevenZFile.read(buffer)) != -1) {
                out.write(buffer, 0, bytesRead);
            }
        }
    }

    public static void main(String[] args) {
        // 指定 7z 压缩文件的路径和解压的目标目录
        String archivePath = "images.7z";
        String destDirectory = "un7z";
        try {
            decompress7z(archivePath, destDirectory);
            System.out.println("7z 文件解压完成 ");
        } catch (IOException e) {
            e.printStackTrace();
        }
    }
}
```

这段代码会将 images.7z 文件解压到 un7z 目录。

16.2.3 设置 7z 文件的密码

Apache Commons Compress 库本身不支持 7z 文件的加密和解密，但可以通过直接调用 7z 命令实现。如下面的代码调用了 7z 命令对 images 目录进行压缩，并设置密码为 1234，然后再使用 7z 命令进行解密。

代码位置：src/main/java/zip/SevenZipWithPassword.java

```
package zip;
import java.io.IOException;
public class SevenZipWithPassword {
    public static void compressWithPassword(String fileOrDirPath, String
outputPath, String password) {
        try {
            // 构建用于压缩的命令
            String command = String.format("7z a -p%s -y %s %s", password,
outputPath, fileOrDirPath);
            // 执行命令
            Process process = Runtime.getRuntime().exec(command);
            // 等待命令执行完成
            process.waitFor();
            System.out.println("Compression completed successfully.");
        } catch (IOException | InterruptedException e) {
            e.printStackTrace();
        }
    }

    /**
     * 使用 7-Zip 命令行工具解压带密码的压缩文件
     *
     * @param archivePath 带密码的压缩文件路径
     * @param password 压缩文件的密码
     */
    public static void decompressWithPassword(String archivePath, String
password) {
        try {
            // 构建用于解压的命令
            String command = String.format("7z x -p%s -y %s", password,
archivePath);
            System.out.println(command);
            // 执行命令
            Process process = Runtime.getRuntime().exec(command);
            // 等待命令执行完成
            process.waitFor();
            System.out.println("Decompression completed successfully.");
        } catch (IOException | InterruptedException e) {
            e.printStackTrace();
        }
    }
    public static void main(String[] args) {
```

```
        // 示例：压缩文件或目录
        compressWithPassword("images", "pimages.7z", "1234");
        // 示例：解压带密码的压缩文件
        decompressWithPassword("pimages.7z", "1234");
    }
}
```

16.3　小结

本章详细介绍了如何使用 Java 进行文件压缩与解压，特别是针对 zip 和 7z 这两种常见的压缩格式。通过 java.util.zip 包和 Apache Commons Compress 库，开发者能够轻松地将文件和目录压缩成 zip 文件，以及从 zip 文件中提取内容。同时，本章还探讨了如何使用 Apache Commons Compress 库来处理 7z 格式的压缩文件，包括压缩和解压操作。

在 zip 文件处理方面，本章首先介绍了如何使用 ZipOutputStream 类创建 zip 文件，并通过 ZipEntry 对象将文件或目录添加到 zip 文件中。接着，展示了如何使用 ZipInputStream 类来解压 zip 文件，包括遍历 zip 文件中的条目并提取文件。此外，还提供了一个示例，展示了如何将目录压缩成 zip 文件，并在解压后恢复目录结构。

对于 7z 格式，本章首先介绍了如何使用 Apache Commons Compress 库中的 SevenZOutputFile 类来创建 7z 压缩文件，并递归地添加文件和目录。然后，通过 SevenZFile 类展示了如何解压 7z 文件，以及如何提取单个文件。本章还提到了如何通过调用 7z 命令行工具来处理带密码的 7z 文件，包括压缩和解压操作。